Praise for Giulio Boccaletti's

WATER

An *Economist* Best Book of the Year

"Boccaletti is a renowned expert on natural resource security and environmental stability, lending *Water* a pressing, historically fascinating, and informative arc. . . . *Water*, the book, is a smart new chapter on the same subject that turned Joan Didion's head toward the Hoover Dam." —Sloane Crosley, *Departures*

"Provides essential reading for those seeking to explore how humanity's relationship with nature has influenced the development of legal and political systems and offers invaluable insights into current debates surrounding climate change and sustainability."
—Lee C. Bollinger, president and Seth Low Professor, Columbia University

"Giulio Boccaletti's book is a remarkable achievement: a readable history of the world, seen through the history of water management."
—Chris Wickham, Chichele Professor Emeritus of Medieval History, University of Oxford

"Giulio Boccaletti makes a strikingly original and persuasive case that the history of human civilization can be understood as a never-ending struggle over water. Boccaletti's command of a vast range of material, across time and space, is astonishing."
—Nicholas Lemann, Joseph Pulitzer II and Edith Pulitzer Moore Professor of Journalism, Columbia University

"A transformative, dramatic, and revelatory tale of how our struggle to master water defines us as humans."
—John Bredar, vice president for National Programming, WGBH

"As humanity strays across planetary boundaries, Boccaletti's political biography of water is essential reading."
—Rachel Kyte, dean, The Fletcher School, Tufts University

"A dazzling tale spanning millennia, geography, science, and human civilizations that is more than the story of water. It is a story of ideas and institutions, of tensions between individual enterprise and collective action, of human needs and planetary dynamics."

—Lynn Scarlett, former chief external affairs officer, The Nature Conservancy

"The breadth and substance of the narrative are outstanding. . . . The book is a tour de force!"

—Michael Hanemann, chancellor's professor emeritus, University of California, Berkeley

"It was an inspired idea to write a 'biography' of water, and Giulio Boccaletti has carried it off in style."

—David Blackbourn, Cornelius Vanderbilt Distinguished Chair of History, Vanderbilt University

"Boccaletti has pinned down our complex relationship with our most vital resource. We live, like the ancients, in a hydraulic civilization—one determined to a remarkable degree by where and when we can find water."

—Fred Pearce, author of *When the Rivers Run Dry*

"*Water* could have no better biographer than Giulio Boccaletti, who takes us on a fascinating journey, telling the story of how humanity's interactions with this most precious resource have shaped our history, our present, and will define our future."

—Eric D. Beinhocker, professor, Blavatnik School of Government, University of Oxford

"An ingenious lesson in geopolitics."

—*Kirkus Reviews* (starred review)

"This biography of Earth's crucial resource covers an expansive time line, from antiquity to the present day. . . . During this time of accelerated population growth, climate change, and political instability, *Water* is essential reading."

—*Booklist*

Giulio Boccaletti

WATER

Giulio Boccaletti is a globally recognized expert on natural resource security and environmental sustainability. He is an honorary research associate at the Smith School of Enterprise and the Environment, University of Oxford. Trained as a physicist and climate scientist, he holds a doctorate from Princeton University, where he was a NASA Earth Systems Science Fellow. He has been a research scientist at MIT and was a partner at McKinsey & Company, where he was one of the leaders of its Sustainability and Resource Productivity Practice, and the chief strategy officer and global ambassador for water at The Nature Conservancy, one of the world's largest environmental organizations. Boccaletti frequently writes on environmental issues for the news media and is an expert contributor to the World Economic Forum. His work on water has been featured in the PBS documentary series *H_2O: The Molecule That Made Us*. He lives in London.

giulioboccaletti.com

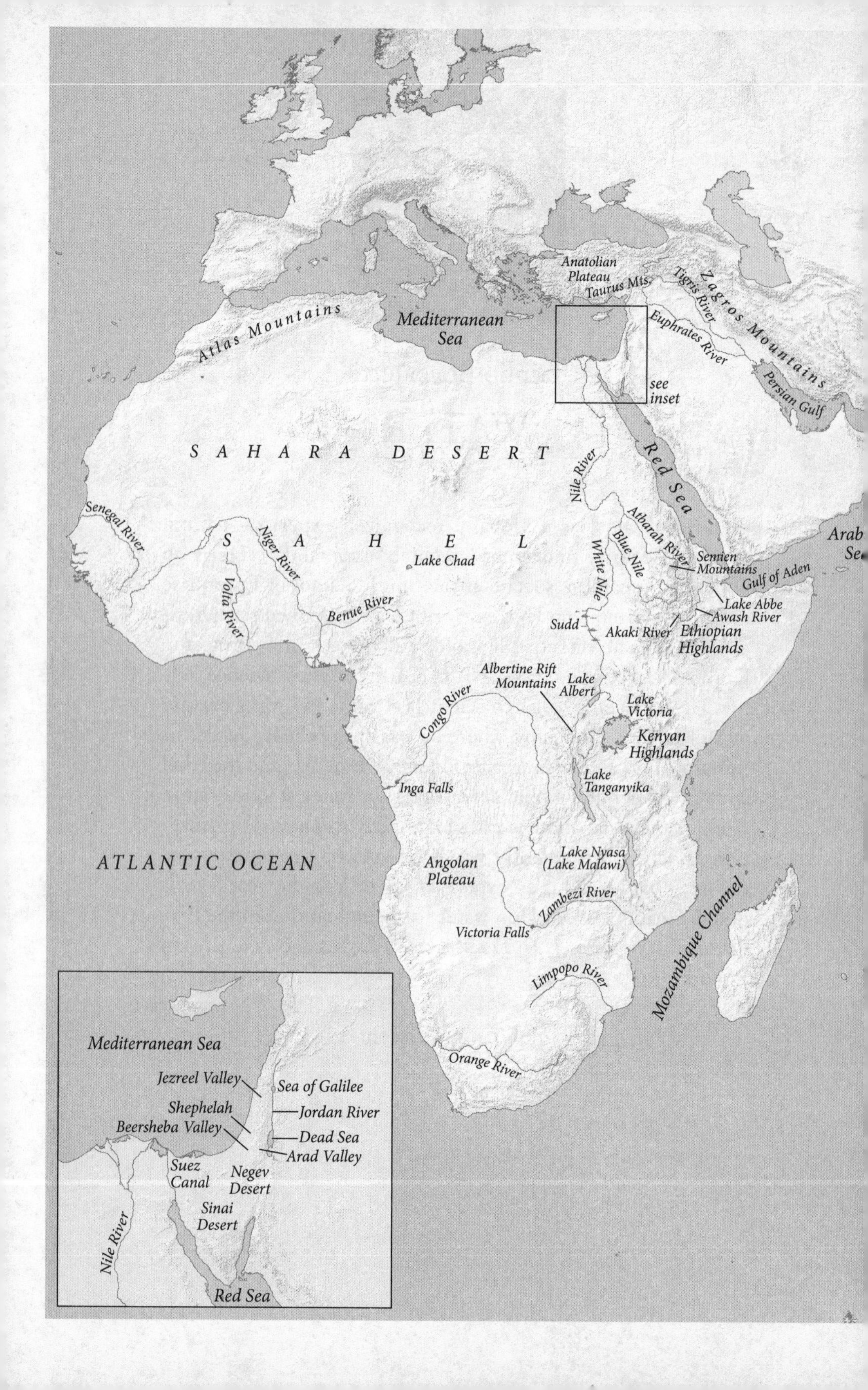

Anatolian Plateau
Taurus Mts.
Tigris River
Zagros Mountains
Atlas Mountains
Mediterranean Sea
Euphrates River
see inset
Persian Gulf
SAHARA DESERT
Red Sea
Nile River
Senegal River
Atbarah River
Arab Se
S A H E L
Niger River
Blue Nile
White Nile
Lake Chad
Semien Mountains
Gulf of Aden
Volta River
Lake Abbe
Awash River
Benue River
Sudd
Akaki River
Ethiopian Highlands
Albertine Rift Mountains
Lake Albert
Lake Victoria
Congo River
Kenyan Highlands
Lake Tanganyika
Inga Falls
ATLANTIC OCEAN
Lake Nyasa (Lake Malawi)
Angolan Plateau
Zambezi River
Mozambique Channel
Victoria Falls
Limpopo River
Orange River
Mediterranean Sea
Jezreel Valley
Sea of Galilee
Shephelah
Jordan River
Beersheba Valley
Dead Sea
Arad Valley
Suez Canal
Negev Desert
Sinai Desert
Nile River
Red Sea

TLANTIC OCEAN
Lake Thirlmere
(Honte) Western Scheldt River
Eastern Scheldt River
The Fens
River Thames
Yser River
River Rhine
Lys River
Marne River
Dnieper River
River Rhône
River Danube
Black Sea
see inset
Sardinia
Attica
Euboea
Island of Samos
Eurotas River
Sicily
Messenia
Laconia
Magoula River
Island of Thera (Santorini)
Crete
Mediterranean Sea
Vajont River
Tagliamento River
Adda River
Piave River
Isonzo River
Po River
Gorzente River
River Aposa
River Reno
River Arno
Tiber River
Corsica
River Aniene
Sardinia
Pontine Marshes

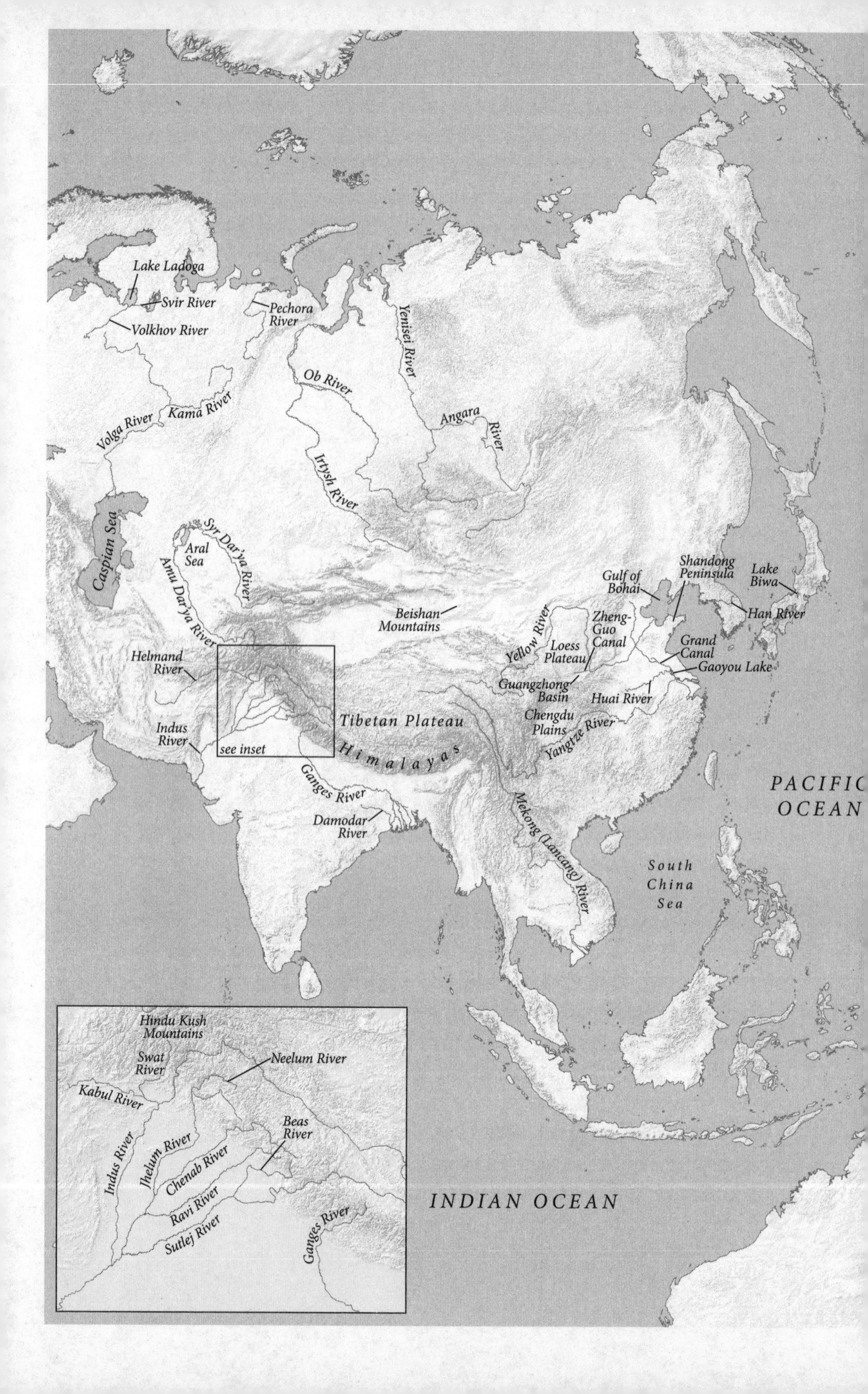

Lake Ladoga
Svir River
Volkhov River
Pechora River
Yenisei River
Ob River
Kama River
Volga River
Angara River
Irtysh River
Caspian Sea
Syr Dar'ya River
Aral Sea
Amu Dar'ya River
Beishan Mountains
Gulf of Bohai
Shandong Peninsula
Lake Biwa
Han River
Yellow River
Loess Plateau
Zheng-Guo Canal
Grand Canal
Gaoyou Lake
Helmand River
Guangzhong Basin
Huai River
Tibetan Plateau
Chengdu Plains
Yangtze River
Indus River
see inset
Himalayas
Ganges River
Damodar River
Mekong (Lancang) River
PACIFIC OCEAN
South China Sea
INDIAN OCEAN
Hindu Kush Mountains
Swat River
Neelum River
Kabul River
Beas River
Indus River
Jhelum River
Chenab River
Ravi River
Sutlej River
Ganges River

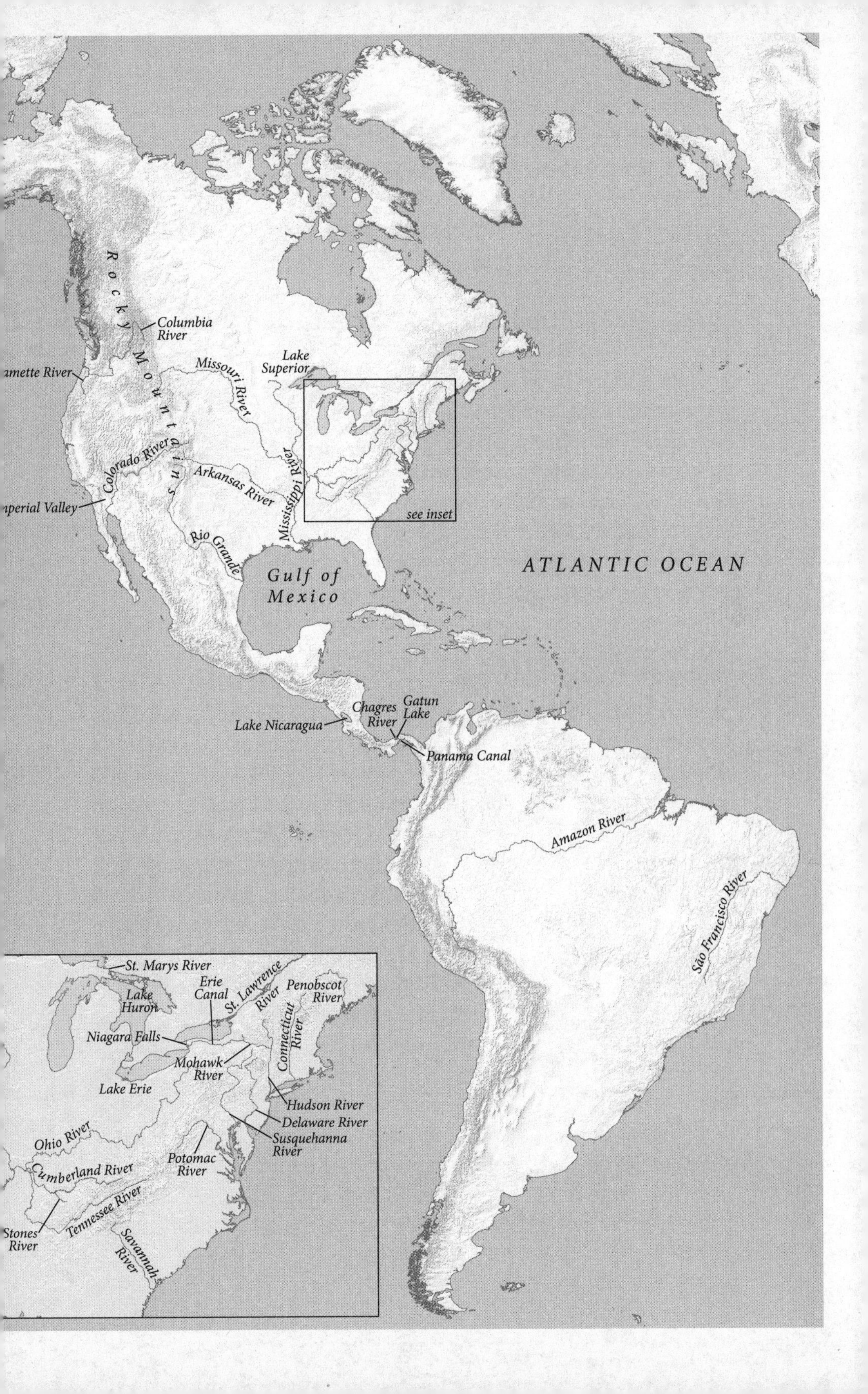
Rocky Mountains
Columbia River
amette River
Missouri River
Lake Superior
Colorado River
Arkansas River
Mississippi River
nperial Valley
Rio Grande
see inset
ATLANTIC OCEAN
Gulf of Mexico
Chagres River
Gatun Lake
Lake Nicaragua
Panama Canal
Amazon River
São Francisco River
St. Marys River
Erie Canal
St. Lawrence River
Penobscot River
Lake Huron
Niagara Falls
Connecticut River
Mohawk River
Lake Erie
Hudson River
Delaware River
Ohio River
Susquehanna River
Potomac River
Cumberland River
Tennessee River
Stones River
Savannah River

WATER

WATER

A Biography

Giulio Boccaletti

VINTAGE BOOKS
A Division of Penguin Random House LLC
New York

FIRST VINTAGE BOOKS EDITION 2022

The Library of Congress has cataloged the Pantheon edition as follows:
Name: Boccaletti, Giulio, [date] author.
Title: Water : a biography / Giulio Boccaletti.
Description: First edition. | New York : Pantheon Books, 2021. |
Includes bibliographical references and index.
Identifiers: LCCN 2020037044 (print) | LCCN 2020037045 (ebook)
Subjects: LCSH: Water—History. | Water.
Classification: LCC GB659.6 .B63 2021 (print) | LCC GB659.6 (ebook) | DDC 909—dc23
LC record available at https://lccn.loc.gov/2020037044
LC ebook record available at https://lccn.loc.gov/2020037045

Vintage Books Trade Paperback ISBN: 978-0-525-56600-7
eBook ISBN: 978-1-5247-4824-1

Author photograph © Andrea Mattiello
Maps by Mapping Specialists, Ltd., Madison, WI
Book design by Maggie Hinders

vintagebooks.com

Printed in the United States of America
1st Printing

For Andrea

Contents

Prologue

On July 19, 2010, a Monday evening, water came crashing down the Yangtze River. Intense rain from the East Asian monsoon hit southwest China. Water poured from the sky. As Monday turned into Tuesday, the flood roared down: every second, seventy thousand cubic meters of water came through, equivalent to thirty Olympic-size swimming pools. In the past, water would have collected in the river, gushing between the rocky banks of three incised gorges in the middle of the mainstem above the city of Yichang. The swollen river would have then overwhelmed the embankments, flooding the plains downstream. Instead, that night the current gently slipped into a wide lake near the city of Chongqing, far above those three incised gorges, oozing out as the crest of the flood dissipated. Six hundred kilometers downstream, the water level in the reservoir rose by four meters, held back by twenty-eight million cubic meters of concrete. Nothing more happened. Three Gorges Dam had passed its first real test.

The plan to build the largest dam in the world had been green-lit in 1992, under Chinese premier Li Peng. The approval had not been without controversy. Li had trained as a hydroelectric engineer in the Soviet Union. He had pushed to get the project commissioned despite concerns for the relocation of one and a half million people, and for the loss of ecosystems and historical artifacts. Eventually, a majority of the National People's Congress voted for the construction. Work began in 1994. Only nine years later the reservoir began filling, ahead of schedule and under budget.

The story of why and how this enormous piece of infrastructure came to be is a familiar one. The high modernist project of the twentieth century was to liberate society from a variable climate, to celebrate the final victory of man in his conquest of nature. Today, everyone operates under the illusion that water on the landscape is, or should be, nothing more than an inert backdrop on the stage of human events. That illusion is created because of the forty-five thousand structures taller than fifteen meters that dam the rivers of the world, a number that grows to millions if all barrages cluttering streams are counted. This enormous stock of infrastructure is capable of catching around 20 percent of the world's annual runoff, the water that collects in rivers and streams across all lands. Modern water infrastructure has replumbed the planet. Three Gorges Dam is one of the latest additions to this vast stock, proof that this modern story of progress has yet to fully run its course. Technology enthusiasts celebrate its achievement, while environmentalists bemoan its impacts. Either way, it is the story of a technological emancipation from nature, in which science and engineering have given humanity, for better or for worse, full control over its own destiny.

This story is familiar. It is also wrong. The story of water is not technological, but political. The impact of water on society must be read through the scars left by a continuous cycle of adaptation. All communities relate to water over time through a process of action and reaction. A levee might protect people settled behind it. A dam might store water for those times when none comes from the sky. But as towns grow and farms expand, people forget why those structures were built in the first place. Society evolves and habituates to its newfound security. Institutions develop in the shadow of infrastructure designed to create an illusion of stability. Then one day, unexpectedly, the levee fails or the reservoir behind the dam goes dry. Loss follows, sometimes catastrophically. People are forced to reconsider their environment, which is no longer the inert scenery to their life. They learn, rebuild, expand, reaching a new level of security. Their institutions adjust, habits change. The cycle repeats.

Technological progress and people's emancipation from nature are a secondary theme in this story. The effects of humanity's ongoing relationship with water are not merely written in rivers. They are etched into the fabric of society, into the beliefs, behaviors, and systems that regulate everyday life. What is most engineered is not landscape, but political institutions.

The central argument of this book is that humanity's attempts to organize society while surrounded by moving water led people to create institutions, which tied individuals together in mutual dependence as they tried to deal with their environment. From countless variations over centuries, the republic emerged as the most successful mechanism to mediate the modern concerns of individual freedom and collective benefit in the face of water's overwhelming force. The argument is not strictly deterministic: water by itself could not have "determined" the form of political institutions. However, institutions did emerge, at least in part, so that society could express its agency over a changing environment. In that sense, the heart of the story of water on the planet is a political answer to material conditions.

Seen through this lens, the roots of modern society's relationship to water go far back in time. The story begins when, ten thousand years ago, humans took the crucial step of becoming sedentary. By then, *Homo sapiens* had already been around for three hundred thousand years, but from a fixed place of observation, the full force of water became overwhelming. Droughts interfered with food production. Storms disrupted people's lives. Floods destroyed communities. Because of water's force, individuals had limited power in controlling their environment. Rather, society as a whole had to learn how to exercise its own power.

Over the course of human history, life on the water landscape forged a social contract. Water is the ultimate res publica—a public good—a moving, formless substance that defies private ownership, is hard to contain, and requires collective management. People developed institutions that required mediating individual desires and collective action in the face of water's force. Those institutions eventually became dominant across the modern world. Legal and political systems, the territorial nation-state, finance, a system of trade, all evolved over thousands of years, while communities tried to ensure they could survive—even harness—the force of water in service of a commonwealth. Without understanding where those ideas came from, and how their development related to water, it is impossible to make sense of why and how the landscape looks the way it does today.

Part I of this book follows the dialectic relationship between the water landscape and human society from the neolithic to classical antiquity, showing how it contributed to shaping statehood. Part II then shows how—over a thousand years—antiquity was metabolized by European nations into the modern state. The legal legacy of Rome, classical republicanism, political liberalism, the sirens of utopianism, all mixed to inspire institutions, from the American Republic to the British Empire, which set the stage for the twentieth century. Part III describes how the power of the modern state and the force of industrial capitalism led to the most radical transformation of landscape in history. Its success was so complete as to make society's relationship with water invisible, hidden under the fabric of modern life, and sowing the seeds of the dangerous illusion that governs the present. Part IV, the last one, describes how, below the visible surface of a society that believes itself separated from nature, the undercurrents of water's agency still flow as powerful as ever.

Such a millennial story is not just an account of events and physical constructions. It is a story of ideas. In fact, it is impossible to explain the former without the latter. Three Gorges Dam, for example, was first the product of Dr. Sun's dream. Sun Yat-sen is often referred to as the father of the Chinese nation. He was a character of extraordinary complexity, a voracious intellect, a lifelong radical, a charismatic leader. Dr. Sun's life exposed him to a broad spectrum of cultures, tracing a path from his birth in a village in Guangdong, to his early schooling in an Anglican college in the kingdom of Hawaii, to his training as a medical doctor in Hong Kong in the 1880s. Along the way, he converted to Christianity. He was a physician, but his gift was revolution.

Dr. Sun was inspired by the profound transformations of the end of the nineteenth century, caught between British imperial aspirations and the utopianism of a modern, industrial society mesmerized by the echoes of classical republicanism. He sought to first reform and then overthrow the reactionary Qing regime in China. During years of exile and failed revolts, his anti-reactionary fervor grew. Like many modern revolutionaries he was intimately familiar with the history of Western political thought. He embraced ideals of emancipation and justice, admiring the French, American, and British constitutional settlements, even while resenting the policies those powers pursued. After the revolution of 1911, the Qing finally overthrown, Dr. Sun became the president of the Provisional Government of the Republic of China. The opportunity to make his dreams a reality had arrived.

Alas, China's first modern republic quickly descended into chaos, as the old military elites turned to dictatorship. Unable to realize his utopian vision for the future, Dr. Sun moved to the French concession in Shanghai and wrote about it instead. *International Development of China* was his blueprint for the country's economic rebirth. His point of reference was America. Sun Yat-sen proposed to "make capitalism create socialism in China so that these two economic forces of human evolution will work side by side in future civilization." His political philosophy required transforming the water resources of China. He compared the potential of the Yellow River to that of the Mississippi, imagining a delta designed to mimic the jetties of New Orleans. He imagined improving existing canals and embankments, constructing new waterways, hydropower, and irrigated agriculture. Then, he imagined a dam in the middle of the mainstem of the Yangtze, at three incised gorges, to "form locks to enable crafts to ascend the river as well as to generate water power." It was 1920.

Dr. Sun was not an engineer, but the interpreter of ideas that stretched as far back as human history. His was the dream of a utopian and a revolutionary. The Three Gorges Dam that stopped the flood in 2010 was not about catching up to the latest technology. The dam was the product of a society that had long chosen to tame the environment on an unprecedented scale. It was the product of a hundred-year-old dream steeped in republican values, one which spoke of commonwealth and progress, of rights of individuals and national aspirations, and which had crystallized long before the modern multipurpose dam had become a common feature of the landscape.

Dr. Sun's dream gave the idea of Three Gorges Dam the strength to persist through time, through Chiang Kai-shek's Nationalist government, through Mao Zedong's era, through Deng Xiaoping's reforms, to, finally, Li Peng's premiership. Once built, the dam seemed to prove that those living downstream could sleep soundly in the knowledge that something powerful watched over them. The significance of that security was in its political intent. Its engineering had become an instrument of the state in creating an illusion of final emancipation from nature in service of a commonwealth. The question is what happens when—not if—the illusion of emancipation is shattered.

At the dawn of the twenty-first century, humanity has become a force on the planet so powerful that some have renamed this era "the Anthropocene." But that has not heralded the conquest of nature. Far from it. The profound modifications inflicted on the planet have tightened, not severed people's relationship with water. The increase in greenhouse gas concentrations in the atmosphere is having a measurable impact on the energy balance of the planet, modifying Earth's water cycle. The extraordinary rainfall of 2010 on the Yangtze basin was a harbinger of much more to come. Changes in the climate system will eventually shatter the illusion of any final emancipation from nature. When they do, what will be concerning, above all, will not be the flaws they expose in the engineered landscape, but the societal response they stimulate.

The success of a republic in managing the tension between individual liberty and collective action rests on fragile and unstable foundations. By destabilizing those foundations, the water events of the twenty-first century could have profound political consequences. What compromises people will be willing to make in order to achieve further security in a newly uncertain world—what sacrifices to individual liberty they will endure and what choices they will make in the pursuit of a collective benefit—will determine whether the unstable balance between freedom and commonwealth can be preserved. That is, above all, what is of fundamental importance to everybody's future.

The questions posed by having to manage the power of water on the planet are not primarily technical, scientific, or even aesthetic. They are fundamentally questions about power, about who gets to decide what happens in everyone's home. The answer is often found in the minds of radical dreamers. Dr. Sun's dream a hundred years ago led to the Three Gorges Dam. Similar dreams created the modern world. To imagine what kind of future current dreams might bring, it is crucial to understand humanity's relationship to water, the most powerful agent of the climate system on Earth. For that, the combined story of people and water—a biography of water—matters a great deal.

PART I

ORIGINS

I

Standing Still in a World of Moving Water

HYDRAULIC CONTROL

Long before Earth ever formed, the subatomic particles that emerged from the Big Bang's first instants formed a plasma of hydrogen and helium. Gravity pulled them together in a nuclear fusion that fueled the first stars, the furnaces that forged heavier elements like oxygen. In the proto-stellar material left by the death of those first stars, hydrogen and oxygen reacted. They produced water.

That is why water is everywhere in the solar system. Saturn, Uranus, Neptune, Mars, Jupiter, and many of their moons were formed from a solar nebula that contained water, the debris left by previous generations of stars. But Earth could not have begun its existence covered in water as it is today. The inner solar system, the portion closest to the Sun in which Earth coalesced four and a half billion years ago, was too hot in its early phase for liquid water to survive on the surface of a planet. So, whatever water is found on Earth today must have either arrived after it cooled (carried by asteroids) or been released as vapor from the planet's interior. Either way, the amount of water on Earth has been fixed ever since.

If distributed on the Earth's surface as a uniform, liquid layer, it would be just over two thousand and seven hundred meters thick. That may sound like a lot, but when compared to the radius of the planet—over six and a half million meters, a thousand times more—

it is hopelessly thin. Today, 97 percent of that water resides in the oceans. Almost all of the remaining 3 percent is divided between ice caps and groundwater. If liquefied, the first would form a layer of about sixty meters, while the second would amount to about twenty. What remains—a tiny fraction, less than a fiftieth of 1 percent—is the water contained in lakes, rivers, and soils, which creates the environment that surrounds all terrestrial creatures, including people. If spread over the planet, it would be less than half a meter thick. The amount of water vapor in the atmosphere, a crucial quantity in this story, is less still, just two and a half centimeters, while the ice and water droplets that form the clouds floating in the sky would contribute a layer the thickness of a human hair.

The amount of water in each of these stocks has changed over the planet's existence, from an ice-covered world to one entirely ice-free, but human beings were not around for most of those changes. In fact, hominids appeared and grew in number during a period of relative stability in the planet's climate, over the last three million years. Still, very significant changes were occurring in the water environment during this time, the most significant of which were the ice ages, periodic changes in ice cover with a hundred-thousand-year cadence.

What drives ice ages are small, periodic variations in the planet's orbit around the Sun and in the tilt of its axis, which in turn modify the amount of energy reaching Earth. However small, the response of the planet is still extraordinary in human terms: twenty thousand years ago, at the time of the Last Glacial Maximum—the peak of the last ice age—ice covered much of the Northern Hemisphere, from Canada to Russia, and most mountains, from the Alps to the Himalayas. In many places, ice sheets were over a kilometer thick. Ice accreted enough water that the global sea level was about 130 meters below where it is today. Why and how relatively small changes in sunlight would result in such an enormous response is still a matter of substantial debate. But in almost all explanations, water itself plays a crucial role. Understanding this role requires appreciating how water interacts with the Sun's energy.

The Sun emits electromagnetic radiation on a broad spectrum of wavelengths, the peak of which is between a quarter and three-quarters of a micron. This is the frequency band which the human eye evolved to interpret as visible light. When this sunlight reaches the Earth's surface, it warms it up. The surface then emits back to space infrared radiation, which is at much longer wavelengths. Oxygen and nitrogen, which account for over 99 percent of the volume of the atmosphere, absorb and scatter visible light (hence the blue color of the sky), but are largely transparent to infrared radiation.

Were the atmosphere exclusively made up of those two gases, almost no heat would be trapped near the surface and the planet would be much, much colder. Water vapor, on the other hand, is largely transparent to visible light, but the water molecule, with its three atoms and slightly bent shape, happens to be particularly effective at intercepting and absorbing infrared radiation. As a result, water vapor acts as an enormous blanket over the planet, trapping outgoing heat: it is the principal greenhouse gas. Of all the forms in which water exists on Earth, by far the most fundamental is water vapor, for it is thanks to its presence in the atmosphere that the planet is habitable.

But water is not just a powerful greenhouse gas. It is also an amplifier of change. The atmosphere absorbs water vapor until it saturates, but the saturation point is itself a function of temperature. The higher the temperature, the more water the atmosphere can absorb: 7 percent more water for every degree of temperature increase, in fact. The more water there is, the more the atmosphere becomes opaque to infrared radiation. The more it is opaque, the higher its temperature. This is called the water vapor feedback. It is a powerful amplifier.

A small change in sunlight such as that associated with orbital changes (or, for that matter, a small change in the concentration of carbon dioxide) would have by itself a commensurately small impact on the temperature of the planet if it wasn't for the water vapor feedback. But because of the latter, a small change in temperature increases the amount of water in the atmosphere, in turn amplifying the temperature change further. Earth's climate is sensitive because there is water in it. Earth's climate is controlled by water.

SAPIENS AND THE GREAT MELT

The first chapter in the story of water and people must describe the role this consequential, ubiquitous substance played in the development of complex societies. As colossal as the peak of the last ice age was twenty thousand years ago, its impacts on people were even more remarkable when ice began to melt. In the Northern Hemisphere, ice sheets began retreating around nineteen thousand years ago. Melting was punctuated by a series of abrupt regional shifts. Between fourteen and eleven thousand years ago, for example, there was a cold spell called the Younger Dryas. It was named after the *Dryas octopetala,* a flowering shrub that thrives in the cold. Fossil traces of the plant had revealed to scientists that the spell had happened. During the Younger Dryas, the climate of the Northern Hemisphere, particularly in Europe, reverted to glacial conditions for a thousand years, before warming again.

As water streamed down from the ice sheets, it shaped the landscape. It ground down mountains, cut valleys, flooded plains, and created coastlines. To be clear, no one would have experienced these phenomena as a sudden shift: at its peak around 12,000 BCE, melting caused sea levels to rise four meters per century, which is four centimeters per year. But changes would have been measurable over the course of a lifetime, particularly a human one.

The population of *Homo sapiens* swelled in Africa around 130,000 years ago, between the last two ice ages. It ultimately replaced all hominid species: *erectus, heidelbergensis,* and *neanderthalensis.* However, any evidence of human culture—anything beyond mere existence, that is—has reached modernity almost exclusively from the last twenty thousand years, right when the planet was leaving the Last Glacial Maximum. By 5000 BCE, sedentary farming was established, various forms of proto-writing had developed, and complex societies were well on their way. So, the years between roughly 18,000 BCE and 5000 BCE were not just years of great change for the water landscape. They were also critical for humans to set themselves up as organized societies.

Early humans emerged from the last ice age still nomadic foragers and hunters, as they had been for all of their existence up to that point. But, as the population grew, megafauna like bison or mastodons, all of which represented a primary source of food, disappeared. Why this so-called Pleistocene extinction occurred at all is a matter of some debate. It may have been the result of the effectiveness of *Homo sapiens* as hunters or of environmental change.

Be that as it may, hunter-gatherers had to broaden their diet to survive, foraging a wider range of food. These early foragers moved around, relying on multiple food webs and highly productive ecosystems like wetlands and forests. Because the productivity of ecosystems waxed and waned with climate, the demographics of these foraging communities followed. Natufian communities in the Levant, for example, thrived during warmer periods, but became more ephemeral as the Younger Dryas set in.

And then, the transition to sedentary farming happened. The first step in that transition was the domestication of plants, evidence for which dates back to the very end of the Last Glacial Maximum in Israel, long before any stable transition to full-scale farming. In other parts of the world, domestication happened later. In north China, for example, millet seems to have been domesticated around 8000 BCE. The simplest hypothesis for how domestication happened is trial and error: given the small number of wild plant and animal species suitable for domestication, it probably took some time for foragers to happen onto the right specimens.

Sedentary farming was the next step and implied creating artificial ecosystems. That is when people locked their journey to water. Any productive ecosystem, natural or constructed, needs enough water to survive. Early communities faced different options, depending on where they were. Dry farming was entirely reliant on rainfall, and would have been the most straightforward form of agriculture. But it was labor intensive and needed abundant rain. Flood-recession agriculture, in which farmers took advantage of the moisture and nutrients left behind by the receding flood of a river, was less labor intensive, but did expose communities to malaria and the destructive power of floods.

Between the ninth and eighth millennia BCE, as it turns out, the first communities to transition to sedentary agriculture relied on dry farming in the "fertile crescent," a rainfed area that traced an arc from central Israel, through Lebanon, along the southern side of the Anatolian plateau and down on the other side of the Tigris River, along the Zagros Mountains. Rather than adjusting to a changing environment, these communities became a force controlling nature in service of the plants and animal species they had domesticated. It was the Neolithic revolution.

Sedentism required human societies to make radical adjustments to their relation with water. Rivers meander, landscapes change, floods and droughts may completely alter the ability of an ecosystem to support a community. While nomads could relocate in the face of these changes, sedentary people were unable to move and had to modify the surrounding environment to fit their needs, or suffer the consequences. That was the true Faustian bargain that society made when it transitioned to stationary farming: it chose to tame an unstable, dynamic environment. The journey of modern man had begun and the distribution of water had defined its starting point.

PRODUCTION BEGINS

Sedentary agriculture changed human society. Most natural ecosystems do not maximize digestible calories for humans, but farming can. The numbers are stark. The basis for agriculture was the cultivation of grains. Unlike other crops—legumes, fruits, or tubers—grains were particularly well suited for constrained landscapes. They were highly productive on a calories-per-hectare basis. A small territory under the control of a sedentary community could yield well beyond subsistence levels. Grains could also be harvested all at the same time, which meant that they could be more easily packed and stored. Ancient dry farming could produce about six hundred kilograms of grain per hectare. With irrigation and multiple cropping, farmers could obtain up to two thousand kilograms of grain per hectare, providing a hundred times more calories than grazing cattle, for example. As a result, farming could support more people than nomadism.

Nomads needed several hectares per head of cattle and had to constantly relocate, so having too many mouths to feed would have strained their scarce resources. Those limits were not an issue for sedentary communities. More children meant more people to feed, but they were also an insurance policy against early mortality. Living in settlements inevitably increased exposure to water-borne diseases like dysentery and cholera. A number of diseases crossed over to humans from the animals that lived with them. As a result, sedentary communities entered a demographic regime of high fertility and high mortality, through which the population developed resistance. This gave sedentary populations a competitive advantage over others. The sedentary population grew.

Between the sixth and the fifth millennium BCE, the earliest settlements were in northern Mesopotamia, far from large, dangerous rivers and near spring sources. Initially, communities were sparse, ten or fifteen people per square kilometer. As the settlements grew, they did so in clusters, surrounded by uninhabited areas. Hierarchies among settlements developed. The landscape began to specialize, with some areas used for grazing, others for growing crops, mostly barley and wheat. Smaller settlements might be one or two hundred people, their size limited by the scale of social interactions. Eventually, larger, walled centers might reach a few thousand people, again limited by the environment and the potential of the local economy. Transport of cereals was limited to three to five kilometers, because everything had to be carried by mule or on foot. This was not an integrated economy.

Rapid expansion of organized society had to wait for the southeastern plains of Mesopotamia. There, richer ecosystems with greater carrying capacity allowed higher concentration of people. But this transition could only happen once the melting world had fully stabilized. At the end of the Last Glacial Maximum, the Persian Gulf was dry from the Strait of Hormuz all the way up to modern Kuwait. Then sea levels began to rise. A few meters' sea level rise per century might seem like a relatively small affair. However, because of the continental slope into the Gulf, such vertical displacement would have moved the waterline back horizontally one or two kilometers. Once the water reached the shelf, which is significantly shallower and with a much lower incline, a ten-meter change in sea level would have moved back the shoreline somewhere between a hundred and two hundred kilometers. This rate of change was too fast to allow any coastal ecosystem to stabilize. Productive wetlands or estuaries could not form along the moving coastline until the sea level rise had stopped.

That happened around 5000 BCE. At that point, the water flowing down from the glaciers and across the landscape brought nutrients to coastal ecosystems, transforming them into remarkable sources of food. Estuarine fisheries alone could yield up to a ton of fish per hectare per annum, which in caloric terms is comparable to the productivity of dry-farmed agriculture. As the coastline stabilized, the landscape of southern Mesopotamia also formed its distinctive structure. During the great melt, rainfall had been high in this part of the world. Full rivers had brought additional sediments, which in turn had lifted the riverbeds. Close to the coast, saltwater intrusions mixed with freshwater to create a vast system of marshlands. Wetlands and marshes, particularly those in brackish waters, are among the most productive ecosystems in nature. They offer a myriad of food sources, grazing for cattle, resources for communities. Lower Mesopotamia turned into an area of unparalleled richness and productivity.

The communities that formed during this, the so-called Ubaid period of lower Mesopotamia, took advantage of being at the interface between saltwater and freshwater, engaging in water transport and irrigation as well as fishing and growing salt-tolerant crops. They acted as an interface between the earliest farming communities of northern Mesopotamia and the rich ecosystem of the Gulf. The explosion of their population led to the specialization of labor. Centers began to grow. Real towns emerged. Eventually a system of interconnected communities gave way to the first city-states of deep antiquity.

LIVING WITH WEATHER

Social complexity stemmed in part from the need to organize to confront water phenomena, whose scale far exceeded the agency of any individual. The scale of those phenomena, in turn, was a consequence of the physical properties of water. In an average year, seventy centimeters of liquid water fall down, which means that the entire stock of water in the atmosphere cycles through it almost thirty times over that period. In that process, water transfers vast amounts of energy from the surface of the planet into the atmosphere, warming it. The energy involved in weather associated with those cycles of water can overwhelm all human activity, even today. If the energy used in the global economy—all transport, power plants, homes, heating systems—were one unit of energy, then the water cycle of an average hurricane releases roughly one unit, the Asian monsoon about ten units, and global annual precipitation several thousand units. Water overwhelms humanity.

The reason for this awesome power is an astrophysical accident. Earth is the only planet in the solar system to have a particular combination of mass and distance from the Sun, which produces average temperatures and atmospheric pressures that keep the planet close to water's triple phase point, the temperature and pressure at which liquid water, ice, and vapor coexist in precarious equilibrium. Because of these conditions, water can experience all phase transitions within a range of temperatures and pressures commonly found on Earth: from ice, to liquid, to vapor, and back.

In those transitions, water has the highest latent heat of any common substance found on the planet. For example, the energy it absorbs in transitioning from ice to liquid is far greater, for a comparable weight, than what iron, gold, or silver need for liquefaction at their respective points of fusion. Similarly, the energy needed to evaporate water is almost six times that of benzene and ten times that of petroleum. If one had to design the perfect molecule to transfer energy on Earth, water would be it. These phase transitions power the weather phenomena that shaped the development of the first fragile, sedentary communities.

For example, the East Asian monsoon is one such powerful weather phenomenon. It is so powerful, in fact, that in the twenty-first century, China built Three Gorges Dam to manage its rainfall, which gives some measure of how disruptive the monsoon must have been to China's earliest communities.

To understand what those communities faced, it is helpful to know how the monsoon behaves. The East Asian monsoon has both summer and winter phases. During summer, tropical storms form a band of rain running along the coast, from the southwest to the northeast of China, all the way up to Japan, bringing the Mei-yu, the "plum rains." These storms release vast amounts of energy, driving the monsoon. Its surface branch draws air across the equator from Indonesia and Borneo, bringing water into China. Meanwhile, above the Tibetan Plateau, a huge anti-cyclone sucks air upwards, generating storms that pour rain into the Yangtze basin. As the whole front moves inland towards the northeast, it switches from the Yangtze to the Yellow River basin, persisting until late September. During the winter phase, a large high-pressure system over Siberia pushes cold air southward, replacing tropical storms, drying and cooling north China. The cold air then crashes into the Tibetan Plateau and into the moist warm air of the southeast, generating winter storms and cold surges over the South China Sea. The cycle repeats.

The distribution of rain and winds associated with the monsoon was at the heart of the development of early Chinese society. The story of China began in the north, where the hydrology of the Yellow River was controlled by this monsoon. Neolithic communities began settling in the mid-basin around the middle of the fifth millennium BCE, supported by a mix of low-intensity farming, grazing, and foraging. Those Neolithic cultures spread to the northeast Tibetan Plateau between the fourth and third millennia BCE, when abundant rain and a warm climate encouraged greater agricultural production.

The heart of Chinese agriculture was the Loess Plateau, a large flat platform the size of France, a thousand meters in elevation, which stretches over part of the mid-river. It is the largest such formation on the planet. The plateau is covered in a few hundred meters of loess, a loose deposit of yellow fine silt, mostly brought by wind from the Gobi Desert to the west, during the winter phase of the East Asian monsoon. When this loose material makes it into the Yellow River, it suspends in water, contributing the color that gives the river its name.

Over the centuries, as the monsoon changed in strength and location, it transformed the environment of the mid-river. Geographical shifts in precipitation, if sufficiently long-lasting, changed vegetation. During much of the fourth millennium the mid–Yellow River was wet, but starting in the third millennium the monsoon moved south. The resulting drought in the mid-river forced populations to move down in the lower valley, while further east, parts of the upper river transitioned back from rainfed agriculture to pastoralism.

The cooling and drying of northern China created a transition zone that cut through the Loess Plateau, from the northeast to southwest, separating grazing in the steppe from sedentary farming. From then on, nomadic people occupied the middle and northern parts of the middle basin, while farmers withdrew to the southeast and kept expanding to the lower river. That boundary has moved throughout the history of China following changes in the monsoon. Wetter climates meant that more of the upper river could be farmed. Drier climates pushed that boundary further downstream.

With each oscillation, the Yellow River saw transitions between Chinese sedentary farmers and the nomadic populations of the steppes. The tug and pull between them at the margin of the monsoon amplified its impact on sedimentation in the Yellow River. As farming moved in and out of the coarser sediment, the amount of mud and silt that monsoonal rains washed into the river changed, altering the speed at which the channel would rise as a result of the deposits. That affected the rate of breaching of natural levees downstream, which pushed the population to engineer artificial ones to protect agriculture. The example of the East Asian monsoon shows the complex route through which large and powerful weather phenomena could both drive long-term changes in the landscape and provoke the response of the population.

What is true for China is true for most societies on the planet. Storms, hurricanes, and monsoons feed off energy released by water vapor condensing into rain or snow. These powerful weather phenomena are capable of transforming the environment through floods and drought, and of doing work which can overwhelm human activities. In response, early communities adapted, beginning a process of developing institutions to manage this uncertain environment, which continues to this day.

MEMORY

The fundamental struggle with water has never really abated since it first began on the shores of the Persian Gulf. The multiple transitions, from nomadism to sedentism, from hunting and foraging to domesticated agriculture, from small rural communities to a productive, specialized, urbanized society, were severe disruptions. But while individuals would have lived through them as gradual, incremental transformations, over the course of *Homo sapiens'* existence, they amounted to shocking events. From the moment *Homo sapiens,* late in its history, decided to stay in one place, surrounded by a changing environment, it began to wrestle with water, an agent capable of destruction and life-giving gifts.

The reason the early story of water and society matters is that it has left deep cultural traces guiding and inspiring human adaptation ever since. For example, given the experience of the early Chinese communities described above, it is not surprising that water myths are abundant in that culture, and have captured the role of the water landscape in Chinese identity. One Chinese myth tells of how the world formed from the body of a giant, whose blood and veins turned into water and rivers. In another, the Jade Emperor, Lord of Heaven, entrusted four great dragons to bring rain to the people. Their names were Long, Yellow, Pearl, and Black. After they disobeyed him, he entrapped them in mountains, so the dragons turned themselves into rivers, becoming the Yangtze, Yellow, Pearl, and Amur Rivers, the great historical sources of water for agriculture. These are the cultural traces of the great East Asian monsoon.

Chinese culture is not alone in recording its relationship with water. Stories from the past reflected the concerns of societies dealing with the overwhelming power of water. For example, floods are exceedingly common as a foundational myth. The Lenape, the original inhabitants of Manhattan, were believed to be descendants of the survivors of a great deluge who had escaped by riding the back of an ancient turtle while following a bird to dry land. The Navajos believed they originated from Insect People, chased away by gods that sent water to cover the land, and who had been guided by a swallow in their escape. Both found in the story of water a powerful source of identity.

When in the sixteenth century Cristóbal de Molina and Sarmiento de Gamboa spoke to those who had survived from Cuzco, they learned of the Unu Pachacuti flood, which covered the land around Quito. Similarly, Maya populations left accounts of a mythical flood that destroyed the race of men and established a new order. In Scandinavia, the Old Norse frost giant Bergelmir was the only one to survive a destructive flood, saving himself and his wife by boat. Even the aboriginal societies of Australia, isolated on the continent for fifty or sixty thousand years, in their ancient songs told of a time when much of the coast was dry land, and of how the waters came to cover it.

It is unlikely that these flood myths all refer to an actual, synchronous memory, for example that of the melting of the Last Glacial Maximum. But such a wide canon of evocative water-related stories is an indication of just how traumatic those adjustments must have been for early societies.

The stories from Mesopotamia, in particular, speak to the oldest memories of humanity's relationship to water. In a famous tablet from Nineveh, a schematic river, the Euphrates, flows through the center of a map, and through Babylon. Below the city, the river ends in marshes, a canal, and, finally, the sea. One image, a few etched marks on unfired clay, captures the significant elements for early societies: the river and the water landscape as whole systems with the life of an organized people.

What propelled the Mesopotamian records like this one to unprecedented modern fame was one tablet in particular. It told an episode of an Akkadian epic in which its protagonist, Gilgamesh, had set out to find the patriarch Utnapishtim, "the faraway." When he found him, Utnapishtim told Gilgamesh of how he had saved himself from a destructive flood by building a vessel, of how he had received a divine mandate to save animals, of the endless floating, of a dove sent out to seek land, of landing, finally, on the mountain in Urartu. It was Noah's story from the Old Testament.

When Assyriologist George Smith broke the news of this discovery in 1872, it hit the public like a thunderbolt. Here was what seemed to the Victorians as independent confirmation of the biblical story, from about a thousand years before the writing of the Old Testament. The great flood, the deluge, the story of the destabilizing havoc that an immense amount of water supposedly wreaked on a prehistorical society. For forty days and forty nights, an implausibly old Noah, his family, and seven pairs of each animal on the planet floated in the ark, as God let water out from "the springs of the deep and the flood gates of the heavens." The Akkadian story of Utnapishtim and the flood mesmerized the world. Newspapers went straight for the sensational. In a moment of lyrical abandon, the *Daily Telegraph* and the *New York Times* both ran this line: "History was jealous of Romance until last week; but then suddenly she gave to the world, by the marvelous skill of a scholar in the British Museum, a fragmentary story far more wonderful and entertaining than any work of fiction."

That societies produced these composite allegories of the relationship between the powerful and unbridled force of the environment and human society supports one basic conclusion: the story of the struggle with water is the root of organized society, and it started when people stopped and noticed that the water around them moved.

2

The Rise of the Hydraulic State

WATER AND THE STATE

From the fifth millennium BCE onwards, the earliest states began leaving consistent traces of themselves. Societies governed by states produced art, science, and writing systems. Those traces are what can be relied on to reveal how the development of the earliest states was shaped by the water conditions they evolved in.

Now, the word "state" conjures modern ideas of political representation or huge, bureaucratic institutions, but the early states in which sedentary communities evolved were not comparable to modern ones in size or scope. Yet they shared with them basic traits, including an army and the ability to collect tributes and redistribute resources. They fought each other, engaged in diplomacy, relied on ideology to reinforce allegiance and on administration to overcome tribal relationships. To be clear, there was nothing inevitable about the rise of states. In fact, many societies remained stateless well into the eighteenth century, surviving in small self-sufficient communities or nomadic tribes. Some are still stateless today. But societies that organized in states eventually outpopulated all others.

The first—the Sumerian city-states, followed by the Akkadian Empire, which loosely coalesced into a Mesopotamian civilization—are often described as a "river valley civilization," because they developed on the banks of the Tigris and Euphrates, just like the Egyptian state would later develop along the Nile, or the Xia and Shang state on the Yellow. However, the narrow view that attributes the origin of the state solely to river management is largely without empirical grounds. In fact, quite the opposite: large rivers are not essential to explain the rise of complex states. River management was not at the root of the Maya forest civilization, for example, nor did the states of northern Mesopotamia have to deal with a river. And, vice versa, stateless communities continued to exist along rivers for centuries. "River valley civilization," it turns out, is a geographical descriptor, not an explanation of why states happened.

But explaining *why* things happen is not the same as explaining *how* they happen. The water conditions in which a state developed interacted extensively with its institutions. Water moves. Actions by those living upstream affect those living downstream. People had to work together to respond to the power of water. In a world drenched with it, sedentism encouraged cooperation, and states, however they arose, had to adapt to it. In this sense, Mesopotamia is indeed where the first hydraulic states developed.

In the story of the Tigris and Euphrates, there is a tight relationship between the nature of the state and how the water landscape evolved. Water obeyed the boundary conditions imposed by the climate system. People, on the other hand, organized into states to exercise collective power and change the environment to their benefit. Environmental conditions did not cause the rise of the state, but they did help to shape it. From the first city-state, to a multitude of competing states, to the very first empire, the story of Mesopotamia is the story of how these two interacted.

MIDDLE EARTH

The land was once known as Al-Jazirah, "the Island" between the two rivers. "Mesopotamia" comes from the Greek *mesos,* "middle," and *potamos,* "river." The name first appeared in a second-century CE history of Alexander the Great by the Greek historian Arrian of Nicomedia.

The Tigris and Euphrates Rivers have been protagonists of the story of the Near East since the Neolithic. They were the setting for the emergence of the first states, and it stands to reason that their behavior would have contributed to shaping them. The sources of these consequential rivers were shrouded in mystery. In the Babylonian poem *Enuma Elish,* the god Marduk undammed the eyes of the slain water goddess Tiamat, unleashing the rivers. But their sources are far less mythical, if equally powerful. The water that feeds them is the result of the complex interaction between the circulation of the atmosphere and the nearby Mediterranean Sea.

The Earth's climate is more or less in what scientists call a state of radiative-convective equilibrium, thanks to water's dual role as greenhouse gas and as a vector for latent heat, which is released into the atmosphere when vapor condenses into rain. Both effects depend on the amount of water vapor in the air column, which is in large measure controlled by how much evaporates from the surface, itself a function of sunlight reaching the ground. Therefore, the temperature of each air column is mostly dependent on the amount of sunlight reaching the surface at that particular point.

Because sunlight intensity is lower at the poles, higher at the equator, the resulting north-south difference drives a circulation in which colder, heavier air from the poles slides towards the tropics, while warmer, more buoyant air flows above it towards the poles. But air moving towards the poles also moves closer to the planet's axis of rotation, accelerating eastward relative to the surface of the planet (the same conservation of angular momentum that a skater relies on when she draws her arms in close to her body to accelerate her spin). The result is a powerful jet stream, at 40 degrees latitude, 10 kilometers or so up in the atmosphere, flowing at about 180 kilometers an hour relative to the surface.

The jet stream circumnavigates the globe and is strongest during winter, when the equator-to-pole temperature difference is greatest. During that season, its core becomes highly unstable, and perturbations grow, forming bands of weather systems, troughs, and ridges associated with the familiar cyclones and anti-cyclones of weather maps. These bands of turbulence and instability are "storm tracks." During winter, storms that originate around the jet stream enter the Mediterranean basin and follow the axis of the sea from west to east. This secondary storm track passes over warm waters, which act as a fuel injector for the storms.

As storms approach the southeastern corner of the basin, those that veer northward draw moisture from the eastern Mediterranean and crash into the Taurus Mountains of southern Turkey and the Zagros Mountains of eastern Iraq and western Iran, releasing their water. The storms that swerve southward move over the hot, dry Sinai desert, bringing hardly any rain at all. As a result, winter rainfall is distributed along a very steep north-to-south gradient across Mesopotamia and the Levant. During summer, these storm tracks weaken. Local weather turns dry as the Indian monsoon generates large-scale standing waves across the atmosphere, including a persistent rainless region of high pressure over Mesopotamia.

The flows of the Tigris and Euphrates experienced by the early communities of Mesopotamia were the result of these annual cycles. They fed on the abundant winter rainfall in the northern mountains and traversed the persistently arid south. Despite their common origin, the rivers were not the same. The Euphrates had a relatively stable flow and was what river morphologists call an anastomosing river: before modern engineering constrained it, it separated into multiple smaller, meandering channels. The Tigris, on the other hand, was bigger by at least 50 percent, as a number of tributaries from the Zagros Mountains added to its flow. Silt suspended in its water was three times that of the Euphrates, making it a dangerous river of mud. Its floods were so violent that they could drown most artificial structures of antiquity. If given the choice, people would have found it easier to take someone else's territory along the Euphrates rather than to dig a canal from the Tigris to move water.

The rivers traveled through a fertile plain. Even with the limited technology available in antiquity, the Euphrates alone could support irrigation on land in excess of a million hectares. The two rivers combined, if tamed, could in principle reach up to three million hectares. But both rivers presented a similar problem. Both had strong peaks in the spring, when the snow left by winter storms in the high mountains melted. As a result, the timing of their flow was wrong. Their peak flow did not correspond to the needs of irrigated agriculture: floods would happen when grains were ripe and ready for harvest. This timing meant that farmers had to do a lot of work to transform the Mesopotamian plain into a viable environment for agriculture. And that is the story on which hinges the evolution of the first state.

GROWING WITH WATER

Below the rainfed north, the Tigris and Euphrates left their respective valleys, entering the floodplain. For every kilometer the rivers traveled towards the Gulf, they dropped by five or ten centimeters at most. The floodplain was so flat that, had farmers attempted to use its incline to irrigate, they would have had to dig canals over forty kilometers long to generate enough pressure. No individual community of the fifth millennium BCE would have had the labor force and territorial control to pull off something like that. But in this case, nature provided the solution.

On the flatlands both rivers slowed. They became unstable, meandering and increasing their length with every bend, frequently flooding their surroundings. Because the amount of sediment that came down from the mountains was substantial, each inundation lifted the banks and bed of the river further. Over time, the rivers began traveling a few meters above the rest of the plain, held in by natural levees. This process created a landscape of elevated banks. In the direction perpendicular to the river, they were much steeper than the incline of the floodplain heading towards the Gulf.

People took advantage of this configuration, particularly settlements on the braided part of the lower Euphrates River. These communities dug short irrigation canals, taking off from the top of the embankment and running downhill across it, generating enough force to irrigate. The canals were about three to four hundred meters from one another. Over time, sediment from the floods filled the space between them, establishing narrow strips of cultivable land, a few kilometers wide, straddling the length of the river. The nature of this landscape shaped agriculture.

The space between the canals defined long fields of about thirty or forty hectares, which formed a herringbone structure along the river. Some of the long fields were divided further into plots of about two hectares each. This field structure could be plowed using large seeder-plows pulled by multiple oxen, digging furrows along the length of the fields and pushing seeds into the soil. The effect on productivity was dramatic: while rainfed agriculture in the north could harvest at most two or three seeds for every one planted, barely above subsistence, the system in the south could produce, under ideal conditions, twenty or thirty to one, a tenfold difference.

The problem was the timing of the flow. When irrigation was needed, the flow was low. It then peaked when it was time to harvest, not to irrigate. If the off-take canals had been dug sufficiently shallow to enable the low flow to spill into the irrigation system at the right time, they would have then easily overtopped when the flood came, putting at risk the entire harvest. The solution was to cut the canals deep enough to contain the flood, and then use temporary structures—dams made out of mud and reeds—to raise the water levels during low flow.

It was a labor-intensive approach. Canals needed continuous dredging. Soils needed care, because water applied to the fields evaporated, leaving behind salts, while the high water table prevented water from leaching the salts away. And so the water landscape shaped several traits of what communities did in southern Mesopotamia: the type of crops they grew, the maintenance of the irrigation system, plowing techniques. These activities in turn required institutions and social structure. The calendar had to be organized around a predictable, time-bound set of tasks. Their variety led to labor specialization, which encouraged more efficient production. Population concentrated, facilitating further specialization. Tax collection and food rationing became easier. Population grew. Villages turned into towns. Towns into cities. Bureaucracies emerged. As the first Sumerian city-states appeared, the river shaped their life.

It is worth pointing out that these processes were not unique to Mesopotamia. On the other side of the world, mud and silt flowed down from the middle Yellow River in northern China, depositing on the riverbed and raising it. Eventually the channel became unstable and the natural levees breached. A new branch of the river formed, while the old one got left behind. The cycle then started again.

As discussed in the previous chapter, the amount of sediment coming down the Yellow River was a function of the conditions under the East Asian monsoon. During the cool and dry climate of the second and first millennia BCE, grasslands kept much of the loess in place and sediments in check. When the monsoon shifted north or south, vegetation on the loess would follow it, changing the sediment yield and sedimentation rates. There too society and the river began a precariously balanced dance from which, once started, neither could be freed.

The complex interaction between the violent monsoon and agriculture transformed the Yellow River, making it far more challenging for the Chinese state than the Euphrates was for Mesopotamian states. Agriculture was supremely vulnerable to the continuous breaches of natural levees, so much so that even the earliest communities had to intervene to manage the river. Although no archaeological remains have been found of the supposed first dynasty of China—the second-millennium Xia—myths and legends survived in the Chinese canon that give some sense of just how difficult the environment was for those first communities. Yu the Great, the legendary founder of the Xia, supposedly drained the North China Plain, excavating nine waterways to the sea, and, according to the later philosopher Mencius, "the water that passed through the land is what is now the Yangtze, Huai, Yellow, and Han rivers." The stories of Yu rooted the management of rivers in legend, reminding everyone that the relationship with water was at the heart of the Chinese state.

The societies living in northern China, like those of southern Mesopotamia, confronted a complex water environment. The configuration of the river did not uniquely determine the path they took. But it did shape their journey.

MOTHER CITIES

The very first state was Uruk, in southern Mesopotamia, a city founded almost a thousand years before the great pyramids of the Egyptian Old Kingdom, two thousand years before the Trojan War, and three thousand years before Rome's republic. It existed for an extraordinary four thousand years, until it was abandoned in the fourth century CE. It was famous for its mighty walls—over six miles long and seven meters high—allegedly built by the great King Gilgamesh himself. Uruk grew out of an initial crop of towns that developed along the lower Euphrates. By the fourth millennium BCE it had become an important religious center, dedicated to the gods Anu and Ishtar.

Water did not just contribute to shaping the state in terms of its activities—calendars, public works, and so forth. Critically, it also shaped the structure of society. Thanks to the levee system, agricultural productivity had made a massive leap. In early Uruk, however, that wealth did not end up in the hands of private individuals. Instead, temples were the principal landowners. As agricultural yields grew, temples grew bigger and began organizing work on the long fields: they drew corvée labor from local communities, paying them grain rations in return. What was not used for rations was collected, centralized, and invested in public buildings or used to pay administrative elites, strengthening the state.

The early temple was a total institution, it was designed to capture most of people's life. It provided storage for food, a meeting place, a place of worship, an administrative center. The temples could exercise authority over people, even commanding corvée labor, because they were thought to intercede with the gods. Even that intercession was related to the water environment.

The Sumerians, the civilization that dominated southern Mesopotamia, had a theocratic mode of experiencing reality. Causes of events were always the realization of the actions of gods, which gave temples authority. This was particularly true for the water landscape. When the spring floods came down the river Tigris, for example, the red, muddy waters of the flood were evidence of the deflowering of the goddess Ninhursag, the "Lady Foothills," by Enlil, the god of the great mountains and of the winds. People went along with corvée labor and paying tributes, because the temple was the center of divine power.

In *Atrahasis,* one of the oldest Akkadian epics discovered among the tablets of the ancient library of Ashurbanipal in Nineveh, a world with no humans saw the gods organized in a hierarchy: lesser gods were forced to maintain canals under the guidance of the god Ennugi, the canals controller. Eventually, the gods, tired of having to do all the work, created man to do the digging for them. In other words, those who wrote *Atrahasis* believed that humans existed for the struggle of managing water. That is what gave authority to the temple.

Over time, surplus production made urban life more attractive. By the beginning of the third millennium BCE, 80 percent of the population was urban, a higher percentage even than today. At its peak, in the third millennium, the city of Uruk covered an area of two and a half square kilometers. Its population was around a hundred thousand, the largest city in Mesopotamia until Babylon surpassed it two thousand years later.

In the levee system, cultivable land was limited to a thin tract around the river. As Uruk grew, it began facing the limitations of its immediate surroundings. Besides, a society of growing wealth and power needed metals, wood, precious stones, all of which could not be found in lower Mesopotamia and needed to be imported. The easiest solution was to trade, taking advantage of fluvial transport. The canal system became a transport network.

Modern ideas of market forces do not easily apply to this early trading system. After all, trade was centralized because the state controlled the surplus to pay for imports. However, there were a few recognizable traits of modern markets. Commerce along the fluvial system depended on merchants, who inevitably escaped centralized control while traveling. At the point of exchange, evidence from stamp seals suggests that they behaved as rational agents in the modern economic sense, sensitive to price and profit.

The management of water had not only enabled the accumulation of surplus to support a state, but had also created the conditions for the rise of the first long-distance markets. As trade inevitably shortened distances between urban centers along the levee system, competition among them increased. During the Early Dynastic period between the twenty-eighth and the twenty-fourth centuries BCE, a series of city-states of comparable size and status displaced Uruk in dominance, to compete with one another over Sumer.

The network of canals around which Sumerian cities were organized shaped their international relations. A state upstream could supply or cut water downstream, allowing it to control its neighbor. The destructive force of water could be used as a weapon. That was the setting for the oldest known state conflict over water, between Lagash and Umma.

Lagash was a state between modern-day Baghdad and Basra, located in the southeast corner of modern Iraq, close to both the Persian Gulf and Elam. Girsu, Lagash's religious center, was caught between Umma, an ancient city to the northwest, and its allies, the cities of Ur and Uruk, to the southwest. The object of contention with Umma was a land parcel of a few thousand hectares called Gu'edena, which translates to "the Edge of the Plain." It was a "beloved field" of the god Ningirsu. A few thousand hectares might seem small, but area is a deceptive measure of importance in this context: on irrigated embankments, the wealth of a state was not the area of land it controlled, but the amount of linear embankment it managed. Umma had access to two hundred thousand hectares, but only twenty thousand were cultivated, because it could not move more than half a kilometer out of the levee. Small plots of land could become significant casus belli.

The conflict played out over several generations. According to Girsu's side of the story, the only one to survive, Umma had leased part of Ningirsu's land, but had failed to pay rent, accruing a stratospheric debt of over four trillion liters of barley. It is hard to imagine this was a real number: it would have corresponded to an annual interest rate of 50 percent over forty years. But the fact that Umma was unable to repay was probably real. To avoid settling its debt, Umma invaded. Because it was upstream, its strategy was to divert irrigation waters from reaching Girsu, starving its agriculture. In response, Eanatum I, king of Lagash, launched the construction of a sixty-kilometer-long channel, taking water "from the Tigris to the Nun-canal." This was probably the Shatt al-Gharraf in Iraq, an ancient waterway that still connects the Tigris and Euphrates today. Eventually Lagash defeated Umma, reclaiming the land.

The conflict had severe, unintended consequences. Eanatum's canal transferred water between the Tigris basin and that of the Euphrates, but the volume of water far exceeded the needs of Girsu. As a result, the water table rose, leading to salinization. Salinization is a process by which the magnesium, calcium, and sodium found in water accumulate in the soil, binding with clay and making the soil impermeable. In those conditions, plants struggle to germinate, and roots fail to absorb nutrients. Farmers tried to leach the soil by pouring more water on it, to grow deep-rooted weeds to dry it out and to drain it, all to no avail. Salinization could not be stopped. The impacts on productivity were catastrophic. Yields in the twenty-fifth century BCE had been over two and a half tons per hectare. By the eighteenth century BCE, they were below nine hundred kilograms per hectare, less than was needed to support a complex bureaucracy. Sumerian control over southern Mesopotamia collapsed.

Water had not only shaped the environment in which powerful states developed and competed—it had also sown the seeds of their destruction.

THE CURSE OF AKKAD

Over time, changes in water conditions could not only provoke an adjustment in society, but the development of new state institutions. Changes in rainfall in Mesopotamia, for example, could be felt over hundreds and even thousands of years. Why the climate system should shift on such timescales is unclear: it is not the result of any obvious external force. As with the Younger Dryas before, it was probably the result of complex interactions between ocean circulation, atmospheric variability, and ice cover. The end result, though, was reductions in rain of up to 30 or even 50 percent, which could persist for decades and even centuries in parts of the Mediterranean.

These changes left abundant evidence in the archaeological record. During Uruk's expansion in the middle of the fourth millennium, colonies of the southern states began growing food in the dry-farmed north. Some even reached significant scale, but at the end of the fourth millennium BCE, a prolonged period of drought caused many of these early societies to disappear. Precipitation in that same region increased again starting in the twenty-ninth century BCE, reaching an apex around the twenty-seventh century BCE. During that time urbanization and political integration really took off.

How water affects the state depends on the state's vulnerability. While states were geographically limited to a single productive community, their vulnerability was largely economic: water's availability and timing was a crucial determinant of how productive agriculture could be and how much surplus the state could rely on; occasionally floods might destroy some of the capital stock the state relied on. But as states grew in complexity, so did the routes through which changes in water conditions could affect the stability of their institutions. Although the initial perturbation could be localized in nature, it could then ignite a chain of events that strayed far from the original problem.

Rainfed agriculture and pastoralism were dominant in the north, but rarely productive enough to support endogenous state growth. Southern Mesopotamia remained the center of wealth. When a northern state arose, as in the case of Ebla in the second half of the third millennium BCE in modern Syria, it was wholly dependent on the existence of powerful trading counterparts to the south. Yet, between the twenty-seventh and twenty-sixth century BCE a number of northern cities did grow, reaching significant levels of wealth. That growth continued during the conflict of Umma and Girsu until, eventually, the first empire in history emerged.

Starting around 2350 BCE, Sargon the Great created the Akkadian Empire. "She conceived me, my en-priestess mother, in concealment she gave me birth, / She set me in a wicker basket, with bitumen she made my opening water-tight, / She cast me down into the river from which I could not ascend." This was not the beginning of Moses's story in Exodus but Sargon's birth legend, from about a thousand years earlier. From the city of Akkad, Sargon embarked on an unprecedented military campaign, during which he conquered much of Babylonia. He then united southern Mesopotamia, defeating Uruk and the once-great Sumerian city-states, finally venturing a thousand kilometers north, up the Euphrates, conquering all in his path, including Ebla.

Sargon's empire was based on trade more than subjugation. He sought access to wealth and had little appetite for bureaucratic control, mostly relying on pre-existing administration. Halfway through the twenty-third century, his successors extended Akkad's domain fully into northern Mesopotamia, incorporating its dry agricultural economy. Sargon's grandson, Naram-Sin, stretched Akkadian dominion over southwestern Iran, northeast Iraq, and even northeastern Syria, standardizing bureaucratic institutions and taxing all reaches of the empire. The network of states under Akkad was integrated by waterways, the connective tissue of this vast empire. The grain trade was centralized from the four corners of Mesopotamia and connected to an international network for metals and other goods that, through the Gulf, reached Oman and even the Indus delta.

The problem with such a sophisticated system, dependent on trade across a vast region, was that it was vulnerable to shocks to any of its corners. In the twenty-third century, about a hundred years after Sargon first established the Akkadian Empire, the northern plains went dry again. As had happened before, changes in rainfall deteriorated the productivity of rainfed agriculture. Controlling northern Mesopotamia, an area of three million hectares, required an army, administration, transport infrastructure. Without surplus production, its complex administration could not be sustained.

The empire unwound quickly. In the Khabur plains of northeast

Syria, for example, villages were abandoned suddenly, constructions left half-complete. In a domino effect, the centralization of grains collapsed. The state spiraled out of control. Because of the high degree of integration, the impacts on Akkad propagated through the empire. The "Curse of Akkad," a legend dating to the subsequent period of Ur, interpreted the events as Akkad being hit by the wrath of the god Enlil. The angry god brought famine and drought to the land: "As if it had been before the time when cities were built and founded, the large arable tracts yielded no grain, the inundated tracts yielded no fish, the irrigated orchards yielded no syrup or wine, the thick clouds did not rain, the macgurum plant did not grow."

What accelerated the demise of the empire were large-scale migrations, which accompanied Akkad's loss of coherence, precipitating even greater unrest. The Gutians, a people from the Zagros Mountains, between modern-day Iraq and Iran, also responded to the drought. They descended from the mountains, moving into the plain and taking over roads and waterways from the Akkadian state, "like small birds they swooped on the ground in great flocks." They recruited local pastoralists, while the rest of the population retreated to cities, hoping to survive on the produce of urban gardens, eventually supplanting the Akkadian dynasty.

The record of the demise of the Akkadian Empire is the first to reflect directly the tight connection between water availability, state control, and migration. While changes in water availability can weaken a sophisticated state, its ultimate collapse is often the result of someone else's vulnerability. In the face of significant drought, a population with less-developed institutions may not just weaken, but fail to survive altogether. At that point, the only alternative is to break the bond with its environment and move. It turns out the most common route for water to have deep, lasting impact is not a flood of water, but one of people. This lesson would repeat itself throughout antiquity, finding its most famous realization in the collapse of the Western Roman Empire.

Over the course of two millennia, several states developed, then fought, and, ultimately, collapsed. Through all those stages, society's interaction with water played a crucial role in their architecture. With increasing political complexity came a more nuanced relationship with the water environment. The geopolitics of water was next.

3

Bronze Age Globalization

A NETWORK OF TERRITORIAL POWERS

When communities became sedentary in a world of moving water they set in motion a series of events that contributed to the creation of the state and to the development of increasingly sophisticated institutions. Along the embankments or over the fertile floodplains of the great rivers of Mesopotamia, periods of growth had been the prelude to conflict and collapse of one sort or another. But the scale and scope of what happened during the late Mediterranean Bronze Age of the second millennium was unprecedented, showing how water could play a crucial role not just in state formation but in the development of regional systems of states.

During the second millennium BCE, a vast commercial and diplomatic network developed between the territorial states of the eastern Mediterranean: Mitanni, Assyria, and Ebla in northern Mesopotamia; Babylonia in southern Mesopotamia; Elam in Iran; Ugarit on the Syrian coast; Mycenae around the Peloponnese and the Greek islands; the Hittites in Anatolia; Egypt. Tin and copper, the main ingredients of bronze metallurgy, pushed people to search for sources wherever deposits were and to trade over long distances.

Reconstructing these trade routes is difficult. Mesopotamia may have sourced its tin from Anatolia, Afghanistan, or central Asia, supplied either by land, through the northern mountain ranges or Elam to the east, or by sea, from the Indus Valley and into the Gulf. This system of trade went well beyond metal. Written records from Ugarit, in modern-day Syria, describe vessels carrying four or five hundred tons of cargo. In one famous case, archaeologists were able to recover a fourteenth-century BCE ship that had sunk off the coast of Turkey. It had been carrying copper and tin, of course, but also an assortment of other goods, including cobalt-blue glass, ebony, resin, hippopotamus teeth, an elephant tusk, tortoise shells, ostrich eggshells, spices and food like coriander, figs, grapes, almonds, pomegranates, and olives.

It was as globalized a world as one could reasonably expect given the technologies of the time. Merchants expanded their reach following coasts, rivers, mountain passes, and trails, whatever routes they could find, to source and distribute raw materials, commodities, and artifacts. As a result, the Near East ended up being part of a vast network that stretched from the Baltic to the Mediterranean, and from the Atlantic to the Black Sea.

Water played a central role in this trade system. Favorable rainfall and skill at managing water resources were principal determinants of agrarian surplus. A water-rich country could support merchants and sailors, a diplomatic class, and, inevitably, an army. It had the resources to develop a ceremonial culture, which required importing rare materials in great quantity. States with limited water resources, and therefore without surplus capacity, were also able to share in this wealth by sourcing metals or other raw materials, or producing sought-after artifacts.

States in this commercial network pursued specialization of labor and land, determined by each society's comparative advantage, partly defined by rainfall. If trade was predicated on the ability of some countries to generate agricultural surplus, it stands to reason that such trade could also be vulnerable to long-term changes in water conditions. Less rain in one corner of the commercial network would decrease crop production. Without adequate storage, this would lead to the collapse of economic activity, which would then propagate through the network, affecting distant places. When that collapse threatened livelihoods it led to migrations, turning domestic crises into international ones.

The complexity of the relationship between water and this international system is best seen through the lens of its natural hegemon:

Egypt. In the late Bronze Age, Egypt was extremely wealthy. Over the course of the second millennium, Egypt was the target of several migrations, provoked by both changes in climate and the enormous disparity between Egypt's wealth and that of other societies. And the source of Egypt's wealth was water, or, more specifically, the Nile.

AN EXTRAORDINARY LAND

At almost seven thousand kilometers, the Nile is the longest river on the planet. While the Mediterranean storm tracks fed the Mesopotamian rivers, the sources of the Nile were watered by much more remote weather in the tropics. This made Egypt a liminal land: abutting the Mediterranean but with deep roots in Africa.

In the tropics, air flows from both the Southern and Northern Hemisphere, converging close to the equator in a narrow band, where it is warmed by the intense sunlight and lifted in tall, convective storms. In those storms water vapor condenses into rain, releasing huge latent energy, which drives the corresponding flow of air poleward at high altitude and into the jet stream. The rising tropical branch of this overturning circulation is known as the intertropical convergence zone, a narrow band of rain clouds that encircles the planet. That is what feeds the Nile.

Everything seemed aligned to obtain the highest productivity out of the Nile. The tropical rainfall band is not stationary. Over the course of the year, it moves regularly back and forth between roughly five degrees north and five south across the equator, following the seasonal movement of the sun at the zenith. Lake Victoria, the source of the White Nile at the equator, receives from it over one and a half meters of rain per year. From there, water then spreads into a vast swamp in modern South Sudan called the Sudd (from the Arabic *sadd*, meaning "obstacle," because it used to be an insurmountable, malaria-infested region) and evaporates. What remains flows down to Khartoum as a steady stream.

As June approaches, the rain band moves north over the Ethiopian highlands, where it drops two meters of water. That water then flows down the Semien Mountains along the Blue Nile and a smaller tributary, the Atbarah River, contributing in one concentrated pour three-quarters of the annual flow of the Nile. At Khartoum, the Blue and White converge into the Nile proper, which then winds its way down through the Nile Valley below Aswan and into the Mediterranean.

The timing of this annual cycle could not have been better for agriculture: the White Nile supplied the valley a steady, low flow during sowing, growing, and harvesting; then the Blue Nile brought the peak flow down a four-thousand-kilometer journey from Lake Tana. By the time the Ethiopian waters flooded the Nile Valley at the end of the summer, replenishing the soils with silt and nutrients, the harvest was done.

When Nile floods overtopped, water would slow down, settling mud and creating new levees. Beyond those, backwaters would collect finer silt and standing water, which percolated belowground, creating new groundwater reservoirs. Those then moved laterally, filling swamps at the floodplain's edge. It was a fertile, wet, dynamic landscape. Unlike the Mesopotamian rivers, the Nile did not require irrigation canals to distribute its waters, because the flood did the job, leaching soils, adding nutrients and organic matter, and naturally preventing salinization. Even fallowing was unnecessary. Land use was relatively simple: crops were sown from November to January, once the floodwaters had receded; they would grow primarily on the natural levees and flats, and would then be harvested before the next flood season. A farmer could almost throw seeds on the field at the right time and wait for them to grow.

Infrastructure was still needed. The draining waters had to be slowed down, if farmers were to control and optimize how fast nutrients deposited. But basin irrigation, as it is called, was a management system that worked with the power of the river, accommodating its waters and adjusting its flow, rather than having to fight and constrain it.

Herodotus had once said that Egypt was the gift of the Nile. And it was indeed a state of extraordinary wealth. Based on simple assumptions about how many calories a person needs, a reasonable estimate is that Egypt could support up to five million people on its landscape without major infrastructure. Five million people in the third millennium BCE would have been somewhere between half and a third of the entire population of the world. That is a colossal number, all the more extraordinary because over 95 percent of the land of Egypt was desert, so all that production could come from a thin strip around the Nile.

A RIVER NATION

What made the Egyptian story extraordinary was the degree to which institutions were adapted to its particular water landscape. A state capable of collecting, storing, and distributing food on the scale of the Nile Valley was also capable of overriding the natural variability of water. Production had adapted to the Nile's peculiar water conditions, with many indications that the agricultural system was geared towards integrated management.

The first piece of evidence is that, in Bronze Age Egypt, emmer was the most common variety of wheat. Emmer was robust, able to tolerate extreme drought and moisture, and could be stored away before the husk had been removed, protecting it from insects during storage. It was far more laborious and less productive than free-threshing wheat—the ancestor of modern durum wheat—so its dominance is an indication of just how important storage and transport were to grain management. Before monetary economies, stored grains were wealth. While storage for small communities simply meant spreading the supply of food from the time of harvest over the entire year, storage for a complex society like Egypt was a form of financing: if production one season was lower than expected, just like a bank account, stored reserves could be tapped to underwrite the population.

The second piece of evidence that the entire river was managed as a system was the nature of fluvial transport. Large storage facilities allowed societies to ride out multiple years of drought, but such underwriting was only possible if the stored reserves could be deployed where needed. The Nile was a navigable waterway from the cataracts to the tip of the Delta. During low waters, it might take two months to travel the seven hundred kilometers that separate Thebe from Cairo, but during high waters it might take as little as two weeks. Either way, road transport would have been far slower. Besides, transporting grain over land would have required feeding animals large quantities of it simply to drag it along. Over large distances, this would have been uneconomic. Water transport was far more energy efficient.

Egypt implemented a fluvial tax system which supported centralized food storage and distribution in case of drought or failed crops. The food security system of ancient Egypt relied on organizing the economy around the river. This fluvial system, in turn, had an impact on the structure of society. With the ability to centralize grains, the royal court was able to feed about 150,000 people in cities. Cities would produce artifacts that were used in the cult of kingship, as well as in trade needed to import precious materials and metals. Therefore, the agricultural and urban landscapes were an integrated whole.

While in the Levant or Mesopotamia, fortified walls were the primary features of urbanization, in Egypt the city was an integrated component of the landscape, truly an agricultural state. The consequences of this were profoundly political. Rather than urban borders, Egypt had national borders, patrolled by the army. The integration of the landscape was so strong that, unlike the city-states of Mesopotamia, the ancient Egyptians imagined themselves as a single nation.

In the *Tale of Sinuhe,* a text from the nineteenth century BCE, Sinuhe, a former servant to Pharaoh Amenemhat I, went into exile at the death of his king. He was gone for many years, becoming a wealthy man and the chief of a Bedouin community. But, in a final, moving invocation in his old age, Sinuhe did not long for a city or a community. He longed for Egypt, the land of his birth. Egypt, the nation. A story that had started from a peculiar distribution of rainfall in the tropics had led a society to develop its highest abstraction: identity.

The story of a hegemon, one supremely adept at taking advantage of its endowment of resources, might give the impression of invincibility. The Nile gave Egyptians power, life, and unity, and that exceptional wealth and cultural supremacy led to an unparalleled status in the region. The kingdom could on occasion enter periods of turbulence if the Nile failed. But that was not where its deepest vulnerability resided. Hegemons exist in a context. They define supremacy in a system where others can be dominated. If water within its borders had given Egypt its power, it was water outside them which threatened to take it away.

FOREIGN RULERS

"Now there was a famine in the Land, and Abram went down to Egypt to live there for a while because the famine was severe." The agrarian wealth of Egypt was an unusual, irresistible regional magnet for neighboring people, particularly those who lived in the semi-arid landscapes of the Levant. In the Old Testament, Abraham had moved from northern Mesopotamia, in the throes of drought and famine, to Canaan, the ancient name of the southern Levant.

Abraham was searching for the Promised Land, but the life he found there was rather precarious, so he pushed on, eventually moving through the Negev—Hebrew for "dry land" or "land of thin soil"—and into Egypt. The Old Testament makes a point of emphasizing the huge difference in wealth between Egypt and the Levant. Exodus, for example, describes all normal Egyptians as owning precious objects. Abraham himself is showered in silver, gold, livestock, and servants by the pharaoh.

There are several biblical references to Egypt's geopolitical role in the region as a consequence of this wealth. The story of Abraham's great-grandson Joseph is a case in point. According to Genesis, Joseph ended up in Egypt, sold by his brothers to a group of Midianite merchants. He was ultimately bought by the household of an official at court. When the pharaoh dreamt something that baffled all interpreters, he sent for Joseph. In the dream, seven fat cows had come out of the Nile to graze on its banks. Then seven other, sickly cows followed, and ate the first seven. Joseph predicted that seven years of abundant harvests would be followed by seven years of failing floods.

To avoid catastrophe, Joseph suggested what would be described today as wholesale nationalization of land ownership. The land would be leased back to farmers in exchange for payment: every year, a fifth of the harvest should be taken and stored away by the pharaoh, so that it could be distributed in time of need. It was a form of taxation to help mutualize the risks of drought over time. Egypt, according to the biblical story, followed Joseph's advice and the country was saved. Joseph was a Canaanite. Those who saved Egypt at this particular juncture were immigrants.

One might think that this was simply the bias of the Jewish compilers of the Old Testament, but the evidence suggests otherwise. Egypt, as often has been the case with rich countries, was attractive, its borders porous to migrants in search of fortune. Towards the beginning of the eighteenth century BCE, the Turin Canon, the primary compilation of Egyptian rulers, lists about fifty rulers for a relatively short period at the end of the Middle Kingdom. The Fifteenth dynasty in particular began in 1663 BCE and lasted until 1570 BCE. To the Egyptians, these unusual rulers were known as Hikau-khoswet, or "rulers of foreign countries." In 300 BCE, the Ptolemaic historian Manetho introduced them with a simplified form: Hyksos.

That the Hyksos were Canaanites is inferred from the archaeological record: their houses looked like those found in Syria at the time; their pottery was similar to what was found in Palestine. Over time, Syro-Palestinian people had moved towards the delta of the Nile in search of a more favorable agricultural environment, and had settled near its easternmost branch, flowing to sea roughly where the Suez Canal is today.

The Hyksos initially came to Egypt as laborers and traders, but through the one thousand years of memories and legends that separated the writing of the biblical text from those events, their story came down as the story of Joseph and his family. A migration which had been triggered by a disparity in resource endowment led to a cultural transformation. The Hyksos left a legacy of diversity and innovation, providing an unprecedented bridge to Asia, a conduit for culture, technology, and ideas. Trade with southern Palestine increased. Avaris, their capital, even had a large port capable of harboring up to three hundred ships. This was a strong trading nation, open to the world.

But as they came, so they went. In the story of the Old Testament, the Israelites grew into a great nation, harnessing the riches of the Nile's eastern delta, until the pharaoh, nervous about their growing numbers, chased them out of Egypt. In fact, the last Hyksos king, Apophis, entered into war with his southern neighbor along the Nile. From the south, the seventeenth Theban dynasty began a campaign to expel the Hyksos kings and reunite Egypt. The campaign was finally completed by the kings of the eighteenth dynasty, and the Hyksos were expelled back to Canaan.

The expulsion of the Hyksos has long been believed to be the memory on which the story of Moses and the Exodus is based. By the end of fifteenth century BCE, the Eighteenth dynasty had started. Egypt entered a golden age of commerce, reaching the height of international prestige and prosperity. But the vulnerability to mass migration must have weighed heavily on the rulers' minds. What had pushed Canaanites towards Egypt had been the harsh conditions of the dry southern Levant and the relative wealth of water-rich Egypt. A few centuries later, a more catastrophic shift in climate conditions around the Mediterranean revealed the extent to which this vulnerability could be disruptive.

THE SEA PEOPLE CAME

Around the thirteenth century BCE, the Near East had reached a pinnacle of social and geopolitical complexity. The Mycenaeans were flourishing all around the Aegean, the trading cities of Canaan were thriving, the Hittites in Anatolia and northwest Syria were projecting their power by land and sea to the north of the Levant, and the Egyptians were at the height of their New Kingdom, with Seti I and Ramesses II. Then it all came crumbling down. In a matter of a few years, all of these powers declined or collapsed. The destruction was so widespread that a specific layer attributable to these events can be found in the archaeological record.

The idea that a climate shift was responsible for this particular series of events has been gaining considerable ground since the 1970s and '80s, when it was first proposed. A common hypothesis is that it all started with a shift in solar output. Over centennial timescales, it is not unusual to have periods of greater or lower insolation. Indeed, it seems that between 1500 BCE and 500 BCE such a reduction correlated with an expansion of glaciers, a general cooling in high latitudes, and a drying of East Africa, the Amazon basin, and the Caribbean. The changes in insolation affected the water cycle. Indeed, the Aegean, eastern Mediterranean, and western Asia all seem to have entered a long-lasting drought at this time. These were not short-lived conditions: they lasted at least three hundred years, with a peak around the twelfth century BCE.

Under Ramesses II, Egypt remained strong up to the second half of the thirteenth century BCE. After the battle of Kadesh in 1274 BCE and the subsequent peace, the Hittites had gained dominance over Syria, but the Egyptians remained in control of Palestine. Then also the Nile shifted to drier conditions, and after the death of Ramesses II, things took a decided turn for the worse. During Merneptah's reign at the end of the thirteenth century BCE, and then during the reign of Ramesses III, Nile discharge dropped further, leading to crop failures and lower harvests than usual.

Meanwhile, the Hittite Empire in Anatolia had also begun disintegrating at the end of the thirteenth century. The drought caused its own domestic food system to collapse, leaving it in the grip of a devastating famine. By then, the Hittites were dependent on naval grain transport from abroad. Letters have survived containing their plea for food. They even asked Egypt for help, despite their competition for hegemony in the region. The fact that, at the end of the thirteenth century BCE, Pharaoh Merneptah made the exceptional gesture of sending them shipments of grains reflects the catastrophic proportions of the famine.

Then, as environmental conditions deteriorated across the Mediterranean, the "Sea People" arrived. Their identity is disputed. The chain of events started with the collapse of agro-pastoralist systems in the north Mediterranean. Nomads and rainfed farmers began moving, chasing more favorable conditions, much like the Gutians had done in Mesopotamia a millennium earlier. The invasions seem to have started by land from the northern Balkans into Greece and across the Bosporus Strait into Anatolia. Mycenae was destroyed around 1210–1200 BCE. That civilization never recovered, and memory of their writing, Linear B, disappeared. Political, economic, and social institutions imploded, as demographics collapsed and settlements shrunk. It was the beginning of the Greek Dark Ages.

From there, part of the invasion turned seaward, while the terrestrial invasion continued across Anatolia and headed towards northern Syria. The sea routes took the invading tribes to Crete, Cyprus, and the eastern Mediterranean. The most vivid testimony of that invasion comes from the kingdom of Ugarit around 1192–1190 BCE, under the reign of Ammurapi.

At the beginning of Ammurapi's rule in 1215 BCE, Ugarit was one of the largest and richest capitals of the Near East, right on the coast of Syria, below Anatolia. Its fertile and productive land was a source of wine, oil, flax, and lumber. It had three ports beyond Ugarit itself, from which it could trade along the Mediterranean coast. It was a center of artisanal production, of dyed textiles, of gold and silver ornaments. Its commercial naval fleet sourced Mycenaean ceramics and distributed them in the interior, and its land caravans reached the interior of Anatolia. It was a multicultural, thirteenth-century society, rich in ethnicities, religions, apparently welcoming of diversity, at the crossroads of the commercial routes of an ancient Near East at a pinnacle of social and geopolitical complexity.

When the Sea People began arriving, the Hittite king pleaded for help from King Ammurapi: "The enemy advances against us and there is no number [. . .] whatever is available, look for it and send it to me."

Ammurapi sent his naval force to face the Sea People in the eastern Mediterranean, while his ground troops were sent to the aid of the Hittite king, leaving Ugarit defenseless. That is when a small advance flotilla took advantage of the situation and attacked. In a letter to the king of Cyprus, Ammurapi wrote: "My father, behold, the enemy's ships came; my cities were burned, and they did evil things in my country. Does not my father know that all my troops and chariots are in the Hittite country, and all my ships are in the land of Lycia? . . . Thus, the country is abandoned to itself."

Ammurapi would be the last monarch of Ugarit. A number of preserved fragments detail the destruction that was brought to the country: food storage looted, vineyards devastated, cities destroyed. This was not a slow collapse following a period of decline. This was an unexpected, sudden event. Ugarit disappeared, so utterly destroyed as to be forgotten for the subsequent thirty-one centuries, annihilated in one historical event. The cities of Canaan were next. The traces of this catastrophic invasion are everywhere. Jericho was completely destroyed. It also heralded a new era for the social composition of the region. Some consider the Philistines who settled along the southern coast of the Levant, in towns like Gaza, Ashkelon, Ashdod, Gath, and Ekron, part of the invasion of the "Sea Peoples."

Then, finally, came Egypt. The kingdom came under attack, both by sea from the north and by land from the west, as the terminus of the invasion. The most direct evidence of the Sea People comes from the funerary temple of Ramesses III in Medinet Habu, which celebrates a great battle at the mouth of the Nile. In an engraved image of a chaotic battle, boats and warriors fight, while the hieroglyphs speak of invaders coming from "islands in the midst of the sea" and were depicted wearing Aegean costumes.

It is possible that Egypt was the final target all along. The archaeological record suggests that multiple waves of invaders were accompanied by families and various household goods, probably because they did not intend to go back. Pharaoh Merneptah had successfully defended a first attempt at invasion from the west a few decades before the main migration. The second invasion, more successful, occurred after the fall of Ugarit during the reign of Ramesses III, who again managed to repel the aggressors. But Egypt emerged from this catastrophic event much diminished.

Despite the sudden nature of these events, the crisis that marked the transition from the Bronze Age to the Iron Age was a drawn-out affair, over the course of three centuries from the sixteenth to the thirteenth century BCE. Throughout that time, invasions, migrations, and climate shifts were complementary causes of collapse, not just sequential drivers of it. The droughts that had pushed people southward and towards Egypt had also weakened the states of the southeast Mediterranean, leaving them far more vulnerable to attack. The crisis marked the end of the Bronze Age world, setting the conditions for the subsequent rebirth. The process of finding a new order probably took as much time as the collapse of the old, from the thirteenth to the tenth century BCE. However, there is no question that this period marked a discontinuity in customs and technology.

The late Bronze Age was a peak time of geopolitical complexity in antiquity. It would take another thousand years to recover a similar level of regional integration. The apex of Egypt's Middle Kingdom had redefined the powerful hydraulic nation. The catastrophe that marked the end of this period of prosperity and regional integration appears to have been a climate-induced migration crisis. It contained a crucial insight that would go unheeded by most subsequent hydraulic hegemons: In the relationship between water and society, the resilience of the state is not just a function of its proficiency in harnessing its own resources; it is equally a function of all others in the system succeeding at doing the same.

4

An Article of Faith

GOVERNING FLOODS IN CHINA

The stories of deep antiquity reveal a generative dialectic between water and society, one that shaped social organization, the state, and even early forms of international relations. Yet, the distribution of water influenced not just formal institutions, but also more intimate, abstract beliefs. Beliefs are important to this story because they outlive infrastructure, and survive changed institutions. They are the most basic building blocks of a social contract. Philosophical and theological beliefs matter to the story of water because they provide people explanations for *why* things happen, and instruction on what people *ought* to do about them. It is easy to see how water would relate to such beliefs in the context of naturalistic pantheism, where environmental phenomena are manifestations or embodiments of the divine. But the vast majority of the world today operates on far more abstract belief systems, so it is fair to ask what relation they might have to water. The answer is to be found once again in antiquity, and China and the Southern Levant are emblematic in this respect.

The Chinese classical tradition of the first millennium BCE evolved alongside the water landscape. From the eighth to the third century BCE, during the Eastern Zhou dynasty, China's political landscape was a fragmented, loose federation of local states often at war with each other. During its first three centuries, the so-called Spring and

Autumn period, a growing number of local lords vied for supremacy, competing on military strategy, defense technique, and diplomacy. It was during this period that the doctrine of the "Mandate of Heaven" emerged, the belief that a ruler held on to power because of Heaven's consent. More often than not, Heaven spoke in natural disasters. In a country whose very heart was the destructive Yellow River, this belief inevitably connected power to water. But those beliefs evolved into far higher order abstractions.

In a seventh-century BCE text, the *Guanzi*, the duke of Huan, who ruled on the Shandong Peninsula, asked the philosopher Guan Zhong where to place his capital. In response, the philosopher argued this: of all the natural disasters a city should be protected from, floods were by far the worst and thus they should determine the location of a capital. Rivers were physical threats, he said. The physical threats would lead to social instability. Social instability would compromise power. The duke could not afford such a risk. Destructive floods, which in the Yellow River basin happened on average every four years, were a dangerous deterioration of the Mandate from Heaven. Of course, the legitimacy of a dynasty did not depend solely on their ability to manage the river—individual breaches did not necessarily result in catastrophic collapse—but uncontrolled failure undermined rule.

Guan Zhong went further. Aside from the inevitable levees and canals to control the river, he also recommended that the duke organize the state differently. He suggested that a bureaucracy of hydraulic engineers, brigadiers, commanders, and laborers should be established and organized by province to maintain the dikes, inspect and repair walls and canals, and generally control the population to maintain the hydraulic works. This wasn't just an institutional response. It was a conceptual leap: the environment required a different approach to the design of a state.

Such a conceptual leap did not imply that environmental conditions uniquely determined a philosophy of the state. But it did mean that any such philosophy would entail an approach to water management. The sixth and fifth centuries BCE were a period of great intellectual ferment in China. Daoism and Confucianism, the two principal systems that shaped Chinese philosophy, were among the many traditions to emerge at that time. These two schools are often represented in conflict: Confucianism as the formal, conformist system, focused on public affairs; Daoism as a private, contemplative practice seeking a balance with nature. These philosophical frameworks translated into radically different interpretations of the relationship between the state and the water environment. The Daoist tradition emphasized adapting to the river, placing embankments far away and making room for its expansion during a flood. The Confucian tradition, on the other hand, was far more focused on powerful levees that could constrain and control the river, ensuring it would be managed into submission.

In fact, both philosophical approaches had very real policy implications for the state. The Daoist approach mitigated the risks of catastrophic floods but created a set of challenging social problems. Floods between wide embankments would inevitably create a vast, fertile alluvial plain, which would then attract farmers. Avoiding a catastrophe under these conditions required, therefore, relocation policies and significant social control. Confucians thought that constraining the river into ever higher dikes would lead the river to flow in place, scouring and deepening the channel bed. But in the case of catastrophic breaches, the damage to the population, which entrusted its safety to solid levees, could be unimaginable, posing substantial risks for the legitimacy of central power. These conceptual frameworks therefore implied a different relationship between people, the state, and the territory.

The Confucian view prevailed and evolved alongside the management of the Yellow River. From the fifth century BCE, agriculture in the middle Yellow River grew, and as irrigation works and farming increased, more and more artificial levees were needed to constrain the increasingly muddy river downstream. It was a Faustian bargain. If the levees withstood floods, the water would scour the channel bed, deepening the channel. This would increase the river's power, with greater chance of future breaches. But if the river did breach the levees, its speed would drop as water expanded and silt and mud would deposit downstream, lifting the channel even further, and once again increasing the risk that the state would have to manage a catastrophic failure in the near future. There was no way out: the river and the state were bound to each other.

Water choices and philosophical outlook were far more dependent on each other than even these risks suggest. Over the course of the fourth century, the state of Qin, at the western margin of the Zhou feudal system, fully centralized its government, enforcing direct taxation and conscription for all, becoming a powerful war machine. Qin had a waged army, which required feeding. The only way to feed it was to wrestle with China's overwhelming rivers. Qin's muscular approach to statecraft translated into some of the great early irrigation projects of China.

Qin invaded the southern states of Ba and Shu, in Sichuan, spilling over from the Yellow into the Yangtze basin to acquire the fertile Chengdu plain. There, in 256 BCE, Li Bing constructed the great irrigation system of Dujiangyan. The system, still in operation today, watered about 670,000 hectares of land. A similar story involves the spectacular Zhengguo Canal, described by the historian Sima Qian. The canal flowed along the foothills of the Beishan Mountains in the Guangzhong basin, the heart of the Qin state, and increased land available for agriculture to around four hundred thousand hectares. Production became so abundant, that, during the subsequent Western Han period, the state ran out of millet storage. The productivity of these lands boosted the power of Qin, giving it enough surplus to support an enormous army, and ultimately to unify "all under heaven." Incorporating water in a bureaucratic philosophy of the state had turned it into an instrument of statecraft.

How a people choose to manage their water environment is tightly bound with how they conceive of their own social contract. The Confucian approach to statecraft stimulated state-led water interventions, which, encouraged by bureaucratic centralization, in turn fueled the unification of China. This story might seem to suggest that the most obvious philosophy to emerge in difficult water conditions must be one that encourages centralization. But on the other side of the world, the water experience of the southern Levant encouraged the development of a radically different set of beliefs.

A KINGDOM OF LIMITS

The Amarna Letters, a series of fourteenth-century BCE exchanges between the pharaoh and Canaanite lords, revealed profound cultural differences between the two. In Egypt, the pharaoh drew power from the divine, controlling the river and its floods. He was a god incarnate and, like all gods, required offerings and sacrifices. The Canaanites, on the other hand, appeared to believe in reciprocity. They may well have accepted paying tribute to a lord, but they expected something in return. In fact, a number of letters suggest that the Canaanite kings were under the impression that the pharaoh would come to their aid because they were paying tribute. This cultural trait, one that called for interdependence and mutual aid, reflected the legacy of the much harsher environment of Canaan. Resource constraints can lead to conflict, but far more frequently they incentivize cooperative behavior.

To understand how water conditions in the Levant might have led to more abstract cultural adaptations, one must start from its economic and political situation. Through the Bronze Age, the southern Levant had been a border land, squeezed between the competing superpowers of Egypt, Mesopotamia, and Anatolia. Levantine farmers were incorporated into a regional trade system. Archaeological remains show a domestic life dependent on a number of imported goods, suggesting that most people had access to markets, not just those for luxury goods. Then, the total collapse of the Hittite Empire, the retrenchment of the Egyptian kingdom, and the contraction of Mesopotamian powers in the wake of the invasion of the Sea People left the southern Levant with the room and autonomy to organize. Even as they grew, though, cooperation and trade remained at the heart of those societies.

Between the ninth and seventh centuries BCE, the kingdoms of North Israel and of Judah reached their respective peaks. North Israel was in the rainier highlands of the north and could have supported almost four hundred thousand people. For statehood to be possible, there had to be enough surplus to also cover military conscription, administration, merchants, and other ceremonial postings. North Israel produced enough surplus to feed about twenty thousand such people. Not a lot—these were small numbers compared to what could be mobilized by the much greater empires of Egypt or Assyria.

The far more arid Judah, in the middle highlands of modern-day Israel, would have been able to support a total of just over a hundred thousand people, which meant even fewer bureaucrats and soldiers. A similar estimate could be applied to the eastern side of the Jordan River. Water and land posed fundamental constraints on the size and power of the agrarian state in the Levant.

The most successful among the Levantine states was the kingdom of North Israel. Its greatest expansion occurred under the Omride dynasty, which ruled the country for much of the ninth century BCE. To the north and east, it expanded beyond the Jezreel Valley, well into Galilee, often encroaching on the territory of the kingdoms of Aram-Damascus and Ammon. To the south, it probably went as far as Jericho. To the west it reached the sea, accessing maritime trade through the city of Dor. It was rich enough to support a monumental culture: the palace in Samaria was one of the largest known in the Iron Age Levant. Not coincidentally, the state's boundaries corresponded to the highest rainfall and some of the region's most fertile land. But even though it was the richest of the southern Levant states, its environmental limitations still required sustained trading.

One famous piece of evidence for this is the story of the siege of Samaria, from the second book of Kings in the Hebrew Bible. Ben-Hadad, the Syrian king of Aram-Damascus, had decided to lay siege to Samaria, probably during the reign of the Omride king Ahab. The city was in the grips of a terrible famine. The prophet Elisha told King Ahab that the following day twelve pounds of flour, a *seah,* would sell for a shekel, roughly twelve grams of silver—a remarkably low price for grain—at the gates of the city. An officer listening to the conversation scoffed. This could never happen, he said, not even if "the Lord should open the floodgates of the heavens"!

The officer's mocking response contains a critical insight. He knew that those prices would only have been possible if the landscape had been drenched with enough rain to dramatically increase supply. It also implied that grain was being sold in a market subject to the rules of supply and demand. Elisha responded: "You will see it with your own eyes, but you will not eat any of it." In the story, as predicted, the siege lifted. People came flooding out of the city gates, famished and desperate, and found grains exchanged at exactly those low prices. The doubting officer barely had time to see those incredible prices before he was trampled to death by the crowd desperate for food. Elisha's prophecy had come to pass. Even for the most successful state in the southern Levant, trade was essential to overcome the water scarcity of the environment.

It is unclear what ultimately weakened North Israel to the point of causing its fall in the eighth century. The biblical texts—in Hosea and Amos, for example—suggest that the farming population was squeezed by the taxation imposed by the elite. A population exceeding the carrying capacity of the land might have put too much burden on the nation, even with trade. In the end, North Israel proved to be an inevitable target for its biggest neighbor, Assyria.

PAX ASSYRIANA

After North Israel fell, the Kingdom of Judah to the south benefited from the Pax Assyriana that followed. Judah was bounded on the east by the Dead Sea, on the south by the Beersheba-Arad Valley. The northern border passed through Jericho, although the latter belonged to the kingdom of North Israel. The western border was trickier to define, but probably cut north-south through the Shephelah.

The very limited rainfall on the Judean highlands, while sufficient for some Mediterranean cultivation, was not enough to generate significant surplus for growth. In the eighth century the population of Judah was around a hundred thousand people. Jerusalem, its capital, initially built around the Gihon springs, had been settled since the Bronze Age. The Amarna Letters mention it as a city-state capital in the fourteenth century BCE, but there is little evidence that it was anything more than a relatively small village. When the kingdom of North Israel collapsed, Jerusalem's population grew tenfold, flooded with refugees who swelled the population to about fifteen thousand.

The economy of Judah was as dependent on the distribution of rainfall and trade as that of North Israel, but domestic production struggled to yield the surplus required to support the large, monumental monarchy its rulers aspired to. The region was transformed when Sennacherib, the Assyrian emperor of the late eighth century, launched a campaign to re-establish control over the region. His campaigns of 701 BCE greatly diminished Judah, which lost some of its territory in the Shephelah to the Philistine coastal towns. Under Assyria, both the Philistine cities and Judah, while more or less autonomous, became its vassals. The Assyrian century, as the seventh century came to be known, greatly encouraged trade, on the back of which many communities prospered. Even relatively inhospitable areas in southern Canaan, such as the Beersheba-Arad Valley, were settled. The coastal plains began to grow.

Judah found itself embedded in a regional system of trade spanning three interconnected zones of production, largely organized around water distribution. The well-watered coastal plains occupied by Philistine cities like Ashkelon had always been a bridge between Mesopotamia and Egypt, and were mostly wine producing. Wine on a per liter basis was cheaper than oil, but on a per hectare basis it was vastly more profitable. Both Phoenicians and Assyrians pursued maritime trade in wine from the coast, as did Philistia. For example, Ashkelon, the Philistine kingdom of King Aga, had a huge winery, probably four hundred square meters, which was big enough to produce for both local consumption and export. However, the city-state was not big enough to also support other production, and oil and grain had to be imported.

Further east from the coast, at a slightly higher altitude in the Shephelah, was a second zone, also occupied by Philistine centers like Ekron, where oil was produced as the next most profitable commodity. Like Ashkelon, in the seventh century Ekron was fortified and was by all accounts a large city, with huge oil production capacity. One hundred and fifteen olive oil installations have been found, which could produce some five hundred tons of oil per year, making it the largest oil-producing center found so far in the ancient world.

Further east still was Judah, the third zone, which specialized in grain. Judah could have grown olives, but then they would have found themselves competing with the likes of Ekron, which sat between them and the coastal ports. Grain was a superior choice, and, in fact, Judah was able to generate a considerable surplus in the seventh century by specializing. Behind that, in the desert areas, they then resorted to herding as a fourth economic activity. Some of those easternmost settlements may also have been about extracting resources such as bitumen from the Dead Sea.

The economic system of Judah and Philistia was integrated in a broader Mediterranean commercial system powered by the Phoenicians, who traded not only with the regional powers of Assyria and Egypt, but also with parts of the Mediterranean, as far as Spain. The distribution of agricultural activities reflected both the geography of the landscape and the economic incentives that surrounded the country. Crucially, the imperative to trade came from the particular distribution of water, and the far more productive results that an optimized agricultural system from the coast to the Jordan River could offer.

This period of economic prosperity came to an end as Assyrian power waned, towards the last third of the seventh century. Judah's King Josiah attempted to unite all Israelites under his rule, reasserting Judah as a regional power. But when Assyrian rule was replaced by the Neo-Babylonians, they adopted a far more aggressive approach to hegemony. Maintaining vassal relationships with small states like Judah had become too much of a risk, given how close they were to Egypt. By 604 BCE the Babylonians were in Philistia. Between 597 BCE and 586 BCE, Jerusalem fell. It was the end. The exile had begun.

THE LEGACY OF SCARCITY IN THE LEVANT

From all these stories it is clear that the societies of the southern Levant exceeded the hydrological limitations of their environment, and therefore scarcity of resources, through trade. Trade in turn encouraged more and more ephemeral transactions: more and more people engaged in cooperative behavior with complete strangers. They developed cultural norms that made social cooperation easier. Those cultural norms were then codified in the stories transmitted through the Hebrew Bible. That is how water conditions contributed to shaping beliefs.

The early books of the Bible were, for the most part, a seventh-century BCE artifact compiled during the last decades of Judah's expansionist efforts or soon thereafter, although Genesis itself may well have been added in the sixth century BCE. Their description of water poverty in Canaan is striking. They describe how, when Abraham returned from Egypt in the late Bronze Age, his nephew Lot could not settle with him in the semi-arid Judean hills because the land simply could not support both families. He had to move to the Jordan Valley, which was "well-watered, like the garden of the Lord, like the land of Egypt." For the same reason, Isaac, Abraham's second son, had to move to the land of King Abimelech of Gerar, in the coastal lowlands of the Philistines just below Beersheba Valley. He relied on wells, but the local Philistines, fearing competition, banished him. The same happened with local herdsmen. The names Isaac gave those wells, "dispute" and "opposition," are a reminder of how contentious water was.

Those were more than just stories. The Torah—the Pentateuch, or, the five books of Moses—is revealed law in the Jewish tradition.

Those stories inspired Jewish jurisprudence to focus on sharing, on the issue of ownership of water.

Through the Prophets, the Scriptures, the Talmud, Jewish jurisprudence accumulated a body of cases which created norms for behavior in a water-scarce environment. For example, in the case of Isaac and the herdsmen of Gerar, subsequent interpretation argued that if the Gerar stream had been supplying Isaac's well, the herdsmen of Gerar would have been right to claim that water. But if the well was fed by independent springs, then it should have been Isaac's. This type of thinking ended up having significant influence on attitudes to water in subsequent legal traditions.

The Torah had several of these cases. They served as allegories to define ownership models. For example, Isaac's second son, Jacob, father to the twelve tribes of Israel, went to Haran in northern Mesopotamia, and found a well at which three flocks of sheep were watered. The well was closed by a large stone, which needed to be rolled aside when all the shepherds were assembled, indicating a form of collective ownership. Later still, in Numbers, Moses asked the state of Moab, in Trans-Jordan, for permission to drink from the river, hinting at government ownership of the water. The biblical examples reflected the fact that water scarcity was an attribute of life for the patriarchal generations. They conveyed a crucial norm that the compilers of the Hebrew Bible wanted passed on.

Case law accrued. Additional jurisprudence on water came from the books of the Prophets and the Scriptures, many of which were written during the exile in Babylon, and from the subsequent Halacha, the Jewish laws. As in the case of China, the environmental threat had become a political issue. The political issue had turned into a legal, and ultimately, a cultural one. Divine law reflected in the Torah had become the basis for a social adaptation that was to be passed on over the course of centuries.

MONOTHEISTIC WATER

The relationship between water and people reached a deeper level of norms and behaviors, leaving profound cultural traces, because it was clearly germane to the destiny of a nation. In Deuteronomy, during the Exodus, God told the Jews: "For the land that you are about to enter and possess is not like the land of Egypt from which you have come. There the grain you sowed had to be watered by your own labor, like a vegetable garden: but the land you are about to cross into and possess, a land of hills and valleys, soaks up its water from the rains of heaven."

Unlike Egypt, people believed that no degree of ingenuity could change the fundamental vulnerability of those living in the land of Canaan. And that, of course, contained a very specific political warning: "Beware, lest your heart be seduced and you turn astray and serve other gods and worship them. Then the wrath of the Eternal will blaze against you. God will restrain the heavens so there will be no rain and the earth will not yield its produce. And you will perish quickly from the good land which the Eternal gives you."

The degree to which the nature of religious beliefs could be shaped by the conditions they developed in is a matter of some debate. There is a long modern tradition of interpreting the rise of religion as satisfying a human need, especially in environments characterized by scarcity. William James believed religion played a moralizing function, while Max Weber recognized the dialectic that led economic and political conditions to define religious traits. The story of the Levant suggests that for the Abrahamitic tradition the causal chain would run as follows: harsh environments tend to promote social cooperation as an important means of adaptation; religious beliefs are a cultural manifestation of that adaptation; moralizing gods—gods that are all-knowing and punitive—help maintain trust and social norms that regulate behaviors in an increasingly complex environment.

There is a further inference that can be made on the relationship between theology and water conditions in the Levant. In most polytheistic religions of the Near East, gods operated in a broad, morally neutral context governed by laws external to the gods themselves. Pantheons had leaders, of course, but Baal, Zeus, Marduk, Anu, or Enlil could not be called supreme or all-powerful in a monotheistic sense. Morality was simply about the will of the most powerful of gods. Natural events were their principal instruments, expressions of their intent, but governed by the rules in which they also operated. But the Mosaic tradition went much further: it demythologized the world.

The God of the Israelites had no broader context that subjected God to laws. God was absolute and sovereign. Historical events were an expression of God's will: the invasions of the Assyrians, Babylonians, and Egyptians, for example, unwitting instruments of God's wrath against his own people. Crucially, God transcended nature. In the Torah, power was not inherent in the material world. This should not be surprising. After all, there was relatively little the people of the Levant could do about their material circumstances.

The function of religion could not be to guide people in what they ought to do with the forces of nature because there was not much they could do. Assyrian imperialism was overwhelming because the carrying capacity of the Levant was too small to afford the resources for a competitive army. Rather, the issue was how to encourage cooperation, for that was the principal route to escaping scarcity. Levantine monotheism, as the philosopher Yehezkel Kaufmann remarked, was a discontinuous adaptation to the singularly complicated conditions that those societies faced. Religion encouraged the Jewish people to make cooperation with other people a norm in an environment of scarcity. When the Jewish diaspora began, once the Levantine states disappeared, this cultural tradition spread across the world.

The stories of China and the Levant show that, while religious beliefs may appear remote from the practical concerns of water management, the moral framework they provide played an important role in the story of water and society. Indeed, they are among the most enduring traces of society's struggles with water in antiquity. The norms those struggles engendered continue to exercise extraordinary power into modernity, sometimes in unexpected ways. For example, the classical Chinese philosophical tradition of Confucius collided with the Western religious tradition in the beliefs of the revolutionaries and reformers of early twentieth-century China. The world seen through the eyes of Dr. Sun Yat-sen, the father of modern China, was an eclectic blend of the two. His nationalism was imbued with the universalism of China's Confucian tradition but also with the anti-imperialism he picked up from his American experience in Christian schools.

Beliefs represent the most profound and long-lasting cultural adaptation to the material conditions of the planet. Beliefs frame values. Values shape choices. Not all choices have to do with water, of course, but deep in the folds of moral norms there still flow traces of society's ancient relationship with that fundamental substance.

5

The Politics of Water

THE POWER OF LANDSCAPE

At the end of the Bronze Age, the cataclysm that had caused the collapse of the great empires of the Near East also hit Hellas, the ancient Greek world. Sea surface temperatures and rainfall dropped. Aridity increased. Dorian tribes swept in to replace Achaean ones. By the twelfth century BCE, the Mycenaean civilization, the builders of great palaces, had collapsed. Population contracted. Agriculture reverted to subsistence. Cities, which depended on surplus production, became unsustainable and Greece entered its dark ages.

Things started to change around the ninth or eighth century BCE, when population grew and agriculture intensified again. In *Work and Days,* written around the end of the eighth century BCE, Hesiod described his small farm as able to support six people, including three slaves and a few animals. These were all symptoms of an agriculture that was beginning to grow and produce more. Similar clues are found in other writing from that period. In Homer's *Odyssey,* which told a Bronze Age story but likely reflected knowledge of Hesiod's period, when Odysseus went to see his father, Laertes, he found a "large, well-tended farm" with many field hands. Laertes was "spading round a sapling," while the slave Dolius had gone off to find stones for a "dry retaining wall to shore the vineyard up." Retaining walls reflected hillside expansion: farming had grown enough to encroach on marginal lands. Greece had re-emerged.

In the relationship that society developed with water, the most important legacy of antiquity was its highest abstraction: the form of political institutions. And those that came from the Greek world created a powerful fragment that persisted and grew over time. The Greek story of water was, above all, a political one.

Both Greece and, later, Rome developed institutions to manage the consequences of a sedentary life in a world of moving water for a society organized around the idea of individual freedom. The first introduced a lexicon and a range of ideas that are still in use today. The second metabolized the intellectual traditions of the first, and realized civic institutions that, for the first time, placed citizens' liberty at the heart of a state's purpose. Their institutions are of extraordinary importance for the story of water. They developed in the relatively benign Mediterranean hydrology, but ended up diffusing across the world, spreading far beyond the limits of their original environment. Today, their traces are found in almost every political system. It is through their institutions—updated for societies that thankfully extend the idea of citizenship to all adults, not just to males, as was the case in Hellas and Rome—that the voice of antiquity is heard today. Their constitutional settlements crystallized formal boundaries between individual liberty and public benefit, establishing the state as mediator between the two. They defined citizenship.

In the story of Greece, power poured down into society for the first time. As it did, the experience of water of individual farmers became politically salient. As power cascaded through institutions of governance, it mapped onto agrarian wealth, cementing the relationship between political power and the economics of agriculture. A farmer was wealthy because of his well-watered lands. He was politically powerful because he could dispose of that land's products as he pleased. This was possible in no small measure because of the nature of the Greek landscape. Not that the latter caused the former—plenty of places had similar conditions but did not develop the same institutions—but it was probably a necessary, if not sufficient, condition.

Greece is at the confluence of three tectonic plates—the Eurasian, African, and Anatolian plates—making it the most geologically active area in Europe. The lifting produces fractures in the rocky landscape of mostly limestone and marble. It was a highly fragmented landscape of carbonates called karst, a fractured terrain in which water runs to sea through ravines, sinkholes, underground channels, and crevasses. Water could move around, making it easy for communities to isolate themselves across valleys. They clustered in towns reliant on cisterns and underground tunnels for water supply. It was the age of the polis, the Greek city-state.

Greek geography was limiting compared to the great fertile plains of the Near East. There were no large alluvial plains, only small, rainfed patches surrounded by hills and mountains. But it was also empowering for the individual owner. The experience of water in Greece was, by definition, small in scale. In the stories of Hesiod and Homer one thing stood out: no state was needed to manage an overwhelming water environment, no canals had been commanded by the gods, no flood was divinely timed to ensure the bounty of the land. The farmer could tend to his land and deal with the particular water challenges he faced without recourse to collective help.

That is not to say that Greece's hydrology wasn't complex. Just that complexity manifested differently. For farmers entirely dependent on rainfall, and isolated from other communities and markets, a relatively small change in the climate could spell disaster. Herodotus described how for seven years rain failed on Thera, the island of Santorini. In the seventh century BCE, people fled and ended up sailing west, to settle Cyrene, in Libya. This was not an unusual story. In fact, the combined effect of population growth and limited agricultural resources pushed many Greek farmers to look elsewhere. Around the Mediterranean coastline, rainfall was typically sufficient to support a mix of dry farming, horticulture, and pastoralism. Karst was not unusual. So, as the Greeks traveled, they looked for familiar geology and climate in other parts of the Mediterranean to reproduce their homes.

Over a relatively brief period, the Greeks spread, exporting their model polis to a vast collection of settlements around the Aegean Sea, from Corfu in the west to the coast of modern-day Turkey, and from the island of Crete to the south, up to the Golden Horn and into the Black Sea, with extensions around the Mediterranean from Sicily to North Africa. As they expanded, they added arable land. Because colonies were dispersed, growth in production went hand in hand with increased trade and specialization.

The Greek world was successful: between the eighth and the third centuries BCE, average consumption increased somewhere between

50 and 100 percent, and population grew tenfold, from half a million to about four million. But its most significant transformation was political. The polis was not just a physical place. It was a collective of citizens. Its moral boundaries were as important as its territorial ones, and crucial for the story of water.

THE RISE OF THE FARMERS

Greek agricultural intensification led to an unexpected evolution in society. The introduction, in the eighth century BCE, of the phalanx, the most effective fighting force of antiquity, completely transformed the role of the farmer in society.

The Bronze Age had been a time of Homeric horse-riding aristocrats leading bands of poorly equipped peasants. The phalanx, in contrast, was based on the disciplined coordination of heavily armored hoplites. These were infantrymen armed with a spear and a large circular shield. Because each hoplite had to provide his own weapons, the phalanx could not source infantrymen from the poorest in society. Metals, which had become available thanks to trade across the Mediterranean, were far too expensive for that. Rather, the military capacity of the Greek state became dependent on a sufficient number of farmers who were wealthy enough to be able to afford the necessary gear. Those were people who had been able to harness the productivity of small patches of rainfed land to produce a surplus.

Farmers had become the principal actors in the military structure of the Greek polis. Such an important military role could not come without greater political power. After all, what made the phalanx effective was that these hoplites were invested in the defense of the polis not just as a place, but as a community of citizens. It was a moral commitment. The hoplite agrarians—those who Aristotle identified as the greatest source of stability for any form of government—had become a political force. The route they followed to gain political power would become the driving force of reform.

In the archaic period, power had been concentrated in the hands of aristocratic families. Without institutions that could mediate their claims, the hoplites became susceptible to demagoguery. Their political demands were intercepted by tyrants, military leaders who promised to wrest control of the polis from the aristocracy in exchange for authority. Seen from modernity this was a familiar process: autocrats rising to power on the back of promises to farmers. And just like in modernity, they often succeeded only to then spend vast resources distracting people with vanity projects and misguided policies.

The likes of Pheidon of Argos, Kypselos of Corinth, or, later, Pisistratus of Athens, were often violent, arbitrary rulers who, after gaining power, curtailed the liberties of free men. They used power mostly to ensure they could hold on to it. For example, because tyrants were often patrons of the arts, people flocked to cities, attracted by the opportunities of a court and of the agora. But with more and more people moving to cities, the problem of supplying them with water became urgent. So, to keep the peace and ensure urban centers remained sustainable, tyrants invested in water supply infrastructure. In Athens, the sixth-century tyrants provided the city with a pipe and fountain system. Similarly, on the island of Samos just off Turkey, the tyrant Polycrates had the engineer Eupalinos construct an aqueduct, a tunnel dug through Mount Kastro, to supply freshwater to the city. This infrastructure was a way of legitimizing central authority.

But they could not deliver a sustainable agrarian model. To prevent the entire rural population from moving to the city, some tyrants extended agrarian loans and pursued public building programs. They expanded manufacturing and overseas trade to keep their constituencies happy. In Athens, for example, production of Attic vases, enormously popular across the Mediterranean, skyrocketed. But their economic development policies, if one can define them as such, were rendered largely ineffective by the needs of an autocratic government. Patronage was costly.

Tyrants had to levy substantial taxes on agricultural production to support extravagant court expenses. Taxes in turn led wealthier landowners to shift from grain production to producing the more profitable oil and wine. But vines and olives required less labor, not more. In Athens, which had less fertile land than others, many smallholders went to work in shipping or services, forcing the state to begin importing grains from the Black Sea. That in turn required a fleet to protect shipping routes, accelerating the transition to a more diversified economy. The centralization of power and increased taxation accompanied economic expansion and diversification, but made the domestic agrarian model less, not more, viable. The pressure on the majority of farmers increased to the point that it became unbearable.

As the first military tyrants were succeeded by their children, whose only claim to fame had been to be born in the right family, the legitimacy of tyrannies deteriorated. None of these dynasties survived more than one or two generations. A different model was needed.

SPARTA'S CONSTITUTION

Cascading political power towards infantrymen required involving them in a political process. The tyrants could not do that. One widely admired city-state, which had never had tyrannical rule and seemed to have been able to develop a robust, somewhat participatory system of government, was Sparta. In fact, the polis had the most durable constitution in all of Greek experience. And it also offered the first model of hoplite political power.

What is known of Sparta's constitution is traditionally attributed to the probably mythical lawgiver Lycurgus. It was a complex system, a mixed government, reflecting elements of monarchy, oligarchy, and democracy. The state was ruled by two hereditary kings, who had military and religious roles. The executive function was played by the ephors, elected by the whole citizenship body, but whose term lasted only a year. Aside from electing the ephors, the assembly of citizens voted on policies but had no right to debate or amend them. Much of the real power resided in the *gerousia,* a committee of twenty-eight elders whose appointment was for life, and who had veto power over much of the policy agenda, giving Sparta a distinctive conservative outlook.

Society was stratified. Spartiates were the full citizens of Sparta. The *perioikoi* were Spartans but not full citizens, economically free but without political power, living in the peri-urban area outside of the city itself. The far more numerous helots lived in servitude and were typically conquered people. The constitution called for nine thousand Spartiates to be drawn as equals, who shared in the responsibility and benefits of being a Spartan citizen. These citizens were subject to a set of expectations that included, of course, military training. In exchange, they received a land allotment.

Much is made of the relationship between the ethics of Sparta, its constitutional architecture, and its longevity, but material conditions were fundamental to its success. Rainfall in Greece has a sharp gradient west to east. As a result, Sparta gets approximately eight hundred millimeters of water every year, twice as much rain as what falls over Athens. Sparta also had access to close to 135,000 hectares of the best agricultural land in both Laconia and Messenia—more than twice what was available in Attica. It was more than enough to support the population of the polis.

Because of the size and productivity of the rainfed landscape, and the particularly stratified society, each citizen had access to substantial resources. At the beginning of the fifth century, while in Attica most citizens owned around one hectare, with a few thousand people owning around five hectares or a little more, in Sparta the average landowner could count on about eighteen hectares, with some reaching fifty. Furthermore, while most Greek hoplites had to balance farmwork and military training, Spartan hoplites were sufficiently wealthy so that they could dedicate themselves entirely to war, while the helots, slave labor owned by the state, did most of the farmwork. This made the Spartan military the most effective of the Greek world, but the cost of this leverage was the constant concern for social stability, which required considerable control.

The Spartan constitution recognized the importance of individual agricultural productivity as a basis for economic—and therefore military—relevance, and rewarded it with some degree of political agency. This introduced a causal chain that, from rainfall to production to military economics, led to politics. But it did rest on one, crucial material condition which was remarkably different from those found near the great rivers of Mesopotamia or Egypt.

Sparta's territory included short rivers—the city was built at the confluence of the Eurotas River and its tributary, the Magoula River—but those rivers were not powerful enough to disrupt society. Individual farmers could manage the variability of water within the economics of their farm. Had management requirements exceeded the resources of individual farmers, had a collective response been routinely necessary to deal with the forces of climate, the farmer's claim to political power would have been severely weakened.

To be clear, the distribution of water did not *cause* Sparta to embrace its particular constitution. After all, the same political experience did not occur in places with similar water conditions, such as the northern Levant: water distribution does not determine by itself the political institutions of society. But society's relationship with water does reinforce and encourage specific traits. For example, if taxation is not needed to support landscape interventions, limiting its scope to administration and military affairs, then economic resources do not need to be centralized as much.

The water environment of Sparta, as that of most of Greece, was benign, so the Spartan constitution was free to accommodate citizens acting as independent individuals, bound together by security concerns and economic aspirations, but not by any need to act collectively to counteract water's influence on the landscape.

ATHENIAN REFORMS

Athens's political innovations were less enduring than those of Sparta, but they pushed the process of mapping power to geography to a far more radical limit: the full distribution of power to the entire citizenship. The initial fuel for these innovations was a crisis provoked in large part by the nature of its geography.

Before the expansion into the Mediterranean, the Athenian polis was a closed economy, in which the price of grains increased along with demand. That in turn maintained some level of access to income for subsistence farmers, even in the face of a growing population. As soon as foreign imports began, though, prices collapsed, creating a social and political crisis. The colonies often had an advantage in the production of grains, so it would have been natural for domestic producers to transition to higher-value crops. Vines and olives were the domain of the wealthy, as they take several years to grow, during which time other production would have to be forgone. It was an impossible transition for those who depended on cash crops.

To cope with the variable rainfall and exposure to international markets, smaller farmers found themselves hedging their bets, growing a bit of everything. Any such form of risk mitigation came at a cost, however. Labor could not specialize because different crops required different skills. Land productivity dropped as farmers could no longer maximize the yields of one crop. They sometimes tried to augment rainfall with water collected in cisterns or the occasional gravity-fed irrigation system. When possible, they even overproduced to store crops for the following year. But given that much of what could be stored might go to waste, this meant forgoing income for inventory. It was a spiral of poverty.

As small farmers faced dropping income, many resorted to trading the following years' production for an advance on food supplies. They resorted to personal guarantees to prove that future production wouldn't be sold several times. It was only a matter of time before rainfall would fail to materialize, the debt owner would call in the debt, and the farmer would lose his freedom. Small freeholders were ruined.

In 594 BCE, Solon became archon, chief magistrate of Athens. Solon cancelled debts and outlawed personal security loans, which had afflicted smallholders. He defined property rights on land, and ensured property boundaries were respected. He gave form to the power of the state as arbiter in the redistribution of resources. In his reforms, the individual Athenian citizen became the center of justice.

His fundamental innovation, however, was to map political power onto agrarian wealth. He created four classes of people, defined in terms of agricultural production: the richest, those who could deliver five hundred wet or dry measures of produce; those who could produce three hundred; those who could produce two hundred; and the poor and landless. This reform separated power from lineage, distributing it instead according to material conditions. This redistribution of power inevitably gave geography a disproportionate role in defining political interests.

The two-thousand-square kilometers of ancient Attica, Athens's peninsula, was mostly mountainous. Only about a third of it was suitable for farming. This geography encouraged partisanship, with the population roughly dividing into three constituencies, mapped onto the environmental conditions they faced: those living in the hills, those living on the plains, and those living on the coast. Those owning lands on the best-watered plains belonged to the higher classes; those living on the marginal land of the hills or coast were in the lower ones.

Having structured political institutions around classes of wealth and, in effect, geography, Solon then distributed political offices and military responsibility accordingly, underscoring once again how central the nexus of military force, political power, and agricultural production was to Greek institutions. The first three classes participated in the election of officials, with only the highest class having the right to vote for the highest offices. The fourth class did not have the right to vote for officials, although it did participate in the assembly and in judicial processes. In terms of military contribution, the top three classes would serve as hoplites, with the first two also providing the cavalry, while the fourth class only provided light infantry or personnel for the fleet.

Solon's reforms introduced the notion that power was based on a measure of agricultural production, not on the traditional birthright of the aristocrats. Agricultural production, in turn, depended on the distribution of rain on the landscape. Power had mapped onto water distribution.

While Solon's charisma was enough to transform the institutions, the full transition to democracy took quite a bit longer. As soon as Solon left Athens, the state was lost to the tyrant dynasty of Pisistratus, leading to a period of greater centralization of power and wealth. But the seeds of geographic localization of power and its relationship to water had been sown.

RADICAL DEMOCRACY, WATER REPUBLIC

If Solon had created the principles, it was Kleisthenes who established a system of radical, direct democracy in Athens. From 508 BCE, when Kleisthenes's reforms were enacted, to 322 BCE, when the Macedonians wrested control from the Athenians, Athens pursued the arguably most consequential political experiment in human history.

Kleisthenes maintained some of the Solonic reforms. The assembly of all Athenian citizens voted on matters relevant to the government of the state, and the council of the five hundred, drawn from the citizenship, managed executive affairs. But he completely transformed the political structure of society. Solon had made representation a function of agricultural wealth. Kleisthenes in effect went straight for geography. He organized the citizenry into a system of ten tribes, each divided into demes, local communities that were units of political representation. Now there was a small geographical unit around which decisions could be taken.

It was not a classification of individuals directly related to production, but one related to their spatial location within each tribe. This transformation had profound consequences for society's relationship to water. Over time, the deme became increasingly the focus of political community. As political units got smaller, their ability to underwrite infrastructure or manage risks—already small given the size of the polis—was further reduced. This territorial fragmentation was only possible because of the benign nature of the water landscape. Interdependence between upstream and downstream users on a large river would have made treating individual demes as independent centers of power difficult if not impossible.

The economic efflorescence of Athens in this period increased the wealth of many individuals. During the fifth century, despite the Persian and Peloponnesian Wars, the Athenian economy grew. The silver mines of Laurion, sizeable even by modern standards, were operating at full capacity, giving Athens the resources to pay for its fleet and for its imperial aspirations. By the fourth century BCE, houses were larger than the average middle-class family home of today: they could have five or six rooms, a courtyard, foundations, plaster walls, stone drains, bathtubs, tile roofs, and gutters.

Great wealth encouraged private solutions to public problems. There are several examples of this. In the fourth-century deme of Kephissia—today a suburb twelve kilometers northeast of the center of Athens—a nameless individual was celebrated in an epigraph for having invested in waterworks: a pipeline and fencing for a springhouse to keep animals out. Even larger examples are found across the Greek world. Around 320 BCE, Chairephernes, a wealthy developer on the island of Euboea, contracted with the polis of Eretria to exploit the land of a malaria-ridden marsh at Ptechai, east of the city. The contract gave Chairephernes the right to use the land for thirty talents and lease it for ten years to recover the contract cost. The drained water would then be used in a large-scale irrigation project.

These examples were a symptom of how far-reaching the Athenian experiment had been in transforming society. Chairephernes was essentially made a partner with the Eretrian deme. A political system with distinctly modern traits of enfranchisement had opened the door to contractual solutions to water problems between the community and the individual.

The Athenian arrangement made dealing with collective problems, such as those occasionally posed by water, complicated. For example, the economic adjustments that accompanied the transformation of Athenian society vastly increased pressure on the environment. The metal refineries, pottery kilns, construction, heating and, of course, shipbuilding required wood. Deforestation on a grand scale followed, which in turn impacted water security. In his dialogue *Critias,* Plato described entire hillsides washing away as a consequence of deforestation. There was no contour plowing or terracing, so, without vegetation, water rushed down the slopes carrying away topsoil. Faced with the problem of governing collective behavior to solve water conditions like this, most agrarian states of antiquity would have resorted to a moralizing ideology. This had been true in the theocratic states of the Near East, in the Confucian and Daoist approaches to water management, and in the Jewish legal tradition. But because Greek political philosophy, at least as Socrates first established it, was an ethical pursuit—individual rather than collective in character—the resolution of these collective action problems could not pass through moral norms alone. The solutions had to be judicial and political.

Even though Plato had argued for radical communal ownership in the fifth book of his *Republic,* he conceded a more realistic world of private property in his *Laws.* In the dialogue between an Athenian stranger and Clinias from Knossos, he described how people in Athens were allowed to draw water they needed provided they did not harm others in the process, a basic sharing rule. Under this legal regime, people could be held accountable for what they did and the impact they had on others. In one famous case, a property owner called Callicles accused the son of his neighbor Teisas of having caused him damage by building a wall along a road that separated their two properties. The road was in fact a dry watercourse, and after torrential rains the wall had dammed up the water, flooding his terrain. Demosthenes, speaking for the defendant, disputed the accusation, proving that Callicles simply wanted the land and was bringing a frivolous suit. This kind of litigation still occurs 2,500 years later and is a symptom of a water environment dominated by private concerns.

Greek philosophy had developed in a benign context, one in which the individual was the center of justice and politics. As a result, only the rules citizens gave themselves could guide society's solutions to collective problems. Aristotle probably captured most accurately this idea, which was at the heart of the Athenian experience. Buried in the folds of his political philosophy was the tension between individual ownership and the communal responsibility to regulate use, between individual rights and collective responsibility. Aristotle believed in a moral purpose for the polis. Citizens did not just pay taxes; they also contributed to a moral project, which in turn tied the life of the polis to the idea of justice and that to individual rights. The polis was at once its individual citizens and their collective purpose. It was this dual nature that led to the primacy of politics. Aristotle saw the state as an expression of individual freedom through civic politics. Therefore politics, rather than religion, was the appropriate forum for defining norms to pursue the collective good. This idea would be fully absorbed in the Roman experience and become central to the republican tradition that reached the twentieth century.

It was a crucial moment in the story of water. In the implied, deep connection between the state and its landscape, between the pursuit of individual liberty and collective good, Aristotle captured the ideal foundations for all republican experiences that followed.

6

Res Publica

A MEDITERRANEAN POWER

The final, most important piece in the layered water inheritance of antiquity is that of the Roman state. The Roman world was vast. At its peak at the beginning of the second century CE, it included all lands around the Mediterranean, extending into most of Europe and Britain, the Near East, and North Africa.

From the third century BCE to the second century CE, the Mediterranean climate was particularly benign, with high spring precipitation and warm temperatures, more so than the subsequent two-thousand-year average. These were conditions favorable for rainfed agriculture, on which Rome depended disproportionately.

But as bountiful as production in the Mediterranean could be in the aggregate, it was unevenly distributed in both space and time. Some regions received more than a meter of water a year, while others barely reached a third of that. In many cases, yields only reached three seeds for every one planted, the bare minimum for subsistence. And even for places with high productivity, a bad rain year could spell disaster. Southern Italy, for example, averaged yields of about six to one. But while a good year could yield eight, a bad one would yield only four.

There was a further complication. Rome itself was voracious. By the beginning of the first century CE, Italy was highly urbanized, with up to a third of the population living in cities. Rome alone housed a million people. The capital needed roughly three hundred thousand tons of wheat per year. Production in the region of Lazio, around Rome, could meet the needs of about forty thousand people. Local rainfed agriculture was never going to be able to meet the demands of such a vast, urbanizing empire.

The obvious answer to this problem could have been greater reliance on irrigated agriculture. But the great regional rivers of Mesopotamia or central Europe were not Rome's working landscapes; they were its borders. And although Octavian added Egypt during the civil wars, by itself the Nile would not have been enough to support an empire of over fifty million people.

The fact is that Rome never centralized the administration of water resources. This may seem counterintuitive for a civilization known for its aqueducts. Yet Rome was mostly a world of small dams, diversions, and tiny settling tanks, all developed by private individuals. Farmers depended on rain. Where geology allowed, they used cisterns to collect runoff. If they could afford it, they would supplement them with wells or small-scale aqueducts. In some cases, they shared water by scheduling its use at specific hours of the day. Even drainage was done on an individual basis. Field drains, whether open ditches or sophisticated tiled drainage—Cato himself described some in *De Agricultura*—depended on individual farmers. And despite the Tiber's frequent floods, the countryside of Rome was left swampy and malaria-ridden, leaving the wealthy to avoid the issue by simply building their villas on hills. Rome never pursued a state-led approach to water infrastructure at a scale commensurate to its size.

The reason for this highly decentralized approach was that the institution of private property was essential to the definition of Roman citizenship. Citizens had the right to life and property, and both were protected from interference by the republican constitution that was at the heart of the Roman state. But water is the ultimate *res publica,* a substance that moves, impossible to confine to individual ownership.

To deal with this tension, Rome had to adopt a precise legal distinction between the role of individual enterprise and that of the state in managing this res publica. Only what could be shared without impact on individual interests could be considered public, meaning that the state owned it. The distinction was subtle. In all rivers, water itself was *res communis,* incapable of ownership. Seasonal torrents and brooks could be privately owned. Perennial rivers on the other hand were public, as were lakes and canals. Navigation was particularly important because of its economic function, so, by definition, all navigable rivers were public and additional restrictions were placed on use of their water. Critically, though, public control did not extend to land in the basin of those rivers.

Any large-scale water resource management intervention could not be based solely on the public ownership of navigable water courses. It was not enough. A state-led approach to water infrastructure would have required embankments, levees, canals, drainage systems, all of which would have needed to happen on the land surrounding the river. Given that most of it was privately owned, one solution could have been expropriation. But this was an unacceptable option: the right to property, as Cicero had written in his *On Obligations,* was sacrosanct, posing severe limitations on what could be considered public property.

If infrastructure on a scale commensurate to the empire was not the answer, then institutions with a comparable scope would have to be the solution.

LIBERTAS ET RES PUBLICA

The most enduring legacy of Rome was a political system that enshrined citizen freedom, the Roman *libertas,* as the governing principle of its existence. How it related to water resources was key to Rome's approach to resolving its need to control the environment.

Freedom in the Roman world had little to do with most common corresponding modern ideas. It was not rooted in natural rights, for example. In fact, the Roman state was an intrusive, moralizing one. Freedom was defined specifically in opposition to slavery—that is, the absence of mastery by someone else. Freedom was a status, a state of being, not simply an ability to act. In this sense it was a profound innovation. Freedom was civic, in that it was derived from being part of a free *civitas,* a community. For Sallust, the Roman Republic evolved out of the need to "preserve freedom and advance the state" and as mitigation against "men's minds growing arrogant through unlimited authority."

Crucial to freedom in Rome was the republican constitution, because it encompassed both the notion of political freedom and the pursuit of a commonwealth. Cicero defined it as "Res publica res populi est," the public good is the property of the people. He used the word *populi,* not the word *civitas,* to indicate that it was really about the Roman people as a collection of free individuals, their legal consent, and the commitment to sharing benefits among them.

Much like early Greece, the deep tension between individual rights and collective action at the heart of the Roman state was resolved by weighting political power by agrarian wealth. Rome did not contemplate equality in the right to govern. Its republic had nothing to do with late Athenian democracy. Rome was a government for the people, but it did not necessarily imply it was by the people. The Centuriate Assembly, for example, the most important legislative assembly in Rome, had been created to give people voting representation commensurate with their military duties. According to Livy, its census reflected what "every man's contribution could be in proportion to his means." The small number of infantrymen who were able to afford a hoplite's armor enjoyed far greater political representation, on a per capita basis, than those who could only afford a sling. Those with no resources were not represented at all. Political power mapped, once again, on agrarian wealth. This was reminiscent of the power distribution in the Greek polis. But the nature of the Roman landscape was rather different.

In Greece, agrarian wealth was limited by the small territory of each individual city-state and its colonies. Rome's resources were far greater and growing. The state intercepted ever more land and water, increasing the political stakes of agrarian wealth commensurately. Rome had to balance an egalitarian commitment to citizenship not just with an uneven distribution of wealth across its social classes, which produced unequal political rights. It also had to do so while undergoing an astonishing growth in wealth itself. It was the cost of a republic seeking to build an empire and the source of political energy that propelled Roman society forward.

Centuries of conflicts followed over the control of resources between plebeians and patricians, which made up Rome's two-class-based political system. At first, while Rome was still confined to central Italy, the pendulum swung closer and closer to the demands of the plebeians, much as it had done in Athens. Plutarch likened the late sixth-century BCE pro-plebeian reforms by Publius Valerius Publicola to Solon's reforms. The third century BCE Hortensian law established that plebiscites—laws passed by the plebeian assemblies—bound all Roman citizens.

But halfway through Rome's republican experience, the tide began to turn. Rome's extraordinary expansion into the Mediterranean propelled wealth accumulation to new heights. It began in the mid third-century BCE, after the First Punic War. The acquisition of Sicily, Sardinia, and Corsica was followed, after the Second Punic War, by the annexation of much of Spain. By the second century BCE, Rome turned its sights on the Hellenistic kingdoms of the east.

The second century BCE was in many ways the apex of the struggle between the rights of owners and the collective good. Between 133 and 121 BCE, Tiberius and Gaius Gracchi—brothers and plebeian tribunes—tried to enact land reforms to redistribute the *ager publicus*, the public lands largely in the hands of patricians. They argued that people were entitled to a share in the common property. Both were killed. After that, the republic turned increasingly aristocratic, dominated by the Senate and the occasional strongman.

As the size and scale of the Roman world increased, the ability of its institutions to mediate the rights of individual citizens and collective action on the landscape strained further. The extraordinary growth in the number of citizens complicated matters. Rome's citizenship had started out as a Greek-style polis, but the social wars of the early first-century BCE had extended citizenship to all people in Italy. This was a crucial step. It eventually led to the transformation of the Roman *civitas* from an identity tied to the city of Rome to a broader political commonwealth across the state. This made navigating an increasing network of claims even harder.

The wars of the second half of the first century BCE brought vast further conquests: Pompeius added much of the eastern Mediterranean, Julius Caesar conquered Gaul, while Octavian added more, including Egypt. With a growing number of citizens and ever-more-concentrated wealth, tensions grew further.

Managing the water resources of such a vast empire through direct state intervention would have required far more aggressive taxation and even the issuance of state debt, neither of which was viable. It is true that so-called publicans, wealthy private contractors, could enter into concession contracts with the state for public works, much like in Greece. But this only limited the size and scope of those works to the publican's ability to underwrite their construction. Unlike Qin or Han China, which were flourishing at the same time, the political and economic structure of the Roman Republic was simply not designed to build infrastructure for water management on a scale commensurate with its landscape.

Then, in 27 BCE, after five centuries of continuous existence, the republic ended, replaced by the principate. For all intents and purposes, it was the monarchy of Octavian, now Augustus, although the constitutional architecture of the Roman Republic would remain in place at least until Diocletian, three centuries later. The irreconcilable tension between individual and collective power that had shaped those five republican centuries remained unresolved at the heart of the Roman state. Rome had to develop an alternative system.

ROME'S GREAT WATER MARKET

The distribution of rainfall over space and time in the Mediterranean basin followed a pattern. Even today, while the amount of water can change by double-digit percentages over a decade, it also exhibits remarkable coherence: for example, when Spain and the northwest of the Mediterranean receive more rain, the southeastern Mediterranean tends to receive less. About a third of the variation in the basin over the years can be attributed to such a Mediterranean dipole. This was a crucial geographical trait, which could be exploited for water security.

When the Roman state became sufficiently vast to extend across the Mediterranean, it also could take advantage of a natural hedging strategy: by sourcing grains from Syria and Egypt as much as from Spain and Tunisia it was able to partly insulate itself from variability across the region. The vast demands of Rome could be met not by centralizing the development of water infrastructure for irrigation and flood control, but by connecting distant sources into a stable network of supply chains.

To meet Rome's food demand, grain came from across the Mediterranean, delivered by a wheat market that connected all the corners of the empire to Rome. The closest source was Palermo, in Sicily, roughly five hundred kilometers away. Madrid was about thirteen hundred kilometers, while the Po Valley in northern Italy was fifteen hundred kilometers away, because ships had to circumnavigate the Italian peninsula (land transport would have been unsafe and prohibitive). Further out still were Turkey, Egypt, and Palestine.

For the most part, wealthy freedmen orchestrated the supply, providing capital and contracting agents. There is even evidence that some of these merchants were organized in companies, similar to the joint stock companies of the seventeenth century CE. This trading system was so efficient that the local price of wheat at point of origin was inversely proportional to distance from Rome: the further away, the more of a discount it had to sell at to arrive in Rome at a competitive price.

The market for grain was not just efficient. It was, in effect, a vast water market. Each seed of grain represented the water it had used to grow. Transporting grain was akin to moving water from different parts of the Mediterranean basin, because it displaced the need to use water at the point of consumption. While the local solution to failures of rain could only be irrigation or storage, in an integrated Mediterranean-wide system, other places would make up the difference provided they were unaffected by the change in rainfall. If the market was big enough to cover locations where variations in rainfall were uncorrelated, then, on average, the supply of food ought to be stable. Indeed, the Roman Empire of the first and second centuries CE was larger than an average Mediterranean storm-track weather system, its scope sufficient to average out over most rainfall variability.

The Roman state still played an important role. First, a private market of this scale required strong legal institutions, because trade needed reliable assurance that the counterpart would be held accountable for promises. For this, there were both private and public courts, and a number of provisions for enforcement. While Roman law did not contemplate a law of agency—no one could enter into a contract on behalf of someone else—there were delegations of authorities, and private maritime trade financing was common. Second, this private system was implicitly subsidized through the patrol of the imperial fleet, which ensured a pirate-free Mediterranean. The state also improved the quality of information on prices and volume across the market by acting as a large buyer. The *Cura Annonae*, the office that administered the in-kind taxation of land, procured grains for subsidized handouts to about 250,000 families in the city of Rome.

The Roman state would never have been able to afford the scale and size of infrastructure required to both control water on the landscape and honor the needs and liberty of its citizens. The alternative had been to become a regulator and an enabler of a market economy of sorts.

But while the state did not underwrite much infrastructure, the emperor as a private, wealthy individual was often a direct investor, and often used his own resources to build infrastructure and increase his own legitimacy. When it came to agriculture for the public, the focus of those resources was not irrigation infrastructure or flood control, but trade infrastructure.

The ships that crossed the Mediterranean carrying grains could be enormous. Roman shipbuilding technology was advanced: Pliny described the purpose-built ships that carried to Rome obelisks that weighed several hundred to a thousand tons. While ships that size would not have been the norm, most of them would have transported a hundred tons of grain or so. The Mediterranean grain market relied on several thousand trips per year. That many ships could not navigate the Tiber to supply Rome. Port facilities had to be developed along the coast, from which smaller rivergoing vessels could transfer grains to the city.

For this purpose, Emperor Claudius established Portus, a massive artificial basin, two hundred hectares of water, seven meters deep, which allowed large ships to dock. A smaller *darsena*, an artificial basin of about a hectare, accommodated smaller boats. The system was linked by two canals, one to the sea and one to the Tiber, from which goods could get to Rome. Despite being far from perfect—the port needed continuous dredging and was subject to severe congestion—every year two thousand boats anchored there.

Thanks to the regulatory function of the state and the investments of the emperor, Rome developed a trade system of remarkable connectivity. It was the mechanism by which Rome could square the circle between a political system designed around the property rights of the individual citizen and a landscape that could only guarantee security of supply in the aggregate. This system was extraordinary, supporting the rise of the most extensive Western state until the nineteenth century. But it also contained a fundamental vulnerability, which would be the engine of its collapse. That too had water at its heart.

A VULNERABLE EMPIRE

One of the crucial sources of Roman demise was tied, ironically, to its proficiency in hydraulic engineering. Marcus Agrippa, Rome's *aedile*, the Roman magistrate in charge of public buildings, marked the new Augustan age by developing a water masterplan of extraordinary modernity. If Augustus was the "architect of the Roman Empire," Agrippa was going to be Rome's "superintendent of construction." Unlike irrigation and flood control infrastructure, urban water supply was a tried and proven route to legitimacy for despots.

At the time of Augustus, Rome already had far better infrastructure than most European cities would have until the nineteenth century. Pliny the Elder thought the Cloaca Maxima, Rome's sewer system, so awe-inspiring that he compared it to the hanging gardens in Babylon. At the beginning of the imperial age there were already four functioning aqueducts. Over the course of twenty years, Agrippa transformed Rome's water system. He doubled supply, adding two new aqueducts. He also followed an ambitious plan, centralizing urban water distribution, scheduling, and allocation.

Rather than just increasing the amount of water to the city, Agrippa ensured that different infrastructure would support specific parts of the city and particular uses. While one aqueduct delivered water mostly for private dwellings, another served public works, fountains, and watering basins. Aqua Virgo, the only Roman aqueduct still in use today, served mostly public works and imperial buildings: Agrippa's complexes in Campus Martius and the baths and buildings on the other side of the river.

Agrippa's hydraulic works were a testament to the sophistication of Rome's engineering. As Roman hydraulics and the confidence in its effectiveness grew, hydraulic engineering spread, with profound economic consequences. It was not just a matter of urbanization for Imperial Rome. It drove mechanization in different parts of the economy.

Water mills, often associated with large aqueducts, were everywhere. The military used them, including on Hadrian's Wall in Britain. Rural communities used them to process flour in all climatic conditions, from arid places, where drop towers, tanks, and pipes increased water pressure, to major cities, where rivers or the aqueduct system itself powered them. One should not mistake this for a modern industrial revolution. The power Romans could extract from water did not come close to what would be obtained centuries later from fossil fuels, nor did these technologies accompany a vast expansion of demand and access to markets.

Nevertheless, water technologies spread widely, occasionally reaching surprising scale. A famous milling complex near Arles, Barbegales, probably from the fourth century CE, was powered by sixteen waterwheels, sequenced along an incline. An aqueduct brought water to power the first wheel, which would then drive the next one, and so on, down a cascade. The complex could produce between four and five tons of flour per day, feeding a population of over twelve thousand people.

The water-enabled mechanization of parts of the Roman economy, such as that of Barbegales, provoked a transformation that would ultimately undermine the empire. The chain of events started with mineral resources. Mining was transformed by mechanization, especially in the gold and silver mines of Spain. Pliny described how water was brought by aqueducts, impounded in large tanks above an open-pit mine, then released from above, using its force to scour the side of the mountain. Once mined, the ore needed processing, which was done, much like flour production, by using the power of water mills. The amount of precious metals extracted through water mechanization was enormous. The mines of New Carthage, for example, produced thirty-five tons of silver per year. Roman mining operations were so big as to leave traces of pollution in the ice cores of Greenland: only the Industrial Revolution would surpass the impact of their emissions in the atmosphere. Eventually, however, this impressive water-fueled productivity became a problem.

During the empire, the state had largely functioned by payments in silver. One of the fundamental strains on the Roman state was the loss of bullion. Silver was routinely lost in circulation, deteriorated in use, or leaked out of the empire, so money had to be replaced. But mechanized extraction of precious metals was so effective, that by the third century CE mines were exhausted, and the state was in the throes of a monetary crisis. Without a steady supply of silver, coins had to be debased, which in turn drove inflation. Paying an army its wages with debased coins was a recipe for disaster.

The strain on the military was compounded by a demographic crisis, initially precipitated by the Antonine plagues at the end of the second century CE, which had reduced the available force. An expanded empire coupled with more limited resources made maintaining an effective military presence harder. The crisis of the third century was aggravated by what appears to have been a drop in agricultural productivity across the empire. It is possible that the benign climate that had supported Mediterranean productivity had begun to turn. The archaeological record shows that agriculture had become highly unstable. Commercial agriculture collapsed, which hit smaller farms and communities particularly badly. Ditches silted up, sites were abandoned, then they were recut and silted up again.

The composite of these events drove a profound transformation of society.

THE END OF ANTIQUITY

The unraveling of the Western Roman Empire is the natural end point of a story about ancient civilizations and water. Its immediate cause began with the *Limes Germanicus,* the northern border which ran along two rivers, the Rhine and the Danube. The rivers were separated by a three-hundred-kilometer land gap between modern Mainz and Regensburg. Once fortified, it created an interminable physical boundary, three thousand kilometers in length, from the North Sea to the Black Sea. Since Hadrian's time, diplomacy had maintained the border's stability. The superiority of the Roman military force and generous policies of inclusion towards the tribes limited most threats.

In the second part of the third century, as the empire spiraled into a crisis of resources, the overstretched army struggled to manage the pressure at the border. Things seemed to improve when Diocletian became emperor in 284 CE. With remarkable skill, he breathed new life into an empire that was on the verge of collapse. His reforms included the tetrarchy, the system of government that divided the empire into an eastern and western part, vastly increasing its bureaucracy, and, crucially, the joint reform of the military and the tax system.

To secure the empire, Diocletian's army grew substantially. If Augustus commanded twenty-five legions of six thousand soldiers each, by Diocletian's time the legions numbered more than sixty, with anywhere from six hundred thousand to a million troops, mostly recruited from Germanic and other barbarian tribes, settled in exchange for recruits. The army was salaried, but, because of the debased coinage, three-quarters of a soldier's wages had to be paid in kind.

Such a large standing army meant huge pressure on land and water resources across the empire. To accommodate it, Diocletian expanded and rationalized the tax system. He built a capillary territorial administration to collect and manage census information so as to adjust who to tax and how much to tax them. The most important change was the standardization of a *iugatio-capitatio* tax. The *iugatio* was a tax on land production, while the *capitatio* was a tax on headcount. Both were mostly levied in-kind. In particular, the *iugatio*, the tax on land, was collected in grain. The basic unit, the *iuga,* was not a standard measure of land area. It depended on production and type of crops, which meant that the tax owed reflected the productivity of the land. It was a tax on rainfed production, effectively a tax on how much it rained.

Diocletian's tax system gave the Roman state unprecedented power. Republican institutions had definitively given way to an authoritarian regime. Control of the treasury gave the emperor an independent source of legitimacy and vast public resources to pay the army. In principle, the tax system should have continued to provide the natural hedge built into the Mediterranean supply network. Sources of revenue adjusted automatically to areas of greater production. But just like food security, tax stabilization critically depended on uniform, aggregate revenues. This locked the state's ability to maintain such a large army to controlling a territory large enough that production failures in one place could be replaced by production elsewhere. It was unstable, vulnerable to changes in material conditions. And conditions did change.

Changes in climate had been noticed as early as the first century CE, when the writer Columella reported a warming that gave Rome favorable conditions. Now, though, the Roman world had begun changing for the worse. Colder conditions, wetter in central Europe and drier around the Mediterranean, hit agricultural production badly. Development in the landscape reversed. Then, in the fourth century, the Huns appeared on the northern shores of the Black Sea. They too were responding to a change in climate.

The Huns were dangerous fighters, effective archers, and extraordinary horsemen. They left no written record of themselves, but one hypothesis is that they were the Hsiung-Nu tribe. The Hun invasion was the unintended consequence of climate events far away, in China. The Western Han dynasty, which had succeeded the Qin in 206 BCE, had overseen a period of extraordinary growth. By the first century BCE, the population in the loess hills alone had increased to three million. Then, a period of climate instability set in and the northern boundary of the Asian monsoon moved south. As nomadic tribes followed the steppes a hundred kilometers south, they entered into conflict with the sedentary population. Pressured by a growing sedentary population behind the Great Wall to the south, nomadic tribes turned westward. That is how the Huns arrived from the east, setting off a domino effect. Their arrival pushed the Goths, who had settled between the river Don and the Carpathians, towards the Roman border, represented by the fluvial frontier.

In the summer of 376 CE, hundreds of thousands of Goths started spilling over the Roman border, looking for safety. As had been common practice, Valens, the eastern emperor, agreed to allow the refugees into the Roman Empire, but this time things went wrong. A combination of mistreatment in the refugee camps and insufficient military presence led to a revolt that overpowered local forces. The insurgent forces then coordinated with those that had been left on the other side of the border, and marched towards Valens. They defeated him at the battle of Adrianople in 378 CE. The emperor was killed.

The loss of border control was the beginning of the end for the western half of the empire. To secure the government, in 402 Emperor Honorius moved the capital to Ravenna, on the Adriatic coast, a city built on piles and sandy islands, traversed by canals and crossed by bridges. Many of its canals and fluvial ways had dried out and silted up during the crisis of the third century, but it was still the most defensible option for a government under siege, surrounded by marshes and difficult to reach by land. Then on August 24, 410 CE, Alaric led his Visigoths to Rome, sacking it. For three days the unthinkable happened. Pelagius, who was living in Rome at the time, said that "Rome, the Mistress of the world, shuddered, crushed with dismal fear at the sound of the Shrieking Trumpet."

After a thousand continuous years of unassailable existence, the Eternal City—and the everlasting empire it once stood for—was no more. Saint Jerome said that "the City which had taken the whole world was itself taken." On September 4, 476 CE, in Ravenna, a hundred years after the first Goth influx, Odoacer—one of the great barbarian generals of the Roman army—deposed Romulus Augustulus, the last emperor of the west. The empire that had once ruled all of Europe, north Africa, and the Near East had ceased to exist, in a small town surrounded by marshes on the coast of Italy.

The journey that started with the first sedentary societies and ended with the fall of the Western Roman Empire encompasses over 90 percent of the timeline of recorded histroy. Its episodes describe cycles of institutional development, abstraction, and diffusion that cemented each society's experience, bequeathing it to its successors. The legacy of that journey spread far and wide, through the Byzantine Empire and the Caliphate, to Medieval Europe and into modernity. Even Sun Yat-sen, when lecturing on liberty and democracy, made ample references to Rome and Greece. Roman republican institutions had been gone for two millennia—Greek institutions for even longer—yet they seemed as politically salient to him as his contemporary political situation.

That is the extraordinary legacy of antiquity. The republican ideals that had emerged in the water geography of Greece and that had seen their full realization in Italy, had become foundational in twentieth-century China. In fact, they would become the dominant idea of state across the world, guiding the development of the twentieth as the hydraulic century. Why and how that convergence happened was itself a process of institutional development, abstraction, and diffusion that took as a starting point the debris left by antiquity and played out over the subsequent thousand years. That is the story that must come next.

PART II

A THOUSAND YEARS OF CONVERGENCE

7

Fragments of the Past

LANDSCAPE DISAPPEARING

From the early Middle Ages up to the seventeenth century, Western societies were rather marginal to the story of human progress, the primary drivers of which moved east, from the Byzantine to the Ottoman Empire, or from the Moghul Empire to the Chinese. Europe, however, inherited from antiquity the institutions that have shaped the modern world. It would take over a thousand years for the fractious, underdeveloped continent to turn that inheritance into a comparative advantage.

For the first six hundred years or so of that journey, a fragmented power structure multiplied the opportunities for conflict and litigation. Then, the development of a common, continent-wide legal system transformed the rules of those conflicts. Its widespread adoption became the vehicle for its most important innovation, that of territorial sovereignty. This was to be the first crucial step in the development of the modern nation-state, and the basis for a new relationship to water.

Before taking that step, the story of water after Rome must begin with what was left. Its traditions, institutions, and infrastructure barely survived in fragments, transmitted intermittently into the Middle Ages. That transmission was all the more extraordinary, given the degree to which the Western Roman Empire unwound. It wasn't quite as fast as is sometimes assumed. While the loss of food sources that followed the Vandal conquest of North Africa was a catastrophic blow, because the complex system of trade that fed Rome was largely based on private enterprise, it continued to operate even as the state dissolved. However, there is no question that the late fifth-century CE marked a discontinuity.

In Italy, the Gothic wars reduced fluvial transport on the Po River, the main Italian river, to a trickle. The Goths had no navy of their own, and little interest in inland navigation. There is evidence that a small amount of trade continued because boatmen from across the valley sought refuge along the coast, which was still under Byzantine control. In a letter from 537 CE, Flavius Magnus Aurelius Cassiodorus, a Latin statesman, arranged for a shipment from Istria, in the northeast of Italy, to the Byzantine royal seat of Ravenna along the Adriatic coast. The shipment was to travel via the Venice lagoon and the Po, suggesting that an integrated trading landscape between the river and the lagoon still existed.

However, even that minimal water transport collapsed after the sixth century, when the Lombards invaded from Pannonia, modern Hungary. The seventh-century Edictum Rothari, the first compilation of Lombard law in Italy, barely mentioned rivers at all. Traffic on the great river, which once connected Pavia, Mantua, Milan, and Turin to Venice, had stopped.

The landscape was transformed. The Roman system that fed the imperial army had been organized around a standard field structure composed of regularly spaced, independently managed drainage ditches in rectangular grids. But without dredging, those ditches quickly filled up. By the seventh century, fields had disappeared, boundaries had been erased. Grass returned, land converted to pasture, and the water table rose, surfacing as marshes in most flatlands. Natural ecosystems returned, intercepting more rainfall, absorbing more nutrients, and keeping sediments in place. Rivers ran clearer and more regularly. Freshwater fisheries—trout, salmon, eel—returned in large quantities.

The transformation of the landscape went hand in hand with an unwinding of the rural economy. The spread of arable agriculture, which had been continuously advancing since the Bronze Age, stopped. Water had taken back the landscape, and society—its power weakened—responded the only way it could: by adapting. Smaller houses replaced larger constructions. Farming became far more localized. Society turned to rivers, swamps, and marshes, for survival. A different economy set in.

The unwinding of Roman institutions was not the only change to affect the water landscape of Europe in the middle of the first millennium. The climate system also changed. During the sixth century, central and northern Europe became wetter and colder, while the Mediterranean region got drier. What caused this shift is unclear. In 538 CE, Cassiodorus wrote to his deputy describing a strange climate, one in which "the Sun, first of stars, seems to have lost his wonted light, and appears of a bluish color."

The most plausible hypothesis is that a series of volcanic eruptions—in 536 CE first, and again in 574, 626, and 682 CE—sent large quantities of sulfur dioxide into the stratosphere, dimming the amount of sunlight reaching the Earth's surface. The effects of these eruptions could last up to a decade, during which winters had fewer storms, and summers little heat. Whether that was the specific cause or not, what is known is that cereal production, which was particularly sensitive to summer sunlight, temperature, and rainfall, was hit hard.

The collapse in food production weakened the population. Then, the Justinian plague of 541 CE, *Yersinia pestis,* the "pestilence by which the whole human race came near to being annihilated" as Procopius described it, ravaged communities across the Mediterranean. During the first of many outbreaks, a significant fraction of the population of Byzantium and Europe died, millions of people. The climate continued to worsen. The last quarter of the sixth century recorded the highest rainfall for a thousand years. Rivers like the Tiber and the Rhône flooded. Gregory, the bishop of Tours, described devastating floods that hit central and southeast France around 580 CE. As people abandoned the increasingly dangerous river valleys, forests regrew to cover 40 to 80 percent of the landscape, especially east of the Rhine and across northern Europe. Open fields became rare.

Then, sometime during the eighth century, temperatures increased across Europe. Climate conditions improved. It is not clear why. One possibility is that the temporary regrowth of forests that followed the fall of Rome had decreased the amount of sunlight reflected by the continent, increasing the amount absorbed by the landscape. The recovery may also have been due to internal dynamics of the climate system. Whatever the cause, the milder climate once again pushed up demand for arable land, and farming expanded, both northward and to higher altitudes. Forest clearing restarted. Yields increased. After a long contraction, in the eighth and ninth century the population began growing. It was time to rebuild.

MONASTERIES CONTROL WATER

The conflicts that would eventually lead to the development of a common legal system were the result of the profound fragmentation in territorial ownership and management that had been left by the conflagration of Rome. With Roman institutions gone, others emerged to govern the landscape. Along with Byzantium in the east, the Church of Rome was the principal inheritor of the Roman legacy in the west. The Christian Commonwealth was not just the spiritual glue that held medieval Europe together. It also became its most developed territorial institution. However, the Church did not exercise territorial sovereignty like a centralized state. Rather, it administered the landscape indirectly, through a vast network of monastic communities.

The story of water from this point onwards was driven by the deep fragmentation of those institutions. That monasteries should acquire that role at all, let alone on a scale sufficient to drive a transformation of the water landscape, may seem at odds with their original pursuit of ascetic hermitage, solitude, and contemplation. When Pachomius, the fourth-century ascetic, had introduced monastic communities in the desert of Egypt, he could hardly have had hydraulic engineering in mind. But by the fifth century, Saint Augustine's *De Opere Monachorum* instructed monastic communities to work the land. When the Vandals invaded, his monastic rules migrated from North Africa to Italy, taking root there. Eventually monasteries took over building the mills, aqueducts, weirs, and drainage systems required to harness the landscape.

While monasteries may have taken over the development of infrastructure, the Church as a whole took a rather distant approach to economic issues, and to those concerning the landscape in particular. Even though Augustine's rule spoke of working the land, from an ethical and theological perspective the Father of the Church believed a

Christian society should not concern itself with the material world. He had been deeply affected by the sack of Rome. If even Rome could fall like Babylon, he must have thought, the contemplation of God should be the sole focus of Christians.

In *De Civitate Dei*—Augustine's magnum opus—there was precious little room for concerns as mundane as the management of water. The only city Christians ought to concern themselves with was New Jerusalem, the city of God. Ethics and theology trumped practical politics. This was an important shift. By "city" Augustine did not mean the physical place but the collection of citizens, the *civitas*. He moved the focus of virtuous citizenship away from a res publica organized around the pursuit of shared material benefits and bound by law, towards a primary concern for the spiritual life.

Because of this shift, the Church's decisions on the material world increasingly shunned the economic and turned to the moral. As the monastic movement spread and as the Church grew its territorial influence over Europe, water, the very agent that had shaped people's environment for the previous five thousand years, became politically invisible. This left a significant vacuum in the architecture of rules communities lived by, making it close to impossible to develop a collective response that exceeded the boundaries of one's property. This was in stark contrast to the Roman world, which had developed sophisticated legal and commercial institutions to balance individual ownership and collective coordination.

There is no better allegory for this shift than that of the life of Saint Benedict, the founder of Western monasticism. In his sixth-century *Dialogues,* Gregory described Benedict of Nursia's life of hermitage in a natural cave above a secluded, bucolic lake near Subiaco, a town east of Rome. The lake was important in the stories of Benedict, mostly because it was a contemplative setting where several of his miracles took place.

But the lake was not the pristine natural setting those stories suggest. It was entirely artificial. Both Pliny and Tacitus had described it as one of three massive artificial reservoirs—700 meters long and 150 meters wide—that Emperor Nero had built to decorate his villa on the banks of the River Aniene. It was held back by a dam, which at 40 meters was probably the tallest in the world until the Alicante dam of 1594.

As an allegory, it is fitting. Roman society had built institutions and infrastructure to bend the landscape, and people's behaviors on it, for its benefit. By the time Benedict came along, a unified state had all but gone, and even the infrastructure that was left was no longer an expression of power, but simply part of the landscape. Managing water in the early days of Saint Benedict was less about infrastructure and more about prayer.

DEFINING A FRAGMENTED LANDSCAPE

The territorial fragmentation that took over the landscape, particularly in northern Italy, led to a myriad of irresolvable small conflicts and competitions between local potentates. Rivers and waterways, whose flow could hardly take sides, was caught in the fighting. Systematic conflicts over water were set up by the political developments of the last centuries of the first millennium.

Inland navigation on the Po River resumed with the Lombard king Liutprand, who awarded navigation concessions to the coastal city of Comacchio in 715 CE. Boatmen were once again allowed to trade salt upriver, in exchange for payment of duty at ports in Lombard territory. Inland navigation turned the Po into a domestic market that stretched to Pavia, the capital of the Lombard kingdom, three hundred kilometers from the coast.

When Charlemagne descended across the Alps and conquered the Lombard kingdom in 774 CE, he annexed northern Italy into the Carolingian Empire, integrating Europe once more into an imperial whole. Charlemagne's political consolidation was extensive, but while he aspired to Roman greatness, his institutions were structurally thin. He did not rebuild a tax system or an administration comparable to that of Rome. Absent a strong administrative state, the Italian landscape ended up in the hands of a network of local institutions, lords and bishops who managed more or less on behalf of the sovereign.

To govern communities in Italy, the Carolingians relied on bishops more than they did on the local aristocracy, so the structure of the Roman state was replaced by the temporal power of the bishops. During this time, monasteries, particularly those of the Benedictine order, also grew as an economic power. By the end of the tenth century, local power in northern Italy came from extensive land ownership along waterways, much of which was under the Church, either directly through its bishops or indirectly through monasteries.

The two institutions that claimed universality in medieval Europe were the Church and the Holy Roman emperors, successors of Charlemagne. This universality did not manifest itself through direct rule over territory, however, as much as through a complicated web of personal and institutional allegiances. Those networks were often overlapping, particularly in the fragmented world of northern Italy. Emperors gave monasteries and bishops concessions to extract taxes from activities on the river, as a means of exercising some control over them. Therefore, the Church and the empire competed through territorial patronage, leading to a further deterioration of any common management of the river.

The case of the Abbey of Polirone, the most important Benedictine monastery in Italy, is emblematic. Its story starts with Adalberto-Atto, the founder of the Canossa family, who was made a count by Emperor Otto I. Canossa chose Mantua as his capital, in part because it allowed control of the family's dominions around the river Po. But the Canossa family also needed to tap into the legitimacy of the Christian network. They did this by founding a monastery.

Adalberto had acquired the island of San Benedetto, between two branches of the river Po—the old Po and the Lirone—at the end of the tenth century. The island was covered in woods, frequently flooded, and uninhabited. It seemed economically worthless, but turned out to be strategically priceless.

In 1007, Count Tedaldo of Canossa, Adalberto's son, converted a small church on the island into a monastery, calling it San Benedetto in Polirone after the two surrounding rivers, and endowing it with half the island. Once the abbey entered the circuit of Cluny in 1077, it became the most important institution for Gregorian reforms in northern Italy. At the beginning of the twelfth century, Matilde, Tedaldo's granddaughter, donated the other half of the island to the monastery and the abbey became the largest Benedictine community in Italy, exercising its influence over a number of affiliated abbeys, priories, and hospitals, and several fortified villages. San Benedetto in Polirone thus found itself a node at the intersection of imperial and ecclesiastical power.

If the complexity of imperial and ecclesiastical power wasn't enough, a third important actor entered the fray. After a period dominated by an entirely rural economy, cities grew in importance in the early second millennium of Europe. They were seats of both temporal and religious power, particularly in Italy. They were also crucial for the agrarian medieval economy, as city markets were the principal venue for the distribution of agricultural production. Cities were the ultimate sedentary structure, dependent on the active management of water and on infrastructure that could convey it where and when needed.

Churches and monasteries were often involved in the development of water infrastructure to serve cities. In 1269, Milan levied a tax to pay for the conversion of its canal system into a fully functioning navigation system. The Cistercian Abbey of Chiaravalle, near Milan, helped by bringing water to the south of the city, enlarging a pre-existing irrigation canal. While monasteries like Chiaravalle adopted an emollient approach to their dealings with cities, others like Polirone were more intent on projecting power over them. The abbey used its wealth to invest in its lands as commercial enterprises, and made claims on navigation rights on the Po in conflict with other centers. In the second half of the twelfth century, Polirone entered into conflict with cities and other monasteries for the control of waterways for the lucrative salt trade to Chioggia on the coast.

There was one final complication in medieval competition over water resources and landscape: the distinction between individual and corporate ownership. This distinction was particularly important for the Church. Members of the clergy did not typically own land or rights to the use of rivers as individuals, but as an institution. This distinction allowed the Christian Church and its monasteries to accumulate wealth while continuing to exercise moral and spiritual leadership. It was a powerful system, which led the Church and the monastic movement to ultimately accumulate close to a third of available land in Europe. But it reinforced the atomization of the political landscape. This extraordinary fragmentation made the medieval world prone to aggressive litigation, which is what drove the next crucial step in the development of modern water institutions.

LANTERN OF LAW

The solution to the fractiousness over the use of water could only come from a common set of rules. After all, everybody's ownership of landscape and access to water resources was tied to having a viable legal system that could establish land title.

The most important system of rules in the European early Middle Ages was canon law. Charlemagne's efforts to establish a Christian empire had fueled intense interest in canonical sources, and in the eleventh century, Ivo, bishop of Chartres, redacted one of the most extensive syntheses of canon law in his *Decretum* of 1096 CE. In such texts were the echoes and fragments of the Roman legal tradition, but the problem was that, much in keeping with Augustine's direction, canon law did not typically concern itself with the "public," the legal responsibilities of the state towards the commonwealth. The Church was really its subject. It was concerned with providing a framework for Christian theology, not for practical management. Water, with all the public issues it posed, was practically absent.

However, at the end of the eleventh century, Justinian's *Corpus Iuris Civilis* resurfaced. The Roman legal code had last been mentioned in a letter from Gregory the Great in 603 CE. It had then largely disappeared from the sources. As soon as it was rediscovered, it became standard curriculum at the newly established University of Bologna, in northern Italy, the pre-eminent European center of learning. Its high priest was Irnerius, a legendary scholar, so influential as to earn the title of *"lucerna juris,"* the "lantern of law."

Irnerius's teachings restored the integrity of Roman law to Europe, revealing a legal system of a lucidity and coherence that had been long forgotten. Because water played a fundamental role in the medieval economy it was inevitable that jurisprudence would focus on it. For example, in December 1125, Irnerius arbitrated over a piece of marshy land near Mantua, between two canals that poured their waters into the Mincio River. The population of the nearby fortified settlement of Casale used the territory, as it was rich in fisheries, woodlands, and cultivable land. The Abbey of Polirone had authority over Casale, and declared sovereignty over it. But so did the abbot of San Zeno, a monastery near Verona. Four jurors were called in to settle the matter—Irnerius was one of them—and found in favor of Polirone.

The law had provided a reliable and acceptable mechanism to settle disputes. Roman law, along with its ecclesiastical counterpart, canon law (eventually summarized in the *Decretum* by Gratian, another scholar from Bologna), came to be known as *ius commune*, the common law. It was the integration of the laws of the two universal institutions: the Roman Empire and the Church. Its teaching spread far. Elites from around Europe came to study in Bologna and, upon returning home, introduced the methods and principles of the *ius commune* there. And it is through that common ancestry that legal approaches to water spread.

The celebrated English Magna Carta had its roots firmly in this legal tradition. It provides the most famous example from this period of how temporal and religious claims could be translated into a system of rules to govern all. The concessions which English barons first received in 1215 from King John on the fields of Runnymede reflected a mix of Roman private property tradition, moralizing canon law, Norman legal traditions, and the realities of medieval power relations in England. It was full of language and references to the *ius commune*. Although today the Magna Carta is remembered for its proclamation of rights, most of it was concerned with administrative issues, including water.

The management of water entered the charter through the concerns of the city. In the complicated negotiations between the barons and King John, the citizens of London had drawn up demands for the king. The first on the list, the most important, was "Concerning the Thames, that it should be absolutely and wholly the city's." London's ability to command the use of the river would have provided a huge economic benefit, ensuring its security. This demand was received in Magna Carta as clause 33, part of a small group of clauses that singled out the city of London.

Clause 33 dictated that all fish weirs should be removed from the Thames and other rivers of England. The fish weir was a large V-shaped weir, made of rows of upright poles stuck in the riverbed, connected by transverse beams, interwoven with flexible branches. Between the two lines of poles, a net caught salmon coming upriver, or eels and other fish as they moved towards the sea. But clause 33 was not really about fishing: the provision was an indirect way of introducing freedom of navigation, by eliminating its most frequent obstruction.

Freedom of navigation would have been known to the compilers of the English statute from both the *Digesta* and the *Institutiones* of the Corpus Iuris Civilis. In fact, the glossator Vacarius, the first teacher of Roman law at the nascent university of Oxford, had studied in Bologna and had referred to those specific provisions in his *Liber Pauperum.*

The complex web of power relations that had emerged out of the conflagration of the Roman order had finally been resolved by the establishment of a "universal" set of rules. While other European medieval institutions with universal claims—the Holy Roman Empire and the temporal power of the Church—would ultimately be swept aside by modernity, the legal framework that emerged out of the Middle Ages not only persisted but spread. This was the route through which Rome made it back into the modern world.

SOVEREIGNTY ON THE RIVER

The re-introduction of Roman civil law did not just provide a body of jurisprudence to settle disputes on the use of rivers. It had profound consequences for the governance of water resources. In particular it provided the starting point for the definition of territorial sovereignty, which would transform the fate of water in Europe and, eventually, the world.

The source of this institutional innovation was conflict over who could claim ownership of the land. The eleventh-century conflict between the pope and the emperor over investitures concerned the right to appoint bishops, and was to a large extent about land. Given the territorial control of ecclesiastical institutions, it is not surprising this was a matter of contention. The full force of the newly rediscovered Justinian code might have seemed to the emperor just what he needed to justify his control over the landscape. After all, Roman law had been developed when one, and only one, emperor was the source of all power. But there was a complication.

When Italian cities, formally still under the authority of the emperor, began asserting their independence, things got substantially more complicated. During the twelfth century, many cities in northern Italy had transitioned to being communes, a form of republican organization that turned them into more or less self-governed corporations. When Emperor Frederick I attempted to reassert control, Italian cities coalesced in the Lombard League and defeated the emperor at the battle of Legnano in 1176. The settlement of the Peace of Constance in 1183 gave the members of the league even further autonomy, accelerating their republican transition.

As it happens, the independence of cities unleashed one of the most fertile periods of fluvial transport in the Po basin. Cities needed to trade with each other and form a market for intermediate and final goods if they were to supply an increasingly connected Europe. The fluvial system had grown so much that it had begun creating problems for investors in hydraulic infrastructure. The abundance of water and the relative ease with which people could dig the clay soils meant that any additional canals were typically a matter of private agreement. But inevitably, once dug, these canals would become part of the navigation system. In Roman law, navigable waterways were public, and the state—therefore the emperor—had jurisdiction over them. Under those conditions, anyone could draw water freely as long as it did not harm others. This created substantial problems to those investing, for while they paid good money to build infrastructure, anyone could then benefit from it without having to pay. The problem was that imperial primacy, as enshrined by Roman law, prevented city-states from ruling on the landscape by themselves. Countless legal conflicts emerged which in principle could only be resolved by recourse to imperial authority. The transition of sovereignty to cities required revising the practice of the Justinian Code.

The fourteenth-century jurist Bartolus of Sassoferrato finally freed cities from imperial obligations. Through his work he defined, for the first time in the post-Roman world, the legal idea of territorial sovereignty. It happened in a set of logical steps.

First, among his many contributions to European civil law tradition was the idea that common practices could eventually constitute the basis for jurisprudence. This gave cities ground to establish a claim. The fact that cities had been exercising sovereignty over a long period of time implied that, even if they could not prove a concession from the emperor, they de facto had one. Then came the nature of their claim. Their jurisdiction applied to territory. What was radical in Bartolus's doctrine was that territory was not just the property of a ruler, but an object of rule, itself subject to jurisprudence. It was, in effect, the end of the universal temporal jurisdiction of both emperor and Church on all land.

Bartolus introduced this innovation in 1355 when, supposedly inspired by a walk along the Tiber, he composed the *Tiberiadis*—also known as *De Fluminibus*. In fourteenth-century Italy, territorial issues on rivers were no marginal matter. Rivers moved, meandered, transformed the landscape. Floods would leave alluvial deposits that changed the borders of the tiny, fragmented city-states. Rivers created and destroyed land. Conflicts between cities arose regularly and the inconsistency of different statutes made resolving them complicated. The question of what law would apply to such deposits was very material. To answer this question, Bartolus moved from arguments based solely on the rights of the owner, to those based on the relationship between people and things. The ownership of new alluvial deposits—farmland where none was available, for example, or a new island—was not just a matter of ownership, but a matter of the relationship between that new land and whatever was around it, and the relationship of the latter to people.

New land might acquire the properties of whatever it accrued to. For example, ownership could be attributed by proximity: unclaimed land should go to the nearest town or village. This seemingly technical point had a profound implication. By focusing on the relationship between the property and the owner, Bartolus had converted an issue of personal rights into a geometrical, territorial problem. He had begun to establish the basis for territorial sovereignty, a legal basis that survived in continental Europe until the Napoleonic Code.

The conflagration of Rome had left a multitude of unresolved conflicts on the landscape. But from the rubble of the Western Empire, not aqueducts or canals, but legal institutions had emerged as its most fundamental water-related bequest to the Middle Ages. Out of it came the rules that would govern the use of water in Europe and, eventually, in much of the world. But the legal problems posed by rivers to an emergent urban economy did more than that. They provoked the development of the idea of territorial sovereignty, which, a few centuries later, would become the basis of another fundamental evolution, again tightly linked to a river: the rise of the modern nation-state.

8

The Republic Returns

A FINANCIAL REVOLUTION

The recovery of republican forms of government from the Roman tradition was a crucial step, maybe the most important one, in the transition to a modern relationship with water. It was through the republic that a society could express its collective agency against the force of water in the pursuit of a commonwealth.

The journey back to republicanism was complex. Republics were not in great favor in mainstream political philosophy in the Middle Ages. The long imperial experience of Rome had overshadowed the original institutional innovation that had fueled its expansion. The persistence of Byzantium in the east, the presence of the Church and of the Holy Roman Empire in Europe, had all reinforced the idea that the true Roman inheritance was that of the Augustan era. Dante's plea for the rebirth of Italy was for a return of imperial Rome, for a universal monarchy, not for its republican predecessor.

Nevertheless, Roman republican ideals made it back into Europe much in the same way that Roman law was recovered. One of the side-effects of the university system, which had started in Bologna and spread quickly across Europe, had been a renaissance in the teaching of rhetoric using Sallust's histories and Cicero's orations, which lent themselves to the format of a lecture. These texts had resurfaced during the Carolingian renaissance of the ninth century, and became powerful vehicles in their dissemination among the elites, celebrating

Roman civic virtue. Cities, not just in Italy, were early adopters of republican government from the twelfth century onward.

There was nothing obvious about the success of republics. Indeed, most of those early experiments failed after a relatively short time. However, the period between the twelfth and early fifteenth centuries witnessed a number of circumstances which proved essential to consolidate republican aspirations in at least part of the European population.

First, the rise of republican governments in cities refocused society from feudal relations back towards citizenship. It gave voice to the entrepreneurs and laborers that were the cities' producing heart. The rise of a commercial class shifted the emphasis from subsistence agriculture to the infrastructure they all needed to connect to a bigger marketplace for their products, which fueled an interest in water and navigation infrastructure. But to build it, they needed finance.

The financialization of water infrastructure by commercial investors was another crucial step which helped consolidate the importance of republican government. During the early Middle Ages money was hardly ever invested in productive infrastructure beyond one's property. Savings had been mostly hoarded. Finance was limited by a fundamental moral constraint. As Thomas Aquinas pointed out, the value of coins remained unaltered as they exchanged hands, ergo a loan of money ought to be returned in the same amount, lest one fall prey to sinful usury. This made paying interest impossible.

The moral limits on finance were in stark contrast to the Roman tradition. While the Roman state could not borrow, it could lend money. In fact, in early imperial Rome loans were common to finance consumption and production, as well as trade. Back then, as a matter of course, farmers knew to include interest of up to 12 percent as part of their costs. Concession contracts that rewarded capital investment were the primary mode of infrastructure development. But in the early medieval world, morality governed market behavior. If financing was going to happen, the moral restrictions on usury had to be overcome.

The first successes in overcoming these restrictions were not in infrastructure but in trade. Traders needed credit to absorb the risks inherent in commerce, and they became skilled at exploiting a gray area: even within the moral confines of early medieval practice, it was acceptable to compensate a lender if a loss had been incurred. If someone had lent a horse which then was returned lame, it was acceptable to seek compensation. To unlock finance, the central question was whether the loss of alternative profits over the time of a loan—the *opportunity cost* of capital—was equivalent to being returned a lame horse. The answer seemed to be yes, if the relationship between lender and borrower could be framed as joint ownership of risk.

The right to property had deep religious roots in all Abrahamic traditions. It also had classical ones in the authority of texts like Aristotle's *Oikonomika.* Because of this, most scholastic theologians accepted that human beings had to satisfy their needs through exchanges of property and money. Even the Franciscans, who had made poverty their core belief, had come to this conclusion. Ownership in a partnership to share risk—a company—meant that payment of interest could be compensation for a risk taken. Companies initially developed around families pooling their wealth. Then distant relatives joined in. Eventually, investment capital from strangers arrived, finally leading to deposits. The modern banking sector was born.

The relationship between finance, water, and the rise of the republic tightened. As the mechanics of finance expanded, they unlocked savings and investment. Money supply, which in Europe had been relatively fixed up to that point, expanded. In search of opportunity, companies shifted from trade to manufacturing, where risks were lower. Doing business in manufacturing required infrastructure, and particularly water infrastructure, both because water was the only real source of energy aside from human and animal force, and because access to markets required inland navigation.

The example of Bologna is emblematic. The city in northern Italy was a hundred miles from the sea, but nevertheless became one of the largest ports in Europe. Bologna was on the river Aposa, a small torrential tributary to the river Reno, which at the time flowed into the Po. The Aposa was too small to support economic activity on any significant scale. Then, in 1183, a group of private entrepreneurs created a company to develop a dam and a seven-kilometer canal to divert water from the river Reno and use it for their production, mostly wheat mills and wool fulling. That masonry dam in Casalecchio, a small town outside Bologna, is still in operation today as the oldest such hydraulic structure in the world.

In 1208, the entrepreneurs of Bologna entered into a contract with the city government, which could take canal water and use it for its own purposes. The city would not have been able to finance the original works, but once the dam and canal had been built, it was able to pay for its ongoing maintenance. It was in effect a concession contract, in which the owners gave water in exchange for the payment of maintenance cost, while retaining use of the water for themselves. With that water, the city was able to feed its *Navile*, a navigable canal on which finished products could then be shipped through the fluvial network to Venice, where they were traded into the global markets. By the fifteenth century the city had become one of the most important fluvial ports in Italy, the fifth-largest trading port by traffic of goods in the world.

The company had become the principal mechanism for landscape development, an instrument for the delivery of public goods. But the introduction of finance did more than make it possible to deliver projects. Republican governments gave greater political power to a larger set of people, which forced a shift in how states, even small ones like Italian cities of the thirteenth century, viewed decisions about the landscape. When, in 1217, representatives of small artisans, notaries, and merchants entered the republican government of Bologna, they enacted significant institutional reforms, including the forced purchase of the water infrastructure operating on the Reno. This expropriation, an act that would be replicated in a number of cities across northern Italy, meant that whatever proceeds came from the use of water infrastructure would now feed the treasury of the city, and that the city could make decisions about that same infrastructure in the interest of the entire population.

The collision of finance and republicanism had led to state-led development strategies that framed decisions in economic rather than just legal terms. This was a radical transformation. When applied to water, these terms change the role of projects in society, from the object of conflicts over their use, to political instruments in pursuit of a public good.

THE END OF UNIVERSALITY

The development of finance and of a commercial class was not enough by itself to ensure the success of any civic project. The rise of republican governments would have been impossible while the Holy

Roman Empire and the Catholic Church maintained strong claims of universality. But those claims disintegrated with growing exposure to the more sophisticated societies of the East. That exposure had a particularly explosive moment during the thirteenth century, which accelerated the end of a view of the world limited to the European experience.

China's fabulous wealth had captured Western imagination even as the Western Roman Empire dissolved and Islam took over the Persian Empire. However, contact between the West and China during the early Middle Ages had been limited to occasional exchanges through Arab traders. That changed in the early part of the thirteenth century.

Harsher conditions in far northeast Asia pushed nomadic tribes southward, where they found wetter, warmer conditions. These conditions also lifted the water table, increasing the productivity of grasslands. Grasses were fuel for horses, the backbone of the formidable Mongol army. Much like the Huns of a thousand years earlier, under the leadership of Genghis Khan, the Mongols followed the steppes westward across Asia and into eastern Europe, entering the plains of Hungary in 1241. They quickly overwhelmed the local population, plundering and killing, even reaching the Adriatic in pursuit of their enemies. The impression they made on Europe shattered any illusion of domestic superiority.

The episode was as intense as it was brief. In the end, the European water landscape, rather than local defense, defeated the Mongols: the exceptional thawing of the spring of 1242 left behind swampy terrain, which made movement by horse difficult. Harvest failures left them little to plunder. By late 1242 the Mongols had withdrawn from Hungary into the steppes of the lower Volga, never to return. But the door to the East had been opened wide.

As European eastward trade increased during the course of the century, contact with China grew. Many Europeans learned of China's great Kublai Khan from Marco Polo, a Venetian who, in 1271, left Venice for China at seventeen and only returned twenty-four years later. Polo and his uncles arrived at the court of the Great Khan in 1274, five years before the fall of the Song dynasty. Having just defeated the northern Jin, the Mongol Yuan built their new capital on the site of modern Beijing—then known as Dadu. The capital attracted an enormous population, too large for local agriculture.

The Yuan dynasty had assumed that the Yangtze Valley, which had fed Chinese capitals since the third century, could once again supply the grains. The problem was how to get them to north China. The Yellow River was no longer a viable option for navigation, and the region did not have other major waterways: the century-long conflict between the Jin and Song empires had destroyed any north-south transport routes. Coastal navigation was the obvious solution, but that route was far riskier than they had anticipated. For example, in 1288 a quarter of the shipments of grain coming up that way were lost to storms. Rounding the Shandong Peninsula into the Gulf of Bohai with a large cargo could be dangerous. Inland navigation was the only real alternative, but a route had to be built.

The fluvial project to connect southern and northern China did not start from scratch. The canals that supplied the old capitals of Xi'an, Luoyang, and Kaifeng were still more or less functioning, although they did not reach as far north as Beijing. So was the seventh-century canal from the Yangtze to Hangzhou. The Yuan began digging. Work started more or less at the same time as Marco Polo arrived in China and was completed in 1293, just after Polo left. By linking old and new canals, the resulting Grand Canal connected Hangzhou to Beijing. At over eleven hundred miles, today it is still the longest artificial canal in the world.

In truth, the canal was far from easy to maintain. When the Ming dynasty inherited it from the Yuan in the fourteenth century, they struggled with the economics of such a large piece of infrastructure and its efficient use. Yet back in Europe, a world of city-states and feudal lords was left bewildered and awestruck by what they heard of China and its canal. For centuries, Western visitors like the Jesuit Matteo Ricci, in the sixteenth century, wrote of it with admiration, impressed by tens of thousands of boats carrying rice and grain. Two centuries after Ricci and five after its construction, Adam Smith was still describing it as a key source of comparative advantage for the Chinese economy, still ahead of its Western counterparts, and still making a compelling case for the importance of state-led water infrastructure.

China had not only shown that the medieval institutions of Europe were far from universal. It had also demonstrated how command of landscape could define the use of water resources in the national interest.

CRISIS

Beyond the mixing of finance and republican government, further fuel for the transformation of the European water landscape came from a crisis. In the summer of 1315, the archbishop of Canterbury, in England, ordered his clergy to go out barefoot, bearing sacraments and relics, in the hopes of convincing people to atone for their sins and appease God's wrath. It did not work.

Torrential rain destroyed the harvest. The beginning of the fourteenth century heralded a period of climatic instability and demographic collapse across Europe. The change marked the shift towards a "Little Ice Age" that would last several centuries. Between 1315 and 1322 across northern Europe, a sequence of unusually wet summers and cold winters damaged up to half of the agricultural production. The climate shift also took its toll on sheep herds wintering outside, while disease ravaged cattle all across the Northern Hemisphere, from Mongolia to Iceland.

The economic consequences of this climate shift were catastrophic. Even salt prices skyrocketed, because the poor weather increased demand for it as a preservative. By the spring and early summer of 1316, famine had struck in northern Europe. By 1317, peasants had exhausted all means of survival, thoroughly squeezed by their lords, and moved to cities. Malnutrition and cramped living quarters led to outbreaks of typhoid and other diseases. To top it all, in 1346 a deadly plague appeared. First it was in Crimea. Then the disease moved clockwise around Europe, spreading rapidly across the malnourished population, ending in Moscow in 1352. By the end of its first cycle, about a third of the population of the continent had died.

All these events led to a radical transformation of European society. Disease reduced headcount while leaving wealth—savings, capital stock, machines—virtually untouched. As a result, per capita wealth increased while total labor force decreased. These two trends tipped the scales in favor of finance, providing a powerful incentive for investments that could increase labor productivity. Water was the sole source of mechanical energy, so the concentration of savings stimulated unprecedented investment in water infrastructure to support manufacturing.

Bologna, for example, saw an explosion in the adoption of hydraulic silk mills for twisting thread. Small derivative ducts known as *chiaviche* allowed manufacturing to abandon the large hydraulic milling wheels that needed open-air canals in favor of much smaller overshot vertical wheels that could fit in a city basement. Taking advantage of its canals, a capillary system of water distribution conveyed water into basement factories all across Bologna. The lighter basement wheels transferred two- or three-horsepower torque to relatively light, mechanized wheels in the floors above, which could spool four thousand threads at a time. Standardization of thread sizes, which also happened at this time, lowered costs further.

By the end of the fourteenth century, there were fifty-two hydraulic wheels in Bologna, powering a complex industrial ecosystem: fifteen wheat mills, sixteen silk factories, and another twenty factories for metal, wool, and other manufactured goods. Investments in navigation infrastructure then connected the manufacturers to the open market, fueling a positive feedback between the demand for textiles in the rest of Europe and the decreasing cost of supply coming from the manufacturing base of the city. It was industrialization several hundred years before the steam engines of the English wool mills, unprecedented in Europe and in the world.

Italy was not the only example of water-led transformation. Intensification in the context of a changing climate happened all across Europe. The Low Countries, for example, where people lived only a few meters above sea level, was another case in point. There, catastrophic floods could wipe out inland communities, so people depended on dike systems, canals, and embankments for protection against the encroaching sea. The Little Ice Age made matters far worse, as higher seas and wetter conditions required greater defenses. The complicated system of infrastructure required to manage the deteriorating conditions was expensive. Most villages did not have the capital, organization, and know-how to construct and maintain it, so water authorities evolved to do so in their stead. Water authorities were a mix of bottom-up cooperation among multiple villages, and benevolent encouragement by lords or bishops, which led to centralized management. This was another Faustian bargain. Successful drainage only made bogs sink further, increasing the risks and requiring even more infrastructure.

As more land flooded, people shifted from crops towards animal husbandry and dairy production. This rural specialization was less labor-intensive and more profitable. As a result, many moved to cities, while the capital of city merchants was invested in farming. Dutch society evolved towards a market economy. Eventually, the windmill made its first appearance. In a document from 1408, Count Willem VI of Holland told the water authority of Delfland to visit Alkmaar, northwest of Amsterdam, where a drainage mill had been built. The idea proved popular. Over the course of the fifteenth century over two hundred of them appeared. To finance them, water authorities imposed annual milling taxes, while landowners provided the capital in proportion to the size of their property.

Because the costs were concentrated disproportionately on some people while the benefits were distributed to all, the introduction of mills eventually led to the creation of embankments called polders, separating areas owned by different investors, so that they could protect their benefits. Management of the drainage system was entrusted to a polder board, professionalized by hiring dedicated personnel, and taxes were increased to pay for it. Polders had considerable control of water flow, and, as they grew, interactions among them could lead to serious problems on a regional scale. If someone lifted more water into the canals than could be drained, someone else might get flooded. All of this inevitably required additional planning. Regional authorities were born.

The stories of northern Italy and the Netherlands show that, during the fourteenth and fifteenth centuries, the prototypes for the institutions of capitalism had emerged around the use of water. The development of the modern company, the exercise of territorial sovereignty by small city-states, and the existence of a legal framework for the enactment of contracts were all essential to allocate risks correctly and get the economics of water projects to work. By the sixteenth century, in the Low Countries the delicate balance between local control and regional coordination that had emerged from the experience of the water system inspired the birth of the Republic of Seven United Netherlands.

THE NEW WORLD

As radical ideas about republican governments gained traction, their scope was further challenged by the dramatic expansion of the European world not just eastward, but also westward. October 12, 1492, was a crucial milestone in the expansion of the medieval world. Christopher Columbus's voyage westward, meant to reach the East Indies, ended instead on the beaches of Guanahani, in the Bahamas. He had found a continent.

With the 1493 papal bull *Inter Caetera,* Pope Alexander VI "gave" the new land to the Spanish Crown in exchange for a promise to evangelize the people there. Over the course of the sixteenth century, the Spanish and Portuguese crowns got to work and conquered most of the Western Hemisphere, with the French and the English not far behind. The rural economy of the European feudal period was a thing of the past.

Merchant groups allied themselves with the state to exploit the possibilities of this expanded world, and the distribution of power shifted decidedly towards those who dealt with money. The resources of the New World completely transformed the power relations in the Old, not only because they further demolished any universality claims of European institutions, but also because the influx of precious metals and the access to immense agricultural potential fueled the rise of an enormously powerful merchant class, whose political claims would propel republican demands.

The discovery of a continent entirely new to Europeans stimulated a reorganization of power relationships in Europe. A measure of just how radical a shock it was is given by Thomas More's description of the perfect society in his *Utopia.* Its fictional protagonist, Raphael Hythloday, had traveled with Amerigo Vespucci to the equatorial New World, only to be left behind and to discover the republic of the Utopians. He imagined a state inspired by the republican ideals of ancient Rome, but actually based on communal, agrarian practices with no private property. More's *Utopia* relied on an extraordinarily benign river, the Anyder, loosely based on the Thames, but whose name meant river "without water."

More's *Utopia* reflected an emerging idea that the perfect society would in essence be emancipated from the vagaries of nature. No floods or droughts threatened the idyllic island, no unmanageable riverbanks or overwhelming storms complicated the lives of its people. It purported to describe a fundamentally agrarian society, but More's *Utopia,* just like Plato's *Republic,* was more concerned with the moral psychology of statecraft than with the material conditions in which society had to operate. More's writing was an indication of where the politics of Europe were headed. Urbanization, the rise of commerce, and the increase in wealth had shifted the focus towards the individual.

In truth, had the societies of the Americas survived the onslaught of European invasion, the story of water might have turned out quite differently. At this point, there was nothing inevitable about the particular relationship that European societies had developed with their water landscape.

Those living in the American tropics had developed comparably sophisticated modes of managing water. Some early explorer accounts provide fragmentary evidence of what that world looked like. For example, Friar Gaspar de Carvajal, who traveled down the Amazon alongside Don Francisco de Orellana in 1541, described a densely populated Amazon, with cities of sorts along the banks of the river and plenty of evidence of trade.

Rainforest societies were able to develop complex food systems and forms of urbanization because they had adapted to the particular hydrology of the rainforest. The dominant source of rain in a tropical forest is the vascular system of plants, which transpires water vapor through the leaf's stomata during photosynthesis. Unlike arid or semiarid climates, how much rain falls down locally in a tropical rainforest is a function of such plant transpiration more than direct evaporation from the ground. The trees act as a vast water pump, in which roots tap large reservoirs of water in the soil and evaporate it in the sky. The water is then replenished in the brief but intense rainy season.

This situation made developing intensive rainfed agriculture similar to that of Europe nearly impossible. Instead, the population of the Amazon did not just domesticate crops or animals as their European counterparts did. They had the far more complicated task of domesticating the entire landscape. Their rivers were highly productive, yielding up to a ton of fish per hectare. People used artificial mounds above the flood level to create ponds that could store surplus fisheries, and used those earthworks to grow fruit plants, complementing the managed wetlands. It was an intensely managed ecosystem, supporting a large population. In the Bolivian Amazon, for example, there are traces of several-hundred-square kilometers of hydraulic infrastructure in the forest, from between the end of the fifteenth and the beginning of the seventeenth century: straight canals and causeways up to several kilometers long and permanent fish weirs.

Several studies suggest a pre-Columbian population of about sixty million across the continent at the beginning of the sixteenth century. The impact of the European arrival annihilated almost all their traces. The combination of previously unseen diseases—smallpox above all—slavery, and forced labor decimated the population and collapsed societies. Within a century, the Amazon appeared as nothing more than a sparsely populated forest, with a few tribes living in relatively simple communities. The population had dropped 90 percent, to around six million people.

Because farmers cultivated tracts of land on scales of a few hectares, the disappearance of the indigenous populations meant that during the course of the century over fifty million hectares were abandoned. Within a few decades, forests and vegetation regrew. The regrowth in forest was so significant that it caused a drop of about ten parts per million in the concentration of atmospheric carbon dioxide over the course of the sixteenth century. That is the same amount of carbon dioxide that industrialization added to the atmosphere over the course of the nineteenth century.

And so what came back from America were not its sophisticated landscape experiences but raw materials. Thanks to the influx of resources, by the end of the sixteenth century, gold had become the basis for the global economy, and the lure of El Dorado, a mythical city of gold hidden somewhere in the Amazon forest, proved irresistible for the many conquistadores who sought its vast riches.

The rivers of the New World became natural highways along which the inland search for gold could proceed, fueling the expansion of European trade around the world.

MACHIAVELLI'S REPUBLIC

The first republican experiments to emerge after the Roman experience did not last. It would take several more centuries for a republic to become once again the stable, dominant form of social organization. For some time still, landscape control continued to be the purview of medieval institutions. And yet, ideas, if not institutions, had begun to permanently shift. Florence was their first testing ground.

Florence had its first republican experience in the twelfth century, alongside many other Italian cities. It too eventually dissipated over the course of the fourteenth century, as power consolidated in the hands of Cosimo de' Medici, the lord overseer of the Italian Renaissance. However, the hold of the Medici family was interrupted by a second, briefer experience at the end of the fifteenth century. It happened to generate an intellectual revolution. The mind behind that revolution was the republic's second chancellor, Niccolò Machiavelli, responsible for territorial affairs between 1498 and 1512.

Machiavelli foreshadowed a profound transformation that was starting to happen throughout much of Europe. The state was becoming a powerful economic actor. The growth in mercenary armies and the revolution in military technology had increased the cost of defense. The example of the English king Henry VIII is famous. To support his wars with France, between 1511 and 1512 he quadrupled state expenditure, which under his father had accounted for less than 1 percent of national wealth. He tripled them again the following year. The shipping trade, and the maintenance of a navy, added further military and administrative costs. Public finances were transformed.

Machiavelli noted that financial resources, however centralized, did not by themselves ensure military power. Rich people have often been conquered by poor ones, he reasoned. The power of the state was in its republican purpose, in being an economic entity that made choices about where the resources went for the benefit of the commonwealth. He believed "the republic should keep the public rich and its citizens poor"; in other words, the state should be willing to redistribute resources in order to pursue the public interest. That had been the fundamental tension at the heart of Rome's republic.

Machiavelli's interest in republican Rome was unusual. In his day, admiration for the Roman world focused on the imperial experience. Machiavelli embraced the inevitability of war, discord, and expansion that had come with the republic, and believed it was in the balancing of the tensions between the *plebe* and the senate over land that the greatness of the Roman republican project resided. He spoke of the "splendid consequences" of those tumults, of the great power in the landscape and of the great virtue of involving those that lived on it. This was one of the reasons Machiavelli worked to institute a civic military force made of conscripted peasants rather than rely on mercenaries. It was in the tension between the wealth of the few and the needs of the many, rather than in the maintenance of any particular order, that one could see the purpose of a republican constitution.

Machiavelli's exposure to water issues was eminently practical, mostly tied to military matters. There are some passing comments about the subject in his book *The Art of War,* which took inspiration from ancient Roman practices of water use in both offensive and defensive military stances. Water projects were relatively common in warfare. The opportunity to apply these tactics came in the ongoing war against Pisa, Florence's close commercial competitor near the mouth of the River Arno. The basic idea was to construct a weir to divert the river into two ditches, and leave Pisa dry. Thousands of laborers worked day and night. The problem was that the Arno, constricted by the weir, began flowing faster, digging itself lower into the ground and below the intake level of the ditches themselves. Eventually, Machiavelli abandoned the project. But despite the failure, the effort to enlist the landscape for military purposes stimulated Machiavelli's realization that the security of the state depended on the many that lived on it.

Just like in republican Rome, farmers needed to be part of a political process—and part of its military defense—if the state was to succeed. Through the active engagement of the state's landscape, economics had replaced law as the basis for political decision making. Machiavelli re-imagined the role of the state as an economic actor in the service of the stability of the republic. Its objective was liberty.

The modification of the landscape to manage water and increase agricultural productivity, or to protect economic activities from floods, was a political act. Machiavelli's conception of the state had a crucial, practical implication: it favored rich and large states over poor and small ones. The feudal state, one with limited control over small resources, was essentially dead. Europe was destined to become a multistate system of nearly equal actors, and they would find themselves colliding more and more in an unstable equilibrium, driven by growing competition on a global stage.

9

Water Sovereignty

DISPUTE ON A RIVER

In seventeenth-century Europe, two crucial political innovations proved to be long-lasting and highly consequential: the development of the modern territorial sovereign state and the rise of liberalism. Both were emblematic of a world that was rapidly veering towards commerce and increasingly global competition. Both were tightly associated with substantial transitions in the relationship between society and water.

Simply put, liberalism is a system of beliefs centered on individual liberty and private property, rather than the commonwealth, as the heart of political action. The conception of liberty that came to dominate the seventeenth and eighteenth centuries had a lot in common with the classical tradition, but it differed in that it implied little about the architecture of the state—it was not, primarily, a civic liberty. Liberty was the freedom to act without impediment, not a state of being or a civic status as it had been in ancient Rome.

The development of a liberal culture in Europe was a drawn-out affair and it accompanied a profound political transformation. Up to the fifteenth century, Europe was mostly a fragmented continent of city-states, feudal lords, manors, and ecclesiastical institutions. Despite the fact that monarchies in many parts of the continent consolidated political and military power all through the Middle Ages, local property rights over water were litigated in the courts of local potentates based on common jurisprudence. But change was in the air.

Over the sixteenth century, empowered by a newfound political agency, merchant groups took advantage of the expanded world. Eastern plains supplied grains to western cities. The New World supplied sugar to the Old. Resources poured in, fueling global commerce. The sixteenth century was a period of commercial growth. Claims of universality, which the Church and the emperor had made by evoking the long shadow of the Roman Empire, were melting away in the sustained encounter with the Far East and the New World. Larger states asserted themselves along new commercial routes. Migration increased, leading to exchanges, including of disease and knowledge.

Traded goods landed at European seaports and made it into the continent through inland waterways. It was therefore inevitable that any new political arrangements would be tested on the rivers of Europe. The pre-eminent northern European route was the River Scheldt, in the Low Countries. The first sixty miles of the River Scheldt lie in what is today France. It then flows into Belgium to Ghent, where it joins with its most important tributary, the Lys River, turning east towards Antwerp, and finally swerving north to enter contemporary Dutch territory. Its mouth has always been the gateway to an enormous landscape: via the Meuse and the Rhine, it provides access to the vast German hinterland.

In the High Middle Ages, two branches of the Scheldt reached the sea: the eastern Scheldt, or the Scheldt proper, and the Honte, which is the name of the western Scheldt. The delta that resulted was a complicated system of natural canals, water expanses, and inlets, protected from the tides by outer islands. At the beginning of the fifteenth century, maritime navigation was still mostly a coastal affair: the river was not navigable by seafaring ships. Bruges, which was closest to the seaward edge of the delta, was the principal commercial center. But over time, the river changed.

One of the side effects of the drainage schemes that had protected the Low Countries from the growing floods of the Little Ice Age was that they blocked many individual waterways in the delta. The tidal currents were therefore forced to flow with increasing force through a decreasing number of bottlenecks in the delta, scouring and widening the delta channels in the process. The biggest change by far was on the Honte. Individual narrow channels merged into a big one, creating and deepening a new continuous waterway between Antwerp and the sea. By the second half of the fifteenth century, seagoing ships could use it.

Thanks to this change in navigability, during the sixteenth century, Antwerp rose to become the most important port in Europe. It was one of the biggest ports ever to be built by any society: no other place has ever commanded as large a fraction of global commerce as Antwerp. In its markets, European merchants exchanged English cloth, Italian luxury items, foodstuffs from the north, German and French wine, Portuguese spices. Individual success was increasingly determined by this global flow of goods, but this new entrepreneurial ebullience triggered a political response.

As transport switched to the navigable western river, tolls on transit, which previously had no reason to be collected, became a matter of conflict between Antwerp and the Low Country province of Zeeland, on the coast. The former sought free passage, the latter revenues. Zeeland claimed it would use the tolls to pay for investments in the maintenance of dikes and embankments that would preserve the river. But from Antwerp's point of view, nature had provided a deepened canal and there was no reason for Zeeland to charge for passage where no charge had been made before.

In the past, these conflicts would have been under the jurisdiction of courts steeped in Roman and ecclesiastical law, which would have declared the river res publica, or owned by the state. The problem, of course, was that for such a solution to work, a sovereign state was required. In principle, the pope and the emperor were still the supreme authorities of the Western Christian res publica. In practice, though, by that point their ability to enforce any kind of rule had long waned.

For a while, the dominant master of both the Scheldt and the lands of the mouth was the Duchy of Burgundy, held by the Hapsburgs. During that time, Antwerp managed to secure its access to the sea without too much interference. But eventually that equilibrium dissolved. From the 1560s onwards, the Catholic kings translated the struggle for territorial control into a religious conflict with the Protestant Netherlands. In 1566, Philip II sent an army to suffocate a rebellion of the northern provinces. That in turn fueled the revolt led by

William of Orange, starting the Eighty Years' War in 1568. The political unity that had ensured that Antwerp could have unfettered access to the Scheldt finally broke in the spring of 1572, when Dutch rebels took control of the mouth of the river. By 1581 the northern provinces, the Netherlands, had declared independence.

The economics of Europe were changing and so was its politics. Both Zeeland and Holland saw an opportunity to take advantage of increasing global trade and established a regime of import-export so as to maintain the traffic on the Scheldt and benefit from tolls and the transfer of goods on smaller fluvial vessels. Antwerp, which relied on direct maritime trade coming to its port, rather than off-loading onto secondary vessels, had been cut off.

This was the Scheldt question, which persists to this day: Antwerp, one of the largest ports in the world, had been separated from its access by a national border with a competing maritime power downstream. Even today, the Belgian port of Antwerp, eighty kilometers inland and close to the heart of Europe, is still the second-largest container port in Europe after the Dutch port of Rotterdam.

CRISIS AND COLLAPSE

The transition to territorial national sovereignty was arguably the most important political development in modern history. It was certainly pivotal for society's relationship to water. The European state became an unusual hybrid, a synthesis of different traditions that had at its core an unresolved tension. Early states of antiquity had been territorial, of course, but in what were mostly theocratic societies. The Greek and Roman states, on the other hand, had emphasized private goods and individual agency within a civic contract, but then the polis or the *civitas* were first a collection of people rather than a territory. Roman law was a law of people.

In contrast, by the seventeenth century, a political unified territorial state was the protagonist of international conflict. At the same time, the rise of merchant and commercial classes, and of a modern banking sector, made individuals—whether alone or in partnerships—the engine of emergent capitalism. To accommodate this social development, the state evolved complex institutions better suited to govern an economy based on property rights and, ultimately, the rule of law. These institutions proved to be particularly inadequate when it came to managing waters that crossed territorial boundaries, but those waters were essential for supporting the merchant economy. That is the context in which the Dutch Republic attempted to assert its independence.

Under different circumstances what happened on the Scheldt might have remained a local development. But the transition to the nation-state happened at a time of generalized crisis. What had been a seemingly endless run of wealth creation in the sixteenth century ended in destruction. If war were a disease, the seventeenth century would have been its period of highest fever. Thomas Hobbes thought that the natural state of human society was one of war and destruction, calling for a Leviathan, the state, to impose order. No wonder: the first half of the seventeenth century must have felt like the end of the world.

During that period people died easily, often violently. Growing competition for territorial expansion led to war. The Church still controlled vast amounts of land and human resources, and entered into a bargain with the Holy Roman Empire to protect its own interests, which in effect led religion to becoming this century's principal mobilizing force. Armies grew. Taxation in the Holy Roman Empire increased to fund wars.

When Charles V deployed Spain against the Protestants, German princes mobilized in response. The Holy Roman Empire lost over a quarter of its population because of the Thirty Years' War and related epidemics. It took at least half a century for the population to recover. Similarly, in Italy population declined, as did that of most of Spain. The proximate causes for these and many other conflicts were varied and typically found in the politics of the day: to wit, the conflict between Protestants and Catholics. But there was also something else going on.

The crisis was not limited to Europe. In the middle of the century, China lost about half of its population in the transition from the Ming to the Qing dynasties, and its cultivated land fell from two hundred million acres to less than seventy million. In fact, some have referred to this as a global crisis.

The first very significant cold phase of the Little Ice Age had been the Spörer minimum, between the mid-fifteenth century and early sixteenth. But worse still was the Maunder minimum from 1560 to 1720. Worldwide, temperatures were about half a degree below the average of the twentieth century. European temperatures were much lower, about two degrees below average. During that period, violent storms flooded coastal Denmark, Germany, and the Netherlands. Between 1620 and 1621 the Bosporus froze. Baghdad flooded in 1630. The Arctic pack ice grew enough for Inuits to land kayaks in Scotland. Snowfall, heavier than ever recorded—before or since—lay on the ground for months. In 1658 water between Jutland and the isle of Funen froze solid, enough so that an army could walk across it. Spring and summer were cold and wet.

The cause of this period in the Little Ice Age is hotly debated. One possible culprit is the fewer spots observed on the sun's surface, which would have correlated with less solar energy emission. Volcanic activity may have also played a role: the volcano Nevado del Ruiz exploded in 1595; Huaynaputina in Peru did the same five years later; up to a dozen volcanic eruptions happened in the middle of the seventeenth century.

Whatever the cause, throughout Europe, farmers had to adapt to a shorter, less reliable growing season. Agricultural productivity dropped. The ratio of number of seeds produced to those planted went from seven at the beginning of the sixteenth century to just over three for most of the seventeenth century. Grain prices increased fourfold. The number of famine years per decade spiked for much of the century, while the average height of individuals dropped by about one and a half centimeters, a sign of malnutrition. Tax receipts collapsed. States weakened. Want, famine, strain, and social unrest followed. If religion, justice, treasure, were lost, crop failure could spark a revolt, and as Francis Bacon put it: "Men had need to pray for Faire Weather."

The political crisis of the seventeenth century was inseparable from changes in environmental conditions. Court depositions of Protestants following the 1641 revolts in Ireland speak of the unusual weather, as do Chinese records of the Ming power transition. To be clear, climatic change cannot be neatly separated from the faults of man. On the Vistula, in Poland, for example, a number of catastrophic floods were recorded in commentary from the sixteenth and seventeenth centuries. But land use around the river had also changed. Increasing population had led to a doubling of cultivated area in the basin from the fourteenth to the sixteenth century. During the wars of the seventeenth century, intentional breaching of embankments was also to blame for catastrophic floods, so the changing climate was not a unique cause of crisis. However, if the arc of the Little Ice Age was to last about five centuries, the first half of the seventeenth century was by far its harshest period.

The strain caused by newly assertive territorial states, the ambitions of a growing merchant and entrepreneurial class, and a rapidly changing environment could not but manifest itself along the shared waters of Europe.

GLORIOUS FENS

The response to the changing conditions meant that Dutch experience in reclamation was at a premium. Everywhere. Dutch-Flemish emigrants brought drainage experience to all of Europe: from Italy, where Dutch experts advised on drainage in Tuscany and in the Papal States, to France, where in 1628 the duc d'Epernon asked Jan Adriaensz Leeghwater, famous for draining Wormer Lake in Holland, to drain the marshes near Cadillac, south of Bordeaux. The private reclamation contracts typical of the Low Countries also spread to a number of rivers in central and eastern Europe. These early experiences led to substantial reclamation in most German states all through the seventeenth and eighteenth centuries.

However, arguably, no Dutch-inspired reclamation was as consequential as that of eastern England. At the beginning of the seventeenth century, the Stuarts, who succeeded the Tudors, inherited huge deficits from Queen Elizabeth's wars. The Crown had to sell most of its land to repay them. Once the land was gone, customs and new taxes had to be imposed to try and make up the difference. Eventually, the Crown turned to the water landscape, where reclamation was seen as a promising instrument for creating more productive land. This became a deeply political issue when the desire to find more land driven by a thirst for returns on capital collided with a different model of ownership based on communal property.

At the core of this issue were the fens, six hundred thousand acres of low-lying, seasonally flooded wetlands in the east of England. The plan to drain the fens had been initially drafted by Cornelius Vermuyden, a Dutch engineer from Tholen, who had mostly worked in Zeeland. In 1621 he was recruited to repair a breach in the embankment of the Thames east of London, and was quickly put in charge of a number of projects. While his direct engagement in the fens project was intermittent, his ideas were influential in its development.

As part of this scheme, the Great Level—around three hundred thousand acres of land—was going to be the largest fen to be drained. The undertaker was the Bedford Level Corporation, a private company founded by the fourth Earl of Bedford, who did the initial undertaking in 1630. Vermuyden was its chief engineer. It was a fairly straightforward private enterprise. In exchange for their money, the investors would receive ownership of a third of the resulting land. Notably, King Charles was one of the investors. He hoped the project would add an enormous new piece of land to the kingdom, an entire new country. Charles even imagined building a new town in mid-fen, named after himself.

The project faced significant technical challenges, but none was more severe than the riots that broke out in the local communities. Despite the fact that they were commonly portrayed as destitute barbarians, the Fenlanders were relatively prosperous, if not exactly rich. They certainly could live comfortably off the fens and their marsh economy. It provided summer grazing, fish and fowl, fuel, reeds for thatching roofs, baskets, floor covering, all managed as a common good.

A common good, what the Romans called *res communis,* is not susceptible to ownership. Communal management of the fens was eminently suitable for an environment whose productivity depended on a live ecosystem. But because reclamation required private investment, it was impossible without some form of ownership, particularly in a country in which the state was bankrupt. The conflict that ensued was not, as is sometimes argued, so much about the abstract fight between the marsh economy and capitalist agriculture—in principle Fenlanders may well have accepted a transition to commercial agriculture—but about who would benefit.

After drainage had made the inhabitants far more dependent on the new infrastructure, enclosures gave exclusive access to parcels of land to a few owners, excluding others who now had to live on far smaller plots of land. In addition, the promised improvements of the drainage system were short-lived. The corporation had committed to maintaining the entire drainage structure based on its revenues, but this proved to be economically impossible. In the end, much of the drained land had been freed of water only in the summer, making it suitable for grazing but not for cultivation.

In an attempt to salvage the project, in 1638 King Charles pushed Bedford aside and transferred the undertaking to himself, bringing back Vermuyden, who had been taken off the project at the first protests. But Charles never had the resources to carry out his enormous plan. The consequent expropriation of property the Crown undertook to find the money fueled tensions further. Opposition grew, at times turning violent. Revolution was on the way.

Radicals inspired by the communal economy of the fens fiercely opposed the enclosures and privatizations that had transformed the English marshes. The story of the marshes and of the many enclosures of the seventeenth century inspired radical utopian movements like the Diggers, who would later become embroiled in the English Civil War. People like Gerrard Winstanley, leader of the True Levelers, pursued agrarian, egalitarian reforms, and a return to nature and to communal living. In the end, the English Revolution cost Charles I his head, and the monarchy made way for the Commonwealth, England's only short-lived republican experience. The Commonwealth's initial success cemented expectations of individual freedom. Its subsequent failure transmitted a partial legacy to what came next, a surprisingly stable blend of republican tradition and liberal thought.

WESTPHALIA AND THE RISE OF THE COMMERCIAL STATE

Back on the Dutch side, the wars of the 1570s and 1580s had destroyed Antwerp's potential. The once-extraordinary port was now isolated on the Scheldt. To continue to operate, merchants had to move to the newly founded Dutch Republic, which was gearing up to become a dominant maritime commercial power.

Because Spanish embargoes disrupted trade and the Portuguese dominated the routes to Asia, the Dutch first looked for alternative routes. Prominent cartographers like Willem Barents and Gerard

Mercator—after whom the sea and the map projection are respectively named—were convinced a northeast passage to the Pacific was the answer. The Dutch exploration of the Arctic gave them Arctic whaling, but not a passage. In the end the Dutch resorted to forming the Dutch East India Company in 1602 to compete directly with Spain and Portugal over the southern Asian routes. The company, alongside its contemporaneous English counterpart, would become the backbone of European mercantile expansion.

Despite the difficult political and environmental conditions, the first half of the seventeenth century was a time of great economic growth for the Dutch Republic, and again the economic efflorescence was reflected in the water landscape. In the 1630s, they established a canal system parallel to that for drainage and fully dedicated to passenger transport. Amsterdam and Haarlem financed the first leg with horse-drawn barges. It became an attractive mode of transport and provided a good return on investment. The model grew.

The "trekvaart network," as it was called, transformed the economy of the Low Countries. It was a system suited for a country focused on commercial activities, in which face-to-face contact and frequent travels were the norm. The system became massive, carrying enormous quantities of people, far more than the expensive and limited coaching business. It was used by the vast majority of the population, increasing access to the landscape for a much larger spectrum of society.

Meanwhile, the eighty-year conflict between the Spanish monarchy and the Dutch Republic finally ended with the 1648 Peace of Münster. The treaty was, in effect, about sovereignty over the expansive canal infrastructure. Several articles spoke to constraints imposed as part of the settlement. Article 14 in particular stated that "the River Scheldt, together with the canals of Sas, Zwyn, and other connecting channels, shall be kept closed on the side of the Lords States." The closing of the Scheldt and its canals sealed the fate of Antwerp. In a nod to the legacy of Bartolus of Sassoferrato, Spain was forced to recognize the sovereignty of the Dutch Republic, and it was on the basis of that sovereignty that the States-General based their right to close the Scheldt, which isolated Antwerp and gave the Dutch Republic a monopoly on access to the sea.

By the end of the 1660s, all major Dutch centers were connected by the canal system, enabling an unprecedented degree of coordination. Cities connected in this network specialized: Amsterdam in trade, Leiden in textiles, Delft in dairy, Rotterdam in shipbuilding. Urbanization grew: Amsterdam, which had a population of ten thousand at the beginning of the sixteenth century, had over two hundred thousand by the middle of the seventeenth. Maritime and fluvial infrastructure had become a seamless whole.

The Dutch managed to improve the modes of travel, especially on water, which made the world far smaller for their commerce. Wind was their fuel, unexpectedly abundant due to the changing climate. Evidence from logbooks suggests that the cooling of the Little Ice Age strengthened mid-latitude winds, reducing ship travel times to Southeast Asia. While at the end of the sixteenth century, it had taken 350 days to sail from Texel, off the coast of the Netherlands, round the Cape of Good Hope, pass through the Indian Ocean, and reach Jakarta in Indonesia, by the middle of the seventeenth century, it was down to about 200 days. It was a 30 percent drop. By the end of the century the Dutch had created a vast, integrated system of seagoing vessels and inland navigation. The ferries of the Dutch inland system, combined with the changing global conditions, had set the stage for the development of the Netherlands as a global trading power.

THE SOCIAL CONTRACT

The transformation of the economy went hand in hand with a radical transformation of political institutions. In the process, society's relationship with water changed. By becoming the dominant unit of social organization, the territorial nation-state became the pre-eminent human institution to wrestle with water's power.

The English experience, which emerged partly out of water by metabolizing basic republican ideas and the liberal response to them, created a modern constitutional architecture largely founded on private property and liberty (at least, for part of the population). The British argued for a long time that it was the modern echo of the Roman state. But its social contract was not based on the idea of a Roman *cives.* It was John Locke's liberal state. The system that emerged from the English Revolution and the restoration of the monarchy placed private property at the heart of the British social contract. The aim of a legitimate civil government was to preserve the rights to life, liberty, health, and property of its citizens, and to prosecute those citizens who violated the rights of others.

In continental Europe, the Peace of Münster, along with the Treaty of Münster and the Treaty of Osnabrück, formed the Peace of Westphalia, ending decades of war on European soil and establishing, in effect, the first European constitution. The separation of international law from any one religion meant that, from now on, republican and monarchical states were on an equal footing. It was also the first large-scale application of a principle of balance of power, an organization of Europe based on territorial states that would balance each other, marking the end of the idea of a universal monarchy. By recognizing each other's right to exclusive authority over a specific territory, they were accountable to each other, not to a supreme sovereign, for peace.

The delegates who negotiated the peace were committed to reconstruction after so much turmoil and emphasized the need to re-establish commerce. Central to this was navigation on the rivers flowing through Europe. The Treaty of Münster between France and the Holy Roman Empire that ended the Thirty Years' War included a set of provisions to address this issue. Article 89, for example, established freedom of commerce on the Rhine and regulated tolling, inspection, and duties.

The peace as framed across the three treaties was a fundamental moment in the building of the modern state. Whether modern sovereignty originated in those particular treaties is a matter of some debate, but there is little question that subsequent settlements—the Vienna Congress of 1815 after the Napoleonic Wars, the 1919 Peace of Versailles after World War I, the League of Nations, or the charter of the United Nations—drew from the principles first established in the Peace of Westphalia.

The Dutch and English experiences converged when, after the English Commonwealth failed and the Crown had been restored, William of Orange came to England in 1688, becoming joint sovereign and embodying the unification of Anglo-Dutch interests in the new mercantile world. This posthumously defined Glorious Revolution represented the full transition to a constitutional monarchy and to the supremacy of Parliament, absorbing liberal values in a medieval institution.

It was the result of a two-century-long political evolution: Machiavelli and Thomas Hobbes had established, in different ways, the basis for a politics that defined sovereignty; Westphalia marked the shift in emphasis from rule over people to rule over territory; the Glorious Revolution introduced limitations to the powers of the state in favor of the wealth owners represented in Parliament and opened the gates to the modern liberal state. The institutions of representative government that emerged from it were a model for many subsequent political systems, including an independent parliament and judiciary.

This political evolution had a profound impact on society's relationship with water. On the one hand, it proved capable of delivering infrastructure on rivers on a grander scale because of the economic forces it was able to mobilize. Indeed, after 1688 in England, river improvement proposals, which had previously gone through the Crown, went through Parliament. Return on invested capital, which was at the heart of economic success, was directly proportional to the security of property rights: the more one could trust that they would be able to hold on to an asset, the more likely they were to take risks and increase potential returns. As a result of the increased security of property rights, the government could borrow far more than before. In the ten years following the revolution, access to funds for river infrastructure increased tenfold, setting the stage for the subsequent unprecedented growth. At the same time, the formal mechanisms to mediate between collective good and individual rights beyond the law were limited. In this context, the state could be a powerful manager but would struggle to be an effective political mediator on the landscape.

In any case, from now on the story of water became a story of nation-states. Their genetic code was republican, even when the formal government was monarchical and its political foundation liberal. A state that identified with its territory would have to confront the issue of water as it moves through the landscape, its control an expression of sovereignty. The state identified with the territory, but was not its principal owner. Collective objectives were pursued through the economic activity of individuals and companies.

The tension between a powerful territorial state, which identifies with the control over a landscape, and an economy that emphasizes private action over direct state intervention was going to be the dominant challenge of water management in the modern world and, largely, still is.

10

American River Republic

A NATIONAL PROJECT

The United States became the modern model republic. The young nation had to strike a difficult balance: on one hand, an extraordinary water landscape, the power of which far exceeded the capacity of any individual to manage it and was unlike anything seen in Europe; on the other, a commitment to a liberal society free of despotism and government interference that called for a radically new social contract, at least among its settlers.

This tension was pervasive. In a letter addressed to the deputy governor of Pennsylvania on behalf of the Pennsylvania Assembly on November 11, 1755, Benjamin Franklin warned that "those who would give up essential liberty, to purchase a little temporary safety, deserve neither liberty nor safety." Those words referred to a territorial security threat posed by incursions of the neighboring Delaware and Shawnee tribes on local communities. They could have equally applied to security in the face of the overwhelming force of water.

From its inception, the United States found itself having to contain the unintended consequences of a state sufficiently powerful and well resourced to deliver water security. Building the nation over the course of the nineteenth century was a remarkable process of land accretion, which shaped institutions and rivers along the way.

The starting point was the thirteen colonies, from Massachusetts to Georgia, which declared independence in 1776. Their coniferous and broadleaf forests covered a relatively narrow territory east of the Appalachian Mountains. They were the source of a number of rivers flowing to the Atlantic: the big north-south New England rivers like the Penobscot and Connecticut; the large, navigable Susquehanna, Delaware, and Hudson, which supported an agricultural economy in the middle colonies; the Potomac and Savannah, in the southern colonies. Their ports looked to Europe, their climate heavily mitigated by the Gulf Stream.

This landscape, crossed by so many rivers that drained west to east, took a new role once the westward expansion started at the end of the Seven Years' War. The 1763 Treaty of Paris had assigned Britain large swaths of territory into what was called the Ohio Valley and Quebec. It was split into three new states, and territories assigned to native American tribes. As part of the settlement, Spain obtained the Louisiana territory, west of the Mississippi, and, critically, the port of New Orleans. After the end of the Revolutionary War, those western territories became part of the United States.

George Washington had speculatively amassed considerable land beyond the Appalachians and was keenly aware of both its potential and its limitations. The Ohio Valley was fertile, but it would have been useless to eastern colonists without a viable transport route to bring goods back east. He knew that without a cost-effective way of trading eastward, all goods would flow through Spanish territory to New Orleans.

After independence, he decided to revive an old colonial idea: to turn the Potomac River into a major transportation canal that could improve the links between the newly independent American states and the Ohio River. That idea, borne out of a commercial imperative, ended up having a disproportionate role in the constitutional journey of the federal state.

Washington incorporated the Potomac Company in 1784, becoming its president. The company was based on the tradition, long-standing by then, of resorting to private enterprise to develop projects of public utility. The capital for the company was raised by issuing stock both in Maryland and in Virginia. Large landowners and wealthy traders invested, many of them Federalists who cared about integration of the United States.

What made this company remarkable is that it was the first interstate project of national significance. For the economics of the resulting stock company to work, the Potomac Company had to operate boats across states, through Maryland and Virginia and back into Pennsylvania, to reach the Ohio. It was the same problem of freedom of navigation that had bedeviled the Scheldt. But in this case, the issue was a domestic one for the newly formed United States. Had all the states, independently, exercised fiscal authority over the operation, they would have made its economics impossible.

Setting rules for interstate trade was going to be crucial for the success of the project. The Articles of Confederation which bound the thirteen states were of limited help. They had been drawn up during the Revolutionary War to create a national government, but had been designed to be the opposite of what British rule had been. The articles did not contemplate an executive or a judicial branch, and the legislative one was limited by the need to get a super-majority of nine of the thirteen states. The federal government had power only over states, not over individuals, for whom there was no protection of rights at the federal level. It could not raise taxes or an army. Its institutional architecture was not designed to address territorial issues. In particular, the national government could not control trade among states.

After the War of Independence, this weak form of association was at risk of dissolving into thirteen individual and separate states, in which case the vast expanse to the west would be co-opted by the Spanish, who offered a far more integrated and functional trade infrastructure through New Orleans. A problem of navigation had turned into a problem of governance.

THE BIRTH OF A CONSTITUTION

The story is well known, although its water association is often neglected. Washington convened a meeting at his Mount Vernon estate, in Fairfax, Virginia. Both Virginia and Maryland sent delegates. The hope was to discuss a regulatory framework that the Potomac Company could operate in. The scope of the negotiation was initially unclear. The Maryland delegates had a broad mandate to discuss navigation issues on the Potomac as well as the Pocomoke and Chesapeake Bay, but the Virginia delegates were mandated to discuss only the Potomac. Despite the initial difficulties, they arrived at a set of agreements, which could be enshrined in the legislation of both states.

The Mount Vernon Compact, as it became known, was a first model for interstate coordination. Once Maryland and Virginia had agreed on a cooperation compact, it became clear that Pennsylvania would also have to be invited into the agreement, as the Potomac stretched back into that state. The problem of regulating the navigation system gained steam. On January 21, 1786, the Virginia legislature passed a resolution to appoint five commissioners to confer with those of other states in the union to consider "how far a uniform System in their Commercial regulations may be necessary to their common Interest and their permanent Harmony."

James Madison followed these developments with interest. The intersection of trade, navigation, agriculture, and national unity was a crucial issue. If the Articles of Confederation of 1776 had represented a commitment to libertarianism against despotism, the Federalist argument was one of republican government and stability against the "abuses of liberty." This translated into a system of checks and balances, with at its heart a national government.

The founders of America were rooted in classical imagery of Roman republicanism. Madison, Jay, and Hamilton even chose to use the pseudonym "Publius" in signing their Federalist Papers, after Publius Valerius Publicola, one of the first consuls of the Roman Republic who had contributed to the Roman revolution of 509 BCE. They imagined an agrarian America, a society fiercely tied to its landscape and resisting the burden of dependence on British rule. The vast landscapes of the continent seemed obviously well suited to become the granary of Europe, a vision that would become reality a few years later, with escalating European demand for grains during the Napoleonic Wars.

In the early days of independence, Madison saw an opportunity to use inland navigation as a practical entry into the broader issue of coordinating commerce and resolving the tension between centralization and delegation. If four states had been able to reach an agreement at Mount Vernon, why not thirteen? He invited all of the states to a conference in Annapolis, Maryland, in September of 1786 to discuss these issues. Not all of them attended, but the report of that meeting reflected the belief that trade issues—inland navigation principal among them—were not just a technical matter. The report noted that "the power of regulating trade is of such comprehensive extent, and will enter so far into the general System of the federal government, that to give it efficacy, [. . .] may require a correspondent adjustment of other parts of the Federal System." The delegates had in effect concluded that a constitutional convention should be held.

The Constitutional Convention of Philadelphia took place from May to September of 1787. Its initial aim was to amend the Articles of Confederation. It ultimately led to a draft constitution. The attempts to regulate inland navigation had set a critical precedent. Navigation came up repeatedly during the negotiations. James Madison brought up the example of the Mount Vernon Compact to show what states could drive individually within the Articles of Confederation. Luther Martin, attorney general of Maryland, used it to show that two states could work out issues bilaterally. Gouverneur Morris of Pennsylvania brought it up again in debating the relative power of the president and of congress to promulgate laws. It was an archetype for what became the complex system of checks and balances that ran through the U.S. Constitution.

The rights of individuals to manage their own land and thrive without government interference found its limit in the problem of coordination along a river. In his impassioned case for independence a few years earlier, Thomas Paine—one of the staunchest advocates for the republican cause—had noted that the scope and scale of an independent American government would have to contend with the particular nature of its landscape. He was right.

BUILDING ON WATER

The new institutional architecture of the United States collided with its water landscape in increasingly complex ways. From the end of the eighteenth century, economic activity boomed. The settlements that crept up the eastern Appalachians were in the grips of a canal obsession, similar to what had taken over England and Wales as they industrialized. Washington's intuition had been right.

Canal and river companies popped up everywhere. These were typically publicly recognized private corporations. While in 1793 eight states had incorporated thirty canal companies, by 1823 New Hampshire alone had twenty. To be useful, canals had to cross part, or in some cases all, of the Appalachians. Most of these companies did not have the capital to build hundreds of locks and tunnels to allow boats to overcome the steep gradient. On top of that, unexpected floods, variations in flow, and constant repair works increased the cost of maintenance above plausible revenues. Yet, because inland navigation was essential for the economy of the states, these companies became too important to lose. If private capital was not forthcoming, the states would have to step in by issuing bonds.

Over time, the tolls paid to the companies became revenue for the treasuries of these shareholding states, linking private and public finances. By the 1830s over half of state debt was locked up in canal and river works. When the panic of 1837 hit, the companies went bankrupt and the state treasuries got caught up in the process. This would not be the last encounter between economic cycles and water infrastructure in the story of America, creating a fault line between the entrepreneurial credo that the United States had inherited from European liberalism and the financial, legal, and social oversight that the federal government increasingly saw as its mandate.

This tension boiled over on the issue of regulating the use of the Erie Canal. While erosion had formed the rivers along the eastern side of the Appalachians, shaping many hard-to-overcome steep valleys and rapids, the Hudson River was different. It had formed as a tidal inlet when ice retreated at the end of the last ice age. The Mohawk River, the largest tributary of the Hudson, formed as the ice scraped the bedrock, scouring the valley and leaving it deeper and wider than anything south of it. As a result, this river was far easier to develop and navigate.

The Erie Canal ran along the Hudson and the Mohawk, connecting them to the Buffalo River, which poured into Lake Erie. It was over 360 miles long, equipped with dozens of locks to overcome the 200-meter height difference between the Hudson and the lake. It was finished in 1825 and was a huge economic success. The route from the coast to the Mississippi River via Lake Erie dropped the cost of transport by a factor of twenty. Cities along it swelled. New York became the principal entry point towards the interior of America, and Erie the artery that pumped the goods of the expanding nation. Just after completion most of the flow of transport was westward, but by 1847, twenty years later, eastbound freight was five times more than that heading west. Until rail took over at the end of the nineteenth century, the canal was the lifeline that connected the heart of the continent to the markets of the world.

If the question of developing canal infrastructure across states had given a first push to federalism, use of the canals soon gave it another. The Erie Canal proved to be an opening shot in the battle for centralized oversight and federal power. This conflict was most evident in the regulation of interests on the waterways that benefited from the increased traffic.

The state of New York had granted a steamboating monopoly to the Fulton-Livingston company, which in turn leased portions of the activity on the Hudson to others. Among them was Aaron Ogden, who had obtained a lease to manage the route across the Hudson into New Jersey. Another businessman, Thomas Gibbons, set up a competing line for the same route. This triggered a legal battle that ultimately went to the U.S. Supreme Court.

In 1824, the Court ruled in *Gibbons v. Ogden* that the Hudson between New York and New Jersey was an interstate waterway because it was part of the Erie system, and that the monopoly granted by New York was, in fact, unconstitutional. Inland navigation had become a fully federal matter. The role of the American federal government in regulating and, ultimately, developing large water infrastructure would grow in fits and starts, and would eventually reach unprecedented levels in the twentieth century.

The conflict between centralized collective responsibility and entrepreneurial execution played out at all scales and levels of government. In the case of drinking-water services, two competing models were at play in the United States. The first, famously championed early on by Benjamin Franklin's Philadelphia, was based on public ownership, financed by the city itself. Most large cities, such as Boston, Chicago, and Baltimore, and a considerable minority of other centers, opted for this. The second model had its most obvious example in New York City, which relied, much like London, on the competition between private companies for the delivery of water supply.

Some of the public debates that ensued on the matter had remarkable historical consequences. The business competition for New York led to the creation of Aaron Burr's Manhattan Company, a chartered water company born to break the credit monopoly of the Federalist New York Bank of Alexander Hamilton. It is still around today, of course. It first merged with Chase to become Chase Manhattan Bank, and ultimately merged with the descending institutions of the J. P. Morgan empire to become the largest bank in the U.S.: JPMorgan Chase.

As far as water is concerned, the public delivery model prevailed after the successive crises of the 1870s and 1890s, when private investments in the sector proved untenable while municipal bonds became a superior, attractive investment.

THE WATER FRONTIER

The Mississippi is a gigantic river basin, covering 40 percent of the modern continental United States, from Montana to New York State, from the border with Canada to the Gulf of Mexico. Its drainage is almost three million square kilometers, comparable to the size of India. Only the Amazon and Congo river basins are bigger. The complex system of tributaries, from the Missouri to the Tennessee, looks like a broad, spreading tree. One of the consequences of such a wide, dendritic system is that the river intercepts many climates, from winter rains, to snowmelt, to summer rains, all happening in different parts of the basin. As a result, the flow in the lower river can be highly variable. Peak floods can carry thirty times the water of low flow.

The western expansion of the United States was a coin with two very different sides, telling a different story on each side. The first is a story of radical nation-building. Until the end of the eighteenth century, the Mississippi River had been the western frontier of the United States. When land west of the Mississippi was returned from Spain to Napoleonic France, the United States government worried that France would block the use of the river. Things seemed to get even worse when, in 1802, Spain revoked American access to New Orleans. Without access to the port, the western part of the United States was largely isolated and commercially limited. Then in 1803, Napoleon made an extraordinary offer: the entire Louisiana territory for sale. Thomas Jefferson seized the opportunity and on December 30, 1803, the United States took possession of a landscape that instantly doubled its size, giving it full control of the entire Mississippi basin. The acquisition was politically and economically transformational for the young republic.

The other side of the coin, the one seen from the point of view of the population that predated the arrival of the colonists, was a story of empire. With the Louisiana Purchase, the United States also acquired the Great Plains, the home of the American prairie, the grasslands ecosystems that for centuries had supported the great nomadic and semi-nomadic societies of the Comanches, the "lords of the plains," the Sioux, the Apache, the Cheyenne, and others. This was the territory that supported an economy based on bison and had been historically inhospitable to sedentary farming. Both stories were linked to the waters of the American landscape. Both stories would also be reflected again and again in the subsequent evolution of the republican project, as the United States began projecting its power abroad.

But for the time being, the westward expansion of the country continued within North America. During the nineteenth century, the United States accreted landmass in the west. Texas was annexed in 1845, all the way to the Rio Grande. Oregon Country was fully in the United States by 1846, which added the Columbia River basin. Mexican secession of the western states by 1848 added the Colorado River. The Gadsen Purchase added Arizona and New Mexico. Alaska was purchased by the United States from the Russian Empire in 1867.

The impact this territorial expansion had on the United States was enormous. For one, the Louisiana Purchase was fateful as it gave renewed impulse to the plantation and slave economy that eventually drove the country to the catastrophic Civil War of the 1860s. Over the course of the nineteenth century the country grew rapidly on the back of its remarkable geography. Agriculture consistently dominated exports. In the early nineteenth century, three-quarters of those exports were agricultural products, and a fifth were the output of forests and fisheries. By the end of the century, the United States had begun exporting manufactured products, but agriculture still accounted for 60 percent of exports, including cotton, tobacco, and animal products. Food alone varied between 25 and 40 percent.

But if the western territories were not going to be simply a colonial, extractive appendix, then citizens needed to move there—the enfranchisement of indigenous populations was out of the question. People had to build a sufficient critical mass to give those states sufficient political weight in the complex republican system of territorial representation of the United States.

To move settlers, Congress passed a number of national settlement policies. The Homestead Act of 1862, the Desert Land Act of 1877, and the Carey Act of 1894 were all attempts to help people settle the western part of the country. The Homestead Act, in particular, was signed into law by President Lincoln in May of 1862. The basic terms of the act were that the government would give settlers 160 acres to set up their home and farm in the west. It was successful, at least for the settlers. Far less so for the indigenous populations, of course.

Arguably it was one of the most consequential policies of the nineteenth century, a vast transfer of territorial wealth into private hands. The intent was to favor the yeoman, the individual smallholder, over any large-scale ranching or farming operation. For the most part it worked. There were echoes of Greece's democratic project in nineteenth-century America and the land ownership of the farmer.

But in this case, the nature of the landscape did not permit it to rely solely on individual agency. The problem was that western land could really not be developed without also developing water resources on an unprecedented scale. Without a comparable federal policy on water resources, the efforts to settle the lands would quickly run into trouble. American rivers were too powerful, too large, and too complex to be tamed by individual farmers. People on the landscape were at the heart of its politics, but engineering as an expression of state power was going to have to transform it.

The U.S. suffered several droughts during the second half of the nineteenth century. The first was from the mid-1850s to the mid-1860s, known as the Civil War drought. Twenty years later, in the 1870s, another drought hit, this time associated with a catastrophic swarm of locusts. Finally, a third in the 1890s was comparable in severity to the Dust Bowl of the twentieth century. The droughts challenged the efficacy of the Homestead Act and other landscape policies.

In 1879, John Wesley Powell, then director of the U.S. Geological Survey, wrote the *Report on the Lands of the Arid Region of the United States*, which, in essence, outlined a development agenda driven by water conditions. He observed that only a small fraction of the government lands of the West were suitable for agriculture. He divided the United States into climatological regions and asked where the government should intervene to help take advantage of the vast country. In the humid region of the Columbia River there was ample opportunity for agriculture, but it required clearing the forest. The arid region of the southwest could not be cultivated without irrigation and drainage systems.

Powell was looking ahead. And what he saw was the need for a powerful, interventionist state transforming the water geography of the frontier. And that is exactly what the country pursued.

ENGINEERING A NATION

In 1802, Thomas Jefferson established the United States Army Corps of Engineers. It was organized on the model of the French engineering schools.

Prior to its own revolution, France had had several engineering corps. Of the two biggest—Les Ingénieurs des Fortifications, incorporated as the French Army Corps, or simply Officiers du Génie, in 1776, and Les Ingénieurs des Ponts et Chaussées, bridge and roadway engineers, which dealt with civil infrastructure and depended on the Ministry of Finance—only the latter survived the French Revolution. The former was too much a vestige of the ancien régime, which rampant nepotism had turned into an aristocratic stronghold. The Ingénieurs des Ponts et Chaussées was generally well respected with strong connections with local authorities.

The Ingénieurs des Ponts et Chaussées was an enterprise inspired by the physiocratic doctrine of Francois Quesnay, the Enlightenment idea that a nation's wealth came from its land. The academy attached to the corps, the École des Ponts et Chaussées, continued to operate after 1789. The Constitutional Assembly reformed it, introducing admission by competitive examination, eliminating fees, and providing students with a stipend. It had a distinctly Jacobin character, adopting a "revolutionary method" of teaching: accelerated, centralized, and for a large number of students. The new breed of engineers it produced were part mechanic, part architect, and part military strategist.

The École Polytechnique, as it was renamed, became a foundation for the subsequent Napoleonic dreams of transforming the natural world in the service of progress. The French had inadvertently introduced an age of meritocratic engineering, which, when transferred to the United States, would go on to reshape the face of the planet.

In the United States, the Army Corps was an institution designed for nation-building. As the country expanded, it found itself at the center of the developing landscape and often caught between battles for sovereignty between the federal and local governments, from the 1849 sale of the Louisiana swamps to fund levees, to the response to the floods of the 1870s and the establishment, in 1879, of the Mississippi River Commission, to the California Debris Commission, set up in 1893 to deal with the Sacramento and San Joaquin Rivers.

For much of the nineteenth century, the principal concern on the Mississippi was navigation. The channel had to be kept deep enough for ships to pass, which originated, among other things, the name "Mark Twain" ("mark two" was twelve feet measured by sounding depth, the depth safe for a steamboat). Once acquired, it became the country's most important strategic asset: the world's largest transport system running through the world's most productive agricultural land. The Army Corps became the principal force for its navigation and, therefore, the most important institution for water management in the country.

The engineering prowess of the growing nation was a matter of great pride even at its darkest hour. On November 19, 1863, at the consecration of the National Cemetery in Gettysburg, just before President Lincoln spoke his famous 272-word address, the Honorable Edward Everett gave a long oration that swelled as he called for unity across the war's divide. Everett spoke of the "bonds that unite us as one People." He drew generously from Pericles's rhetoric of a constitution "called a democracy because we govern in the interests of the majority, not just the few" and of laws that gave "equal rights to all." But it was in the landscape that his oratory soared the highest, speaking of "the mighty rivers that cross the lines of climate, and thus facilitate the interchange of natural and industrial products, while the wonder-working arm of the engineer has levelled the mountain-walls which separate the East and West."

When Alexis de Tocqueville visited America in the 1830s, he was convinced that the model of the American democracy would sweep

Europe. He was wrong. In the middle of the nineteenth century the fate of the republican project remained largely in balance. Niccolò Machiavelli had warned that size and liberty seldom go together: "Expansion is poison for republics," and, given that by the 1860s the only large republic of the modern era was stuck in a destructive civil war that might herald its demise, he seemed to have been prescient. But the country survived that catastrophe. America was not just singular because of its revolutionary origin or political philosophy. Its exceptional size, geography, demographic complexity, and determination to tame the environment in service of an ambitious republican project drove its expansion.

By the time the nineteenth century was over, the United States had set itself up to become the most radical architect of water geography in human history. Its mix of institutions, law, and natural resource endowment transformed ideas about what was possible in water security, unlocking dreams that would have seemed science fiction for most of human history.

The Reclamation Act of 1902 promoted the development by the federal government of infrastructure to harness water resources. With that, the twentieth-century development of the West had begun. By that point, American federal republicanism was over a century old; it had survived the Civil War and inspired thousands of liberals and radicals across the world to challenge political forms that had been assumed permanent. The challenging water conditions of the United States simply became the challenge of the westward frontier and the driving force behind the expansion of federal intervention.

The great American rivers—the Mississippi, the Missouri, the Columbia, and many others, which constitute the underlying natural architecture of the country—would become the basis for the rise of the American republic as the dominant economy of the twentieth century.

II

Global Water Empire

THE RETURN OF EMPIRE

While republican principles were enshrined in the constitutional language of the American and French revolutions, the diffusion of liberal thought in eighteenth- and nineteenth-century politics, fueled by the growth of the capitalist economy, encouraged the introduction of another important legacy of antiquity on the international stage. It was to be one that would shape much of the nineteenth and twentieth century: the empire. While they never quite acquired the traits of ancient hegemons, modern imperial nations became a powerful force in international relations. In many ways, they still are.

The British created a global empire, briefly the largest the world had ever seen. By the second half of the nineteenth century, it accounted for a quarter of world output. Through their network of territories and colonies, the British spread their belief in the supremacy of individual rights, laissez-faire economics, and the power of markets across the planet. Of course, if belief wasn't enough, the British navy and chartered companies, whose armies mostly did Britain's bidding, ensured their dissemination.

Britain's constitutional architecture had enshrined the protection of private enterprise and property rights into a Lockean liberal state. But its most important contribution to the story of water was the definition of landscape imperialism for the modern era, enabled by a pursuit of freedom for individual British citizens, while ignoring the problem of defining a citizenship for all the people across its vast empire. The British experience turned the modern empire into a paternalistic institution, a managerial state intent on not just extracting resources from the places it controlled, but on disseminating ideas about the environment. British engineering produced the first universal theory of landscape. It was not the last.

The empire's crown jewel was India. In the eighteenth century, the East India Company had built a fortune exploiting the textile industry of Bengal, and pioneering the development of private and third-party companies for the development of water infrastructure. In 1858, the philosopher John Stuart Mill, who worked for the East India Company all his life, wrote a petition to Parliament, describing the company's achievements. The company's pride was the many canals, waterworks, and navigation routes that made up the irrigation system of India. Mill largely ignored pre-existing water infrastructure, describing instead the company's "exemplary" activities: the East and West Yamuna Canals, the Ganges, Punjab, and Sind Canals.

These projects supported settler colonization, which according to people like Mill delivered the universal British benefits of civilization, peace, and prosperity. In truth, despite Mill's praise, the projects were not always that successful. The Madras Irrigation Canal Company, for example, was supposed to construct and maintain irrigation canals for about half a million hectares of land, but it was mostly a failure. The revenue for the project was supposed to come from selling water at a fixed government price while charging for other services, including transport. As it turned out, the financial engineering was better than the actual engineering. The planning and execution were poor, and expenditures went well over budget. Besides, the business had been predicated on a switch to wet crops like rice. Alas, that did not happen because cotton prices boomed when supply from America collapsed, due to its Civil War. The incentive to switch evaporated. The project was eventually transferred to the government.

In 1858, after the Indian Mutiny, rule was transferred from the East India Company to the Crown, establishing the British Raj. British imperial policy in India focused on agriculture. Most capital investments were geared towards the large public projects that could improve productivity. The Indus River was the principal target of policy. What followed was the most substantial replumbing of a natural system the world had ever seen, producing what is to this day the largest contiguous irrigation system in the world.

In the lead-up to the American Civil War, U.S. investments had become unattractive, so financiers turned to India. In 1859, the Madhopur Headworks—a large barrage that cut the Ravi River high up, near Lahore—began feeding the Upper Bari Doab Canal, a key piece of infrastructure for the Indus basin. That was the opening shot of an unprecedented development. Barrage followed barrage as water was forced into ever bigger channels, innervating the Indus landscape. In the late sixties, the huge famines in Orissa and Rajputana gave the public the rationale, if not the economic incentive, for the Raj to create a department for irrigation and accelerate the canalization of the landscape. Investment in water infrastructure in India soared.

The development of the Indus River irrigation system was not simply a feat of engineering. It was an enabler and a response to profound shifts in the economics of the empire. Breaking with its eighteenth-century protectionist tradition, from the 1860s Britain pursued an aggressive free-trade policy, entering into a series of bilateral treaties to reduce excise taxes on imports. Most British treaties contained a "most favored nation" clause, which meant that the best terms available on any treaty had to be made available to all signatories, pushing production towards the cheapest, most efficient conditions. The country bought goods everywhere, accounting for a third of the world's imports.

British belief in the power of free markets was dogmatic. So strong was its hold on the ruling class that the overconfident country even entered into unilateral free-trade agreements. Unilateral free trade had devastating effects on British domestic agriculture. The United States, under President Ulysses S. Grant, was coming out of the Civil War. It was once again attractive to British capital, and had benefited from investments in transport infrastructure and the Homestead Act. Unlike Britain, America had no qualms about using protectionist policies to support domestic production. In the early 1870s, the British market was flooded with American cereals. Prices collapsed by up to 50 percent. British farmers were unprepared for the competition.

To add to the crisis, the 1870s had experienced disastrous weather conditions for British agriculture. Great Britain had been battered by constant rains: 1872 was the wettest year since records began in 1766. By December over a meter of rain had fallen on England and Wales. Rain crippled the main cereal-producing regions of the country. Yields dropped by a quarter, sometimes by half. Pastureland was waterlogged, which led to disease outbreaks in livestock.

When in 1879 the government set up the Royal Commission on the Depressed Condition of the Agricultural Interest, it concluded—somewhat fatalistically—that the depression was due "to a succession of unfavorable seasons," but the reality was that British agriculture had been destroyed by weather and trade policy. Indeed, agriculture in Britain would not recover at all until the twentieth century, when the blockades of the world wars forced the island to return to its land.

All that did not matter at the time, though, because by the 1870s the agricultural focus of British investors had turned conclusively east. Water resource development of India accelerated. Canals and other water projects were particularly attractive for investment because they increased the productivity of receiving countries, reducing the cost of imports for the British consumer, while providing a return on capital to the financial heart of London.

In 1873, the British passed the Northern India Canal and Drainage Water Act, which established the state's absolute claim on all irrigation waters. Engineers could allocate water on "scientific" principles, and farmers would have to pay a rate for that water, not based on how much they took, but on how much production came from the irrigated land—an unintended echo of Diocletian's tax reform. Given the limited innovation in Indian agriculture at the time, it was an incentive to grow the command area, the amount of irrigated land.

In the 1890s, canal building in the Indus basin reached industrial scale, as perennial canals supplanted previous, smaller seasonal canals. The scale of investments was astonishing. In 1800 about eight hundred thousand hectares were under irrigation in India. By 1900 the Indian peninsula, including Bangladesh and Pakistan, had thirteen million hectares of land under irrigated cultivation.

Water was central to the economic vision that the British had for India. As far as the British were concerned, the hero of their story in India was the irrigation engineer. Statues of Arthur Cotton and other nineteenth-century engineers still stand in Hyderabad and in Andhra. They represented the mix of social philosophy and ideas about landscape, engineering, and power that was at the heart of the British imperial project. In truth, it was a symptom of a hegemon that had made transforming the physical landscape a central part of its mission.

OPIUM AND WAR

The impact of British imperial hegemony on water was not limited to its direct investments. The British trade system was so large that its effects were felt well outside the borders of the empire.

During the second half of the nineteenth century, power shifted on the global stage. While Britain expanded, having won its battle against Napoleon and fully taken control of India, China found itself in a seemingly unstoppable decline, unable to keep up with the Industrial Revolution. As recently as the 1830s the balance of trade with the rest of the world had still been in China's favor, although the finances of the country had already been stretched at the beginning of the century. But by the 1840s, the situation had reversed. Opium from India, which had contributed to the vast agricultural expansion of the East India Company, flooded China as silver poured out of the country to pay for it.

The story of opium is a crucial one in the argument linking water and society. Although its physical footprint was not as substantial as that of grains, its true significance stemmed from its politics. It was the first modern instance of a globalized commodity that bound water management through international trade to the pursuit of war. One of the reasons opium was important was that, since the time of Warren Hastings, the first governor-general of India in the 1770s, opium revenues were treated as excise or tax funds, rather than as private commercial profits for the East India Company, so they were an important source of revenue for the British treasury.

To be clear, this was not a new business. The East India Company had adopted the opium and salt monopolies that had first been established under the Moghuls. But the British turned the opium trade into a mass-market business. The British government provided capital, technical advice, and seed, and in return purchased opium at a fixed price. "Bengal" opium came from the eastern valley of the Ganges, where over a million peasants received government licenses to grow it. In these fertile plains alone, some 250,000 hectares of irrigated land were cultivated with poppies.

The production of opium grew. In the 1830s opium was about 6 percent of the annual revenue of the Indian government. By the 1870s, when the revenue of the Raj was around fifty-one million pounds, opium contributed over eight million, or 16 percent.

Opium was a critical crop in the economics of the empire. Britain imported enormous quantities of tea from the Celestial Empire. Without some counterbalancing export, Britain would have hemorrhaged silver bullion very quickly. Opium was the answer. A "triangular trade" developed between India, China, and the United Kingdom. About twenty million pounds' worth of opium and cotton went from India to China. In return, China paid in silver as well as exporting to India products like sugar and silk. From India, England imported thirty-three million pounds' worth of raw materials—cotton, jute, indigo—and exported to India textiles and machinery. Of course, all of this trade went through British firms—agency houses, banks, and shipping firms, especially after the monopoly of the East India Company was lifted in 1816—and transferred a significant amount of bullion from China via India.

The opium trade was disastrous for China. Both India and China had had a drug culture long before the Europeans ever decided to sell it to either of them. The plant was not indigenous to these countries—it probably originated in western Asia and was likely brought to India and China by Arab traders. But the British opium trade fueled mass adoption, leading to a damaging opioid crisis.

When the Chinese emperor attempted to outlaw it, the British used it as an excuse for the First Opium War. The subsequent Treaty of Nanking forced open the Chinese market to foreign imports at punitive conditions. Things got worse. The war, the opioid crisis, the competition of foreign industry, and the loss of silver all resulted in aggressive increases in domestic taxation which put further pressure on the Chinese population. Karl Marx, writing about it a few years later, thought that the Opium War of 1840 had "forced the Celestial Empire into contact with the terrestrial world," ending the isolation that China had been operating under. He was convinced this would be its demise, "as surely as that of any mummy carefully preserved in a hermetically sealed coffin, whenever it is brought into contact with the open air."

The sad closing episode of the Opium Wars was the 1860 burning of Yuanming Yuan, the great "Summer Palace or the Garden of Perfect Brightness" that had inspired Coleridge's "Kubla Khan" at the end of the eighteenth century. The Chinese state had been made to fully "open up" to international—and specifically British—commerce.

Poppy cultivation in China remained illegal throughout the nineteenth century, but recognizing that it simply could not prevent opium from entering the country, the Qing government chose to encourage domestic production in western China. The Chinese peasantry might not stop using opium, but they should at least use a domestic product. This was a purely economic decision, to attempt to halt the leakage of silver to British India. China's production grew to twice to four times the amount of opium imported from India, but this came at a huge price. Addiction is hard to estimate within China, but there are claims that 4 to 10 percent of the Chinese population smoked the drug by then.

China's ability to resist the assault was further weakened by an unfortunate series of climatic events. In the spring of 1876 a catastrophic couple of years began. The provinces of Hebei and Shandong on the coast, in the lower reaches of the Yellow River just south of Beijing, received half the normal rainfall. During the summer of that year, the drought spread westward to the prosperous Shanxi and Shaanxi provinces, the grain-producing regions of northern China and the breadbasket of the country.

In the spring of 1877, things got worse. This time, the drought first hit Shaanxi, on the west side of the Yellow River, then spread eastward to Shanxi, finally reaching Shandong on the coast by autumn. Rainfall dropped to less than a third of normal. Drought conditions persisted until the spring of 1878. It was a disaster. As harvests collapsed, rice prices shot up tenfold.

Depending on estimates, sixty to eighty million people were affected, and on the order of ten million people may have died. The population of Shanxi province was almost halved. Hundreds of thousands of people moved south towards Sichuan in search of food. The imperial state, already struggling to hold on in the face of international pressure and financial overextension, was unable to mobilize the relief necessary to confront the crisis. Private relief efforts, both domestic and international, stepped in. Chinese philanthropists started a fund-raising campaign, producing a pamphlet with graphic wood-cut pictures that famously could "draw tears from iron" to convince people to give money. In 1878 finally the rains returned. China's power, however, had been greatly diminished.

An empire which had been able to harness its water geography for the better part of two thousand years had been defeated in a quest to open its markets to the very products of water and irrigation that came from British India.

This approach to empire was novel. Water infrastructure investments had become instruments of imperial power and geopolitics. The empires of antiquity had of course intervened in their conquered territories, but their aims were purely extractive. They did not invest in the productivity of the conquered state for the purposes of re-engineering their own economy. Grains had been central in supplying Rome, but for the most part the republic first and the empire later limited their interventions to the acquisition of rainfed land, leaving any investment to private developers. The Qin had transformed the landscape of their conquests to better harness their resources, but they were limited by the isolated terrestrial geography of China.

Britain transformed the water landscape of India to feed the opium trade with China, ultimately leading to conflict, mainly in the service of London's financial sector. In the twentieth century this dynamic would play out at a different scale with another crucial commodity: oil.

DOWN THE NILE INTO AFRICA

Colonialism arrived late to Africa. By the time it reached the continent, it had come and gone from South America. There had been several obstacles that had spared Africa from the worst early on: disease made it hard to reach the interior of the continent; military resistance of local populations had proven too hard to overcome for what was perceived to be a limited prize; sparsely populated, enormous landscapes were going to be expensive to administer directly. European powers had mostly nibbled at the edges of the huge landmass, relying on indigenous overland trade for goods and slaves, which they then shipped elsewhere. Inland Africa was neither reachable nor particularly appealing.

But all this changed in the second half of the nineteenth century. British prospectors kicked off mineral explorations across the continent, fueling an imperialist frenzy. The tone was set in the 1870s by characters like the British colonial secretary in South Africa, Lord Kimberley, and Cecil Rhodes, the owner of the De Beers mining company. Their diamond and gold discoveries propelled a flurry of European and American assessments to determine the distribution of "commercial geology."

Explorers like David Livingstone forged relationships with local African leaders, which then facilitated Britain's entry into the continent. For the same reason, Henry Stanley, who had put the matter of the sources of the Nile to rest in 1876, ended up working for King Leopold in the Congo, and Samuel Baker, the first European to visit Lake Albert, for the khedive of Egypt. Often couched as scientific activities, these explorations were nothing more than expressions of interest and precursors to predatory extraction. Suddenly, Africa was well beyond appealing.

Much of the early colonial experience in Africa was conducted on and around rivers. Stanley and Livingstone had proven that the waterways of central Africa were navigable. In 1873, the Royal Geographical Society sent an expedition led by Verney Lovett Cameron. The expedition started on the coast opposite Zanzibar. The original intent was to reach western Tanzania to find Livingstone.

Having reached halfway across the continent, Cameron decided to continue and arrived on the coast of Angola at the end of 1875. He was the first European to cross Africa, traversing the watershed that divided the Congo and the Nile, as well as finding the sources of both the Congo and the Zambezi. He reported of extreme riches and proposed that an inland navigation system should connect the Zambezi and the Congo, creating an unprecedented commercial waterway.

Seeing the trading opportunity, King Leopold II of Belgium promoted the creation of the International African Association, under the pretense of humanitarian and philanthropic objectives. It was a private association, which Leopold financed personally. In *Heart of Darkness*, with some irony, Joseph Conrad called it the "International Society for the Suppression of Savage Customs." Leopold wanted to claim a huge segment of central Africa, between the Sahel in the north and an imaginary line from the middle of Angola to Mozambique in the south. It included the Congo basin as well as the great African lakes, Tanganyika, Victoria, and Malawi.

Competition among European powers escalated. The French quickly struck a deal with Leopold for the rights to the Congo. The British and Portuguese, who had been negotiating treaties for the mouth of the river, found themselves cut out from the interior. Then the Germans recognized the claims of Leopold over the whole of the heart of Africa. Rules of engagement became urgent to avoid conflict.

In November 1884, German chancellor Otto von Bismarck convened the Berlin Conference of European powers to settle the issue. Even though the conference was focused on the Congo and West Africa, the final document of 1885, the General Act, set the international architecture of European engagement on the whole continent. Leopold was the clear winner. The king of Belgium had secured for himself—not for his country—ownership of the Free State of Congo, a massive landscape, part of a vast free-trade zone which included the Congo-Zambezi watershed. The heart of the matter was navigation on these vast rivers of Africa. Chapter IV of the General Act was an Act of Navigation on the Congo.

The modern idea of freedom of navigation, which had originated at the Congress of Vienna of 1815, was applied in full. When the structure of Europe was re-established in the wake of the Napoleonic Wars, the Congress of Vienna defined the regime under which the Danube, the Rhine, the Elbe, and the Oder could be shared among riparian countries. The Berlin Conference simply adopted the same principles as those of Vienna. The General Act established a free-trade zone that encompassed the Congo basin all the way to the Indian Ocean between five degrees north and the mouth of the Zambezi. There would be no transit tax on ships, nor transit dues on goods, and river infrastructure was to be accessible to all nations. Riparian states were able to charge for certain services on the river—like the maintenance of quays or storehouses.

The General Act also established the International River Commission, an authority to oversee the functioning of this regime, police the river, regulate the pilots, and manage quarantines. It was also to plan the works required to ensure the river continued to be navigable, and it even had recourse to the military vessels of the participating powers in case of need and could enter into loan agreements to finance its works. Nineteenth-century globalization was going to travel on the rivers of the great continent.

A WATER LEGACY

The legacy of the British Empire was not just economic or political. The reason rivers are considered central to the historical development of civilization has its roots in British attempts at justifying their own imperial project.

Britain's state religion provided some moral justification for the imperial project. The Church of England adopted Augustinian ideas of a Christian Commonwealth, which nicely fit with the empire's aspirations of universality. The problem was that Augustine's *De Civitate Dei* and Edward Gibbon's long cautionary tale *History of the Decline and Fall of the Roman Empire* bookended a thousand years of concern about how an imperial legacy could dissipate. Religion mattered, and the deep past held both justification and worrying omens for Britain.

Science was not on their side in the nineteenth century either. Darwin's *Origin of Species* disposed of the need for higher intelligence. In particular, natural selection appeared to provide a plausible explanation for different human phenotypes, at a time when biblical polygenic theory—the idea that humans belonged to different races as they descended separately from Noah's children Shem, Japhet, and Ham—was the dominant, slavery-supporting theory. If the morality of empire was to survive the onslaught of Darwinism, physical evidence for biblical narratives was badly needed. Interest in biblical confirmation, the practice of trying to identify ancient ruins to validate and ground-truth the biblical account, grew for much of the nineteenth century.

Near East exploration, from which much of the biblical confirmation was to come, was conducted by diplomats and members of the military. Major General Henry Rawlinson was a classic product of imperial geopolitics. Early on, he had become an army officer with the

British East India Company, but had developed an interest in Persian language. This eventually led him to be stationed in Persia to train the troops of the shah.

In Kermanshah, in the Kurdish mountains, he encountered cuneiform inscriptions for the first time. He visited the rock of Behistun, a multilingual inscription from the time of Darius the Great, and set out to transcribe and decipher the text. This proved to be the Rosetta stone of cuneiform language. Access to Near East history had opened up.

Rawlinson returned to the region in the 1840s and '50s, this time stationed as a political agent for the British in Baghdad. From Nineveh he retrieved thousands of artifacts, which went to fill the halls of the British Museum. Near East archaeology exploded. In Victorian Britain, the substantial stream of objects fed a powerful public interest in everything ancient and vaguely Assyrian. The British Museum was flooded with visitors keen to see the famous winged bulls of Assur. Even the Sydenham Crystal Palace, which had become an amusement park after the Great Exhibition of 1851, housed an exhibit of a great Assyrian court.

It is through this interest that the story of the great "river civilizations" made it into the mainstream. The British aspired to transform the rivers of India and soon of Mesopotamia and Egypt into breadbaskets for a new universal empire. What better support than the evidence that this was precisely what had happened in biblical times? The press latched on to this, driving further political support for Assyriology.

A similar fate awaited Egypt. The British had competed with the French since Napoleon's daring conquest of the country in 1798. In 1799 Napoleonic troops had found the Rosetta stone, a piece of granodiorite with an inscribed multilingual decree from the Ptolemaic period. From it, Jean-François Champollion deciphered hieroglyphic script in 1822, providing a key to ancient Egypt that had been lost since the early Middle Ages. Then in 1881, a tomb was discovered near Thebes and given the unremarkable code name DB320. In it were fifty preserved mummies of kings and queens, among which was the mummy of Ramesses II. The biblical connection was enough to spark a race for Egyptian antiquities. A year later, Egypt had become a British protectorate, itself entering the imperial system.

The British quest for justification in the depths of the archaeological record and their efforts to root their approach to the water landscape in the imperial tradition of ancient river valley civilizations introduced a story of water that would inspire generations to see in the water landscape the principal path to development.

THE WATER LANDSCAPE ENTERS MODERNITY

Empires are extractive, they bring vast resources to the center. But they are also fragile. They offer at best ambiguous solutions to the question of what social contract—the set of beliefs, rules, and organizations that mediate between individual power and collective action—should govern society. These challenges manifest themselves on the water landscape, which is the most physical and basic stage on which that mediation happens.

In the case of Britain, if water was an instrument of empire internationally, domestically the fight for landscape was over the impact of modernity. In London, during the summer of 1854, five hundred people died from a severe outbreak of cholera near Golden Square in Soho. In one particular house, 40 Broad Street, a family had been stricken by tragedy: a two-year-old had died from diarrhea. The mourning mother had washed the child's soiled diapers, throwing the dirty water down the poorly constructed house drain. The drains did not isolate the waste stream from the nearby drinking well, which, in turn, connected to a pump in the street, a pump famed for the quality of its water and from which all the neighbors drew. The result was catastrophic.

The virus quickly spread from the water supply to all nearby households. The outbreak followed. At the time, people assumed cholera was an airborne disease, spread by the "nauseous fumes" emanating from the poverty-infested slums. John Snow, a local physician, realized fumes had nothing to do with it. He suspected the disease was carried by water, so he mapped all the victims of the disease on a local street map and showed that they all clustered around that single pump. The supply was promptly shut down by removing the pump's handle and the epidemic ended. Snow's insight gave birth to modern epidemiology and convinced people that the source of urban disease was polluted water.

The unsanitary conditions highlighted by Snow convinced city administrators across England that modern, industrial urbanization was not sustainable without developing safe sources of water. These could not be imported from across the empire, no matter how rich and powerful it was. The race was on to find new natural sources outside cities in the domestic landscape.

In 1866, the Royal Commission on Water Supply was charged with inventorying such sources of water. Large impoundments and aqueducts followed, as large-scale environmental and hydraulic engineering transformed the landscape of the kingdom. The water supply of Manchester, the heart of British industrialization, was the archetype of these projects. There too, deteriorating living conditions in the city required clean sources of water.

In the 1870s the city of Manchester bought Thirlmere, a lake a hundred miles northwest of the city, to turn it into a reservoir for the city. These developments were entirely consistent with the anthropocentric philosophy of the British liberal elite. In 1854, John Stuart Mill had written that, however one defined nature, it did not make sense to allow it to simply follow its course. Man should be "perpetually striving to amend, the course of nature—and bringing that part of it over which we can exercise control more nearly into conformity with a high standard of justice and goodness."

But development of the landscape in a system of increasing political power could not happen without conflict. Thirlmere, the heart of the Lake District, was one of the most iconic English landscapes. The Victorians thought of it as supreme natural beauty. William Wordsworth and Samuel Taylor Coleridge had their meetings on its shores, memorialized in the poem "The Waggoner." In fact, the landscape had been heavily modified by man over the centuries, but for romantics like Wordsworth the historical interventions had been part of nature (much like Subiaco had been a pristine lake, as far as Benedict was concerned).

The development of the Thirlmere scheme led to one of the first explicitly environmental fights in modern history, as local communities fought to avoid it. It led to the establishment of the National Trust and to the development of the conservation movement. Ultimately conservationists lost those battles, and the landscape developed across Britain.

Achieving water security in nineteenth-century England wasn't a race for technology. It was a battle for identity in the pursuit of the ideal community. It was the beginning of a modernist project that would find fulfillment in the twentieth century: the progressive conquest of nature.

12

The Great Utopian Synthesis

WATER AND INDUSTRIALIZATION

After a failed insurrection attempt in Canton in 1895, Sun Yat-sen left China for what would turn out to be a sixteen-year exile. In those sixteen years, Dr. Sun turned into the revolutionary leader imagined by his persecutors and celebrated by posterity. He traveled from Japan to Hawaii, from San Francisco to New York, finally arriving in London in September of 1896.

After a botched attempt by the Chinese Legation in London to kidnap and deport him back to China, Dr. Sun stayed in London for eight months, a guest of Dr. James Cantlie, his former mentor at the Hong Kong medical school. During that time, he joined in discussions on current events typical of late Victorian England, from the wars of Rhodes and the British South African Company to the concerns of Irish nationalists.

He spent his days studying at the British Museum Library, where he read the works of Karl Marx, Henry George, John Stuart Mill, Montesquieu, and many other philosophers who had contributed to seeding the intellectually turbulent nineteenth century. Republican aspirations, liberal beliefs, the end of the old order, and the rise of the industrial age all contributed to a complicated ideological context. For a revolutionary in the making, exposure to the intellectual environment of London, with its diverse population of political exiles, was formative.

Towards the end of the nineteenth century, grasping for narratives only in the deep agrarian past had become increasingly difficult in an industrializing European world. Those who lived a rural life, even if the majority, were receding into the background. The last subsistence crises of the Western world—the potato crisis and famines of 1845–47, for example—were symptoms of politically diminished, vulnerable farming communities. The economy was changing. Industrialization had forced many into urban squalor. Labor relations were strained and poverty endemic. The American and French revolutions had shown old orders could be toppled. Tensions were rising. Revolt was in the air.

Marx had died in 1883, thirteen years before Sun's arrival in London. He had been the pre-eminent interpreter of this fundamental social shift. His analysis posited that material conditions were endogenous to society; they were determined by the combination of technical and productive capacity—the means of production—and the social relations that supported production, including the structure of property rights. In response to those material conditions, societies evolved through different modes of production, ultimately embracing communism.

In this process, which Marx called dialectic materialism, water resources and the natural environment had an ambiguous role at best. Marx recognized that the natural world was an important means of production in the early stages of a society's development. In fact, Marx went as far as to say that humans first defined themselves as a species in the modification of nature. But as technology became more important to human activity, nature—Marx thought—would no longer really be in a dialectic relationship with society. It would simply become a reservoir of resources to be extracted. He argued for a progressive alienation from nature, as class structure and industrial production took over in defining the human experience: only relationships between people—in fact, between classes—would define the mode of production.

That an urban, middle-class philosopher who focused on the industrial working class would assume that the natural world had little to contribute to the evolution of modern society may not be all that surprising. But the problem was that it seemed to miss large parts of the world.

His theory was designed for a society seemingly subject to powerful historical forces, in which conditions for entire classes of people were changing over the course of a lifetime. In contrast, China's despotic regime seemed immutable, fixed in time. Marx knew that China was still the pre-eminent water country of the time. The philosophers and intellectuals that inspired his work had also observed as much.

Montesquieu had used China as a counterpoint to his theory of the state in *The Spirit of the Laws,* arguing that its particular climate, in the face of limited agricultural resources, led to a constant cycle of dynastic regimes and revolutions. Adam Smith had also observed that China "seems however to have been long stationary. Marco Polo, who visited it more than five hundred years ago, describes its cultivation, industry, and populousness, almost in the same terms in which they are described by travelers in the present times." Smith implied that the exceptional economic conditions of China—cheap labor from its vast population, and cost-competitive production thanks to its extensive fluvial network—must have played a role in this stasis.

The struggle to incorporate water and climate events in a theory of modern society reflected in part the fact that they operated on scales of time and geography that seemed to have little to do with the industrial working day, or the scale and scope of market supply and demand.

To address China's exception, Marx introduced the idea of an "Asiatic mode of production," a mode which did not appear to follow his theory of social evolution: slavery to feudalism, to capitalism, and—finally—to communism. In this mode, agriculture and manufacturing were inseparable within small subsistence communities. It was an agrarian society under the control of a despotic regime, which exercised overwhelming control of the means of production, in part through the command of hydraulic infrastructure.

Marx did not spend a lot of time elaborating on this mode of production. China's anomaly seemed largely irrelevant. After all, it was still a pre-capitalist society. Marx's gaze was fixated on the British industrial proletariat, for that is where he and Engels believed the revolution would come from. For the same reason, Marx largely ignored Russia, which he saw as a nineteenth-century imperial power akin to Britain, but one that had so far avoided industrialization and that was, for that reason, an unlikely candidate for revolution.

It is therefore all the more ironic that in both China and Russia, not in Britain, his revolutionary ideas would find real-world grounding, and one that was largely obtained on the back of water infrastructure.

VERA PAVLOVNA

At the British Museum Library, Dr. Sun did not just spend time reading. He also engaged in conversation with a number of radicals, particularly those from Russia, who frequented the institution. Among them, Felix Volkhovsky was a leader of the populist Russian movement. He befriended the Chinese revolutionary.

Volkhovsky had been persecuted for his liberal views and imprisoned in both St. Petersburg and in Siberia before he was able to escape and reach London. He had been part of a group of Russian intellectuals that had planned to have Marx's work translated into Russian. Volkhovsky also promoted the distribution of another crucial writer for Russian radicals: Nikolay Chernyshevsky.

Chernyshevsky was one of the founders of the Russian populist movement. He had also been imprisoned for his liberal views. Volkhovsky and others worked hard for his liberation. It is possible that he introduced Chernyshevsky to his friends in Britain, including Sun Yat-sen. Chernyshevsky's novel *What Is to Be Done?* was written in the early 1860s while he was in prison awaiting trial for his revolutionary radicalism. It enjoyed a great deal of success, crucially among radicals and anarchists, including the young Stalin, Trotsky, and Lenin.

Vera Pavlovna, the protagonist of Chernyshevsky's story, was an independent-minded young woman from St. Petersburg who left her family to pursue a revolutionary life. Through her, Chernyshevsky revealed what a rural utopia for Russia could look like. Amid a rather convoluted love triangle, the pursuit of empowerment, rural cooperation, and industrialization, Vera imagined Russia's future in a dream. That future was full of modern constructions of glass and aluminium, suspiciously similar to London's Crystal Palace. But it was also a future in which water infrastructure had allowed people to tame nature and turn it to their service. The desert had been engineered into a wonderful, fertile landscape. Machines, canals, irrigation, and weather control had all been summoned in the service of a rural revolution. And people transformed along with the landscape: they had to be taught, enlightened in the ways of scientific agronomy.

The dream of a scientifically engineered landscape was in part a reflection of a hard reality: while vast and rich in water and land, Russia's water geography was, at best, inconvenient. The United States was blessed with remarkable congruence between fertile land and distribution of water. Russia was not so lucky. Its natural resources were not laid out in a way that made them easy to use.

Russia was also much further north than the United States, so about a third of its land was covered by permanent ice and frozen terrain. Agriculture could only really happen further south, where there was adequate insolation. But sunlight mostly lit the arid landscapes and steppes of southern Russia, Ukraine, and central Asia, where there was limited water. With the exception of the Volga River, over 80 percent of the water of Russia flowed in the large Russian rivers that emptied into the Arctic Sea. Siberia and northern Russia had plenty of water. Seventy percent of people and economic activity were further south with less than 20 percent of the water resources of the country.

Chernyshevsky was writing in reaction to the hopeless backwardness of Russian reality. Russia had been poised to be a significant exporter of grains. In the 1860s, as part of its globalization and push for free-market trade, Britain had repealed its corn laws, eliminating its last protectionist barriers. Russia was one of the first countries to export wheat into the newly integrated international market. The problem was that tsarist Russia had had relatively limited experience of infrastructure development. It oversaw a highly controlled, centralized, and inefficient agrarian economy. Towards the end of the nineteenth century it focused its modernization efforts on the Volga, which had been a critical northwest-to-southeast trade route since the eleventh century. But pursuing industrialization would have required unprecedented investments, far greater than what the tsarist regime had been able or willing to pursue.

The relative stasis of the regime, surrounded by the economic and social ebullience of the second half of the nineteenth century, created the space for radical dreams.

Chernyshevsky's hopes for Russian modernization remained unfulfilled under the tsars. Electrification, which had started to spread in the rest of the world thanks to the rise of hydropower at the end of the nineteenth century, had been controlled and slow. History could have taken a very different turn had Russia been able to use its access to the European market to modernize, much as the United States ended up doing as it developed its own agriculture for export.

Alas, because of protectionist barriers in other countries during the thirty years leading up to the First World War, Russia struggled to join global trade. When the First World War broke out, the German blockade cut Russia out of supplying Europe, pushing the United States to make up the deficit instead, and leaving Russia isolated.

Marx was long gone by then, but his ideas were far from dead. When in February 1917 the tsarist government dissolved and the Duma-based provisional government took over, the government tried to accelerate a number of projects that had been stuck in the bureaucracy. But before any intervention could be implemented, history took over and the echoes of nineteenth-century communism reached the twentieth.

Decades after Marx and Engels wrote their *Communist Manifesto,* with their eye firmly on the proletarian classes of the West, it was Russia that would first embrace their revolutionary call. They judged the capitalist transformation of the nineteenth century to have been the "subjection of Nature's forces to man." The revolution would in turn subject them to the service of the proletariat. Lenin saw it done.

MEGAPROJECTS AND UTOPIA

In the nineteenth century, Chernyshevsky's water utopia was far from unique. Revolutionaries like Dr. Sun imagined a future inspired as much by the political philosophy of the long Western tradition as by the aesthetic response to the modern world. It seems that novelists had succeeded at sensing where modernity was headed far more than political economists and philosophers.

The utopias of literary circles were of course political in nature. Edward Bellamy wrote *Looking Backward,* in which he imagined a technological state-run utopia. In his world, nature had been tamed, so much so that when it started to rain, automatic awnings would be rolled out over the street so that people could continue walking about their business without any worry of getting wet. In 1890, William Morris, another socialist utopian, wrote *News from Nowhere,* a science fiction response to Bellamy, describing a time traveler who visits the Thames of the twenty-first century, finding a libertarian socialist utopia in which people owned all means of production, organized largely around the river.

Water and its power were central to these imagined futures. Sometimes, people even tried to make them happen. In 1902 Theodor Herzl published his utopian novel, *Altneuland*. It was a vision for a new Jewish state, which he would then pursue with all his energy in the spring of the following year. The protagonist of the novel, a young, desperate Jewish lawyer called Fredrich Loewenberg, ended up visiting a fictional Palestine, finding a developed coastline where previously there was only desert; everything electrified; a new society scientifically planned, efficient, entirely based on cooperatives—a modernist dream at the beginning of the twentieth century. The energy source for this fictitious model state was a canal that connected the Mediterranean to the Dead Sea, and hydropower installations on the Jordan and the canal itself. Herzl imagined Altneuland independent of coal and able to develop irrigated agriculture from the Jordan River waters.

Technological utopias were not developing in a vacuum. These water-led ideas had been stimulated by the great water projects of the end of the nineteenth century. The Suez and Panama Canals had convinced an awestruck public that, indeed, man's command of nature could be total. The Frenchman Ferdinand de Lesseps was at the heart of both.

De Lesseps's story of daring enterprise to make the Suez Canal is quite extraordinary. While a diplomat in Egypt, he had become familiar with a group of the Comte de Saint-Simon's followers, social utopians who believed—in the greatest French engineering tradition—that public works and science would transcend national boundaries to lay the foundations of a new world order.

Saint-Simon himself had predicted that the canals across Suez and the Isthmus of Panama were critical steps towards that new world. The group had given themselves the name of Société d'Études du Canal de Suez. Their leader, Prosper Enfantin, had zeroed in on a vision for the Suez Canal that enthralled de Lesseps.

After leaving the diplomatic service, de Lesseps focused single-mindedly on making the canal happen. He established the Compagnie Universelle du Canal Maritime de Suez, incorporated in France with operations in Egypt, and sold its shares to thousands of French subscribers. He managed to convince the new viceroy of Egypt, Mohammed Said, to concede a lease for the gigantic project (that is how Port Said got its name). Against all expectations, the canal turned out to be a huge economic success.

By the mid-1870s, the value of the Suez stock had quadrupled, yielding dividends of 17 percent to its shareholders. Even the British, who had initially judged the project ludicrous, decided to invest, ultimately taking the controlling stake and making it the lifeline of their empire. It was a miraculous story of entrepreneurship: an international venture that, against the odds, had bet on the ability of one man to connect the seas. It fed into the belief that all was possible.

When the time came to attempt Panama, de Lesseps decided to follow a similar recipe. The French company entered into a concession agreement for the development of a canal with the Colombian government, which had sovereignty over Panama. De Lesseps raised the initial funds from several hundred rich French investors, in exchange for shares in the company. He then went on tour, with the assistance of the French financial establishment, to raise the remaining funds. Nationalistic propaganda and a fair amount of corruption in the French media resulted in hundreds of thousands of French people investing their savings in the company. Work started in 1881.

De Lesseps's considerable charisma and his track record at Suez forced a consensus in favor of a sea-level canal, the same basic structure of Suez, rather than the lock canal that was ultimately built across the isthmus. It turned out to be a catastrophic decision. De Lesseps had vastly underestimated the risks and costs of such an effort. The rock he had to cut through was nothing like the sand of Suez. The tropical forest proved a difficult, unwelcoming environment. Endemic yellow fever and malaria decimated his workforce.

By 1889 the company had run out of resources and was bankrupt, as were a number of its investors. But it was far from the end for this modern conquest of nature.

AMERICA TURNS TO EMPIRE

After the disaster of de Lesseps's project and the international humiliation of France's great expansionist aspirations, the ball passed to the United States. By the end of the nineteenth century, long gone were the ruralist ethos and classical republicanism of the Founding Fathers. The commercial expansion that had followed the Civil War had propelled capitalism to the fore. America had embraced the utilitarian ideals that were sweeping the nineteenth century.

The French canal company was at risk of never recovering even a small part of its capital. Its investors were keen to offload both the Colombian concession and the old machinery to the United States. The United States government for its part had needed a transcontinental canal for much of the nineteenth century, because westward expansion had opened both coasts to commerce. The expansion into the Pacific, Hawaii, and the Philippines had strained America's ability to project power. Then, the Spanish-American War of 1898 had made plain the full implications of having to sail warships from the shipyards of the East Coast around Cape Horn to reach the West Coast in time for a conflict. A faster route was urgently needed. The problem was that the U.S. vastly preferred a canal across Nicaragua as it already accommodated most of the transcontinental traffic with a mix of rail and navigation through the country's eponymous lake.

One of the original investors in the canal company, Philippe Bunau-Varilla, engaged a New York lawyer, William Nelson Cromwell, to help generate enthusiasm for the Panama project in America. Cromwell, proficient in the arts of lobbying and public relations, succeeded. In the end, in the liquidation of the de Lesseps canal company, the United States government paid $40 million to acquire its assets and concession contracts.

Who exactly received those funds and what degree of speculation and political corruption was involved in that transaction was never fully established, but the less charitable view—put forward in a series of articles in October 1908 by Joseph Pultizer's *New York World*—is that the United States bought itself into the project to satisfy the speculation of a few American private investors.

Be that as it may, the most remarkable aspect of this saga was yet to come. Colombia had sovereignty over the land of the isthmus and seemed intent to negotiate hard on the conditions under which the concession could be passed to the United States. The difficulty in reaching a negotiated agreement led to unrest and ultimately revolt in Panama, which declared independence in 1903.

The minister plenipotentiary representing the newly independent Panama in the negotiations with the U.S. was now none other than Philippe Bunau-Varilla himself. The treaty that emerged—the Hay–Bunau-Varilla convention—was largely favorable to America, giving it sovereignty over the territory of the concession, which became the Canal Zone, and therefore ownership of the canal, a situation that persisted until 1977, when a new treaty was negotiated for the transfer of the Canal Zone back to Panamanian control.

On February 19, 1906, President Theodore Roosevelt wrote to both the U.S. Senate and House of Representatives. His letter accompanied a long report from the Board of Consulting Engineers for the Panama Canal. Roosevelt noted that the board could not come to a unanimous agreement on what design the canal should be. The European consultants, who saw in Suez the most obvious analogue, favored a sea-level canal similar to the Egyptian one. That was the majority opinion on the board.

The minority, however, made up mostly of American engineers, preferred a lock canal similar to the canal on the St. Mary's River at the border with Ontario, the anglicized "Soo Locks," which connected Lake Superior with Lake Huron. Along with the Erie Canal, it was one of the great infrastructure projects of the United States. During the summer months that passage accommodated an enormous volume of traffic, three times that of Suez. With the endorsement of Roosevelt, Secretary of War Taft and the chief engineer of the project itself went with the minority view. The lock design was approved, and the biggest infrastructure project ever attempted on the planet began.

The final design of the Panama Canal was due to an American, chief engineer John Frank Stevens. It was inspired by a well-known nineteenth-century structure: a system of locks connected to a navigable lake, high enough to allow for lifting ships over hills. The design had been used by Thomas Telford to connect the lochs of Scotland through the Caledonian Canal.

Everything depended on one river, the Chagres. There was no Lake Nicaragua to start with. To overcome the impenetrable Continental Divide, a huge earthen dam at Gatun dammed the river, creating the largest manmade reservoir in the world. The locks of Panama were the most innovative aspect of the canal. Unlike many other such systems, there were no pumps to move water in and out of the lock chambers. Gravity moved water from the river into the lake and to the locks, through eighteen-foot culverts that ran the length of the sidewalls. Valves along the tunnels opened and closed to fill the chambers, releasing water once the ships were through. Everything ran on water. Gates and valves operated thanks to the power of the hydroelectric plants on the river.

On October 10, 1913, United States president Woodrow Wilson hit a telegraph key in the Oval Office of the White House. It was 2:00 p.m. The signal traveled from Washington to Galveston, Texas. It continued to the port of Coatzacoalcos in Mexico, crossing to the Pacific coast at Salina Cruz. From there, it went to San Juan del Sur in Nicaragua, finally reaching Panama and its destination. It took only a few seconds. Four hundred powerful charges detonated, four thousand miles from Washington. The deflagration blew up Gamboa dike, the last earthen obstacle left.

The rather elaborate stunt released water from Gatun Lake into the Culebra cut, the artificial valley dug through the rock of the Continental Divide that runs along Panama. The Chagres River, the source of water for both Gatun and the locks, had once flowed only to the Pacific. It was now going to drain into the Atlantic too, the only river in the world to do so. The Panama Canal had connected the oceans.

PREPARING FOR THE FUTURE

Panama was the most extraordinary culmination of an era of unprecedented construction. It inspired a sense of unending possibility. When in 1914 the New York painter William Van Ingen produced four murals to adorn the rotunda of the Administration Building of the Canal Zone in Balboa, he represented the construction as a "Wonder of Work."

One of the four murals shows the Miraflores lock, which controls the first two steps of the canal on the Pacific side. The image is reminiscent of the building site of a New York skyscraper, the iconic imagery of the roaring century that was beginning. At its center is a section of a culvert buried under a stepped, concrete sidewall, on which Van Ingen painted workers for scale. The wall on the other side of the canal was only partly visible as background amid the cranes, while the scene was enveloped in smoke and dust. In the very foreground, emerging from the lower-right corner, a huge steel boom reminded the viewer of just how deep the site was.

According to Van Ingen the locks brought to mind "thoughts of the Egyptian pyramids." The Egyptian architectural revival was in full swing, inspired by British imperial enthusiasm. In visiting the Gatun lock, Van Ingen saw "a range-light tower, with the architectural details of the Roman period; the instant I saw it the thought flashed over my mind; why was not the form of the obelisk used?" It was a new time of empire, and the control of water held center stage. Gustave Eiffel's bridges and celebrated tower, Baron Haussmann's Paris, the Suez Canal, and countless other engineering feats had shaped perceptions of what it meant for the landscape to be modern during the nineteenth century. And now it was Panama.

The story of all those accomplishments is often told as one of engineering and design, evidence of an unprecedented industrial age. But behind the technical accomplishment were older, more fundamental ideas. Fin-de-siècle Europe and America were societies in the throes of a sophisticated second industrial revolution. The Panama Canal convinced many that humanity had fully conquered nature, that the only problem left to solve was simply finding man's place in it.

In its final design, the Panama Canal was an extraordinary integration of human infrastructure and nature. But it could not work without adequate territorial control. The choice of a lock canal, rather than a sea-level waterway, meant that the entire system was dependent on the water supplied by the Chagres River. It committed the canal to proactive management of land well beyond the Canal Zone. Most of it depended on acts of sovereignty. The creation of Gatun Lake—the largest man-made lake in the world at the time—required expropriating land that was going to be flooded. The forests surrounding the lake, which turned the intense and short tropical rainfalls into slowly released flow into the reservoirs, became integral to the infrastructure.

In *The International Development of China,* which Sun Yat-sen wrote in 1920, he spent considerable time explaining how he imagined the Yangtze was to be transformed in service of the new Chinese republic. He included several detailed maps of different stretches of the river and how they should be modified. At the heart of this complex project, which included transforming the estuary with vast land reclamations and cutting enormous canals to straighten the river, was the impoundment that would eventually become Three Gorges Dam.

Inspired by the great projects of the nineteenth and early twentieth century, he wrote of cutting through rock and blasting obstructions. Throughout the text, Dr. Sun was at pains to argue that the project was economic, estimating both the cost of it and the potential value that came from the economic activities it enabled. He ended his detailed plan by saying that "this project will be more profitable than either the Suez or Panama Canal." A benchmark for development had been set.

Translating a utopian vision of this scale required a sufficiently large, powerful agent that could command the resources and territorial control needed to turn that vision into reality. It was no coincidence that the private enterprise efforts of the nineteenth century had failed. That agent was the twentieth-century state, a state that could pull off something as daring as what just a few decades earlier only dreamers like Chernyshevsky and Herzl could imagine.

But the involvement of such a state had an unexpected effect. For every step that people took to control nature and express their ever-more-powerful agency on the world of water, they were tightening their relationship to it. The more intrusive the intervention and the more definitive its effect, the more society was locked in to the illusion of control it had created. This was an incremental process, in which the dialectic between the power of man and that of water transformed the landscape. The Panama Canal was not the only contribution to that transformation, but it was one of the largest. It echoed the island of Utopia in More's story, artificially separated from land by a fifteen-mile-wide canal that had been an imagined feat of engineering.

The nineteenth century had been the pinnacle of private enterprise, a time of concession contracts and of companies, the natural evolution of medieval principles and their classical inheritance. But now, a new, powerful state was going to be the principal developer of the landscape. The loose coupling between human society and the water environment, one based on variability of the latter and adaptation of the former, one that had lasted for the better part of eight thousand years, had come to an end. From now on, human society was expecting predictability, stability, and control as a fundamental trait of the res publica. The illusion that society had finally decoupled from the forces of nature required the state to absorb its force. The great projects had set a new standard for what the water landscape of a great nation should look like.

PART III

THE HYDRAULIC CENTURY

13

Setting the Stage for Revolution

CENTURY AT THE OPEN

The nineteenth century had led to a remarkable degree of international convergence on liberal recipes that manifested themselves in water. For example, the "conventional basin" of the Congo, the huge landmass that included the Congo basin and that stretched all the way to the Indian Ocean, precisely defined at the Berlin Conference of 1884–1885, is a case in point. Only a hundred years earlier no European had ever stepped inside most of this vast territory. By 1900, it was a free-trade zone subject to river navigation principles that had been agreed to in Berlin. Colonization was not just about European people. It was about their institutions. The inheritance of the Iuris Civilis, classical republicanism, and the Treaty of Westphalia spread far. Sun Yat-sen's vision for China, for example, was one of economic development and prosperity, which pursued a very specific relationship between citizens, the state, and the landscape.

As the twentieth century opened into its first decades, Dr. Sun's particular brand of nationalism, democracy, and socialism, just like everyone else's attempts at governing society, had to contend with a momentous transition. In the early part of the century, Dr. Sun could only intuit the true scope of the changes that were afoot. But like all committed revolutionaries he picked up on the winds of change that were sweeping the world. He knew that a country as populous—and as poor—as China would have to define its own future, or have it defined by the gathering storm. All around him, as he traveled the world in exile during the first decade of the new century, were the symptoms of unprecedented change.

Three trends emerged at that point, which ultimately transformed humanity's relationship to the planet and to water. The first was demographic. There were going to be far more people than there had ever been before: that much was clear. For close to two thousand years, human population had grown at an annual rate below a fifth of 1 percent, gradually increasing from fewer than two hundred million at the start of the Roman Empire to just over six hundred million people in the early eighteenth century. Then, growth accelerated. Two hundred years later, at the beginning of the twentieth century, there were well over one and a half billion people. Those numbers would become a demographic explosion just a few decades later. Indeed, by 2000, over six billion people lived on the planet, a fourfold increase in a single century.

A second, crucial trend was industrial. Capitalism had created vast inequalities. The palliative for those inequalities was consumption, the widespread purchase of mass-produced cheap products that would give everyone access to a slice of progress. But to satisfy the demands of mass consumption from population growth, energy use—a crude metric of production—would grow tenfold over the century. Energy mattered. Water was at the heart of this energy revolution, at least in its early stages. In fact, one of the final gifts of the nineteenth century had been the hydropower technology that enabled people to harness rivers to produce electricity. The growth in hydropower would dominate the story of water for much of the twentieth century.

The third trend, maybe the most important for the story of water, was political. The territorial nation-state had been the dominant political organization of society since the seventeenth century, but it wasn't until the twentieth century that it also became its most powerful economic actor. Engineering had given people unprecedented power over the natural world, supporting a singular modernist idea: that the development of water resources could be an instrument of nation-building in the hands of a powerful territorial state. Replumbing the planet on a scale commensurate to the Panama Canal required enormous amounts of finance to be locked up for decades in hard infrastructure. No private investor could spare that much for that long without explicit or implicit government guarantees. This reality had already become apparent in the provision of clean water for cities: after the financial crisis of the end of the nineteenth century, the only way to finance its development was through municipal bonds.

The ability and drive of governments to finance and deliver a hydraulic transformation of the landscape was going to be the central story of the twentieth century, a collective investment in a future res publica. How these investments were guided was a function of the form and purpose of the state. When investments were pursued as a matter of true public benefit, they intersected growing political enfranchisement. The two decades leading up to the First World War saw broadening suffrage, more formalized labor relations, and occasional hints of a welfare state. The tension between the sovereign state, people's engagement in a collective project, and the mobilization of unprecedented public resources led to multiple attempts at realizing the modern state, from the liberal and social democracies of the West to the Soviet Union and the fascist regimes of the first part of the century. All shared different strands of premodern DNA. All were responses to the trends that were shaping the twentieth century. All contributed to redefining society's relationship with water.

FARMING THE WORLD

The first major trend that was transforming the world was the relentless increase in the number of people. Demographic growth put enormous pressure on most of the world's agricultural systems and water resources from the very early stages of the century. In response, wealthy countries were enthralled by the potential of hydraulic engineering to satisfy their political imperatives and meet the needs of a growing consuming population. Their enthusiasm magnified their ambition to reshape the landscape on a scale commensurate to the extraordinary projects in Suez and Panama.

For example, confident in their abilities, thanks to their Indian experience, British engineers turned into keen developers of the Nile. Cotton, a key raw material, was the crop of choice for the region and a crucial ingredient to the mass consumption economy. But because cotton grows in the summer and needs year-round irrigation, the Nile's floods were no longer sufficient. Cotton needed perennial irrigation. The amount of land that could be irrigated was directly proportional to water storage. So, in 1902, British engineers decided to build Aswan Low Dam, the largest masonry dam in the world.

Then, in 1910, Murdoch MacDonald of the Egyptian Ministry of Public Works published the first full assessment of infrastructure potential, aptly named "Nile Control." For a brief moment, the British held vast mandates that covered most of the basin—the Protectorate of Egypt; Sudan, which Lord Kitchener had reconquered on behalf of Egypt in the 1890s; Kenya, Tanganyika, and Uganda. They could entertain holistic, system-wide solutions. The most famous was the Equatorial Nile Project, later the Victoria-Albert-Jonglei scheme. Dams would store water in Lakes Victoria and Albert, thousands of kilometers south of Egypt, where evaporation was sufficiently low to hold the volume from one year to the next. The water would then bypass the Sudd through an enormous canal from Jonglei in South Sudan to the White Nile further north. The ambition of this integrated plan bordered on the delusional. The project was never carried out in its entirety—the idea of a Jonglei canal survived well into the 1970s, though, when the Sudanese government began its construction, and was finally abandoned in the eighties, a few years into the political instability that took over Sudan—but it reflected the desire to exercise sovereignty over the landscape at a scale commensurate to the emerging democraphic forces of the twentieth century.

While the British fantasized about their pharaonic developments on the Nile, on the other side of the Atlantic the United States was equally enthusiastic about water development. The Progressive Era had kicked off with President McKinley's moralizing government. That experience heralded a period during which the federal government was going to take a far more active role in the life of its citizens. Despite the success of the nineteenth century's Homestead Act, in 1900, 90 percent of the population of the United States still lived in the eastern half of the country. Unemployment had reached 20 percent because of the depression of the 1890s, and millions of poverty-stricken European immigrants, carrying their suitcases full of radical socialism and class warfare, risked turning the industrial heart of the United States into a powder keg. A rural push westward would be a safety valve to reduce population density in the east and defuse tensions with the working class.

President Theodore Roosevelt signed the Reclamation Act on June 1, 1902, as a means of sustaining the expansion of the western frontier. The Bureau of Reclamation would use the funds appropriated by the act to build irrigation projects, dams and canals, with a specific focus on the sixteen dry—and still largely empty—westernmost states. The expectation was that irrigation projects would be initially funded through the sale of public land, and then become self-funding, as settlers paid fees to cover the cost of development. But these projects could not succeed without public subsidy. Farming techniques from the east were inadequate for the harsh environment of the west, so the Department of Agriculture had to fund the development of new agronomic approaches.

Over time, reclamation became the central policy instrument to transform the landscape, shifting the focus of the federal government from navigation to farming. With the 1909 Enlarged Homestead Act, the homesteading terms were doubled. The act was accompanied by a propaganda push for a return to agriculture, to the family farm, and to a mythical rural America as the cornerstone of healthy individualism and prosperity in the face of rapid industrialization. In the first two decades of the twentieth century, settlers moved west towards the Great Plains in droves, setting the stage for the troubles of the thirties. Water had become an instrument in the management of increasing demographic pressure.

Britain and the United States were not the only countries to engage in systematic water-led development of their rural landscape at the beginning of the century. Italy, which had played such a central role in the development of institutions that now were at the heart of the global order, also engaged in rural modernization to adjust to the expectations of the modern world and accommodate its growing, young population. It faced an unusual situation: after its unification in 1861, it was the only major European country for which malaria was not a colonial but a domestic issue. The parasite *Plasmodium vivax,* a relatively mild strand of malaria also found in North America, was endemic to the flood-cultivated rice fields of the north. The south of the country hosted *Plasmodium falciparum,* the same strand present in tropical Africa. Malaria had a huge impact on labor productivity.

In 1878 three-quarters of the railway workers in Sicily had contracted the disease, slowing modernization of the island.

Malaria was still believed to be what the Italian name of the disease suggested: bad air—"mal aria"—emanating from swamps. Reclamation was again thought to be the answer, but it only took off when the government realized this was a public health intervention. The Beccarini law of 1882 subsidized it. By 1915, 770,000 hectares had been reclaimed, with as many in the process of being drained.

In all these examples the state had a growing role in driving a transformation of the landscape in the service of national objectives related to population growth and productivity. But, if the rural landscape was still the state's central concern, the growing role of industrial production in the consumer economy was rapidly driving a second, major issue at the heart of the state's water agenda: power generation.

POWERING A HYDRAULIC NATION

If state intervention in the rural landscape accelerated in the twentieth century, the energy transition was without doubt its most striking evolution. With the benefit of hindsight, one can see just how remarkable the transition was. In 1900, a third of all energy came from sheer human and animal muscle. Fuel wood, mostly for heating, did much of the rest. A century later, human work accounted for only 5 percent, as the total energy use grew tenfold. That energy transition was a story of water. Its early stages had already been tightly bound to water, from mine drainage to canal transport in support of the coal economy, and water resources development would also later play a role in the rise of oil. But at the beginning of the century, the story of energy and industrialization was one of hydropower.

Hydropower generation was a late nineteenth-century technology. The very first installation on a river had been in a private home in England, in 1878. The first commercial installation was in Wisconsin just a few years later. America was among the first to pioneer the use of rivers to produce industrial electricity. It all happened rather quickly. Edison developed the commercial light bulb and electric distribution in the 1880s. The introduction of the alternating current, first shown at the Frankfurt exhibition of 1891, brought electricity to industry, allowing for transmission over some distance without significant losses. At that point, hydropower had become a viable generating technology for production, and between the end of the nineteenth century and the first few decades of the twentieth, it spread everywhere.

Japan was a prime example of how water resource development and the new technology of hydropower could propel a country towards modernization, no doubt inspiring modernizers in the region to imagine what was possible (for example, Sun Yat-sen spent his first years of exile in Japan, a base for a number of Chinese revolutionaries). At that point, Japan had been on an accelerated journey to modernization, becoming the leading Asian economy and fostering a commitment to pan-Asianism.

Japan's complicated water landscape had been the protagonist of its rebirth. In the seventeenth century, the Tokugawa shogunate had encouraged trade, particularly with Vietnam, and had turned Osaka into a strategic port to support this international trade. To connect Osaka to Kyoto, then the country's capital, a wealthy entrepreneur, Suminokura Ryōi, invested in navigation infrastructure, clearing rivers, diverting waters, and creating channels for flat riverboats to transit between the two. He paid for the investment with the monopoly concession for shipping rights he had been awarded by the government. This infrastructure, developed in a manner no different than what was happening at the same time in Europe, created a platform for industrial development two centuries later.

In 1868, sixteen-year-old Emperor Mutsuhito recentralized power away from the shogun into his own hands and prepared for modernization. The 1867 Great Exhibition in Paris had fueled huge commercial interest in everything Japanese, and Kyoto's textile industry was poised to supply global demand, provided it could expand its transport infrastructure. Lake Biwa, a four-million-year-old tectonic lake near Kyoto, could provide water to increase the navigability of Suminokura's waterways. The idea for a canal from Biwa had been discussed since the twelfth century, but the inability to muster the resources and coordinate the development of such a large project had always prevented its realization. At the end of the nineteenth century, its time had come.

The project was led by Tanabe Sakuro, a local engineer. In 1888, he had visited the new hydropower station in Aspen, Colorado, in the

United States. Crucially, this trip exposed him to the new, emerging world of river power. Inspired by what he had seen, he proposed a canal that would connect Lake Biwa and Kyoto, incorporating the new hydropower technology. The canal was an extraordinary piece of infrastructure, traveling through three dug tunnels, one of which, at two kilometers, was the longest in the world at the time. The project generated enough electricity to bring streetlights and streetcars to Kyoto, powering its mills. Wooden canal boats transporting rice and other goods came down from the lake. Out of the last tunnel, the boats were attached to a cradle with steel wheels, which carried them down a six-hundred-meter incline in less than a quarter of an hour. The boats were then pulled up again by steel rope, wound around drums placed at the top and bottom of the incline.

Tanabe's system was a modern engineering marvel, fit for a country poised to lead its region in industrialization. It was also testament to the new role that public finance would have to play in engineering the modern landscape. The project was expensive. Its financing stretched the resources of all involved and had to be pieced together from multiple sources: a third was financed with a gift from the emperor, a quarter from the central government, while the remainder was to be raised by local taxation. The new economy, powered by rivers and canals, and with its need for collective commitments, set the stage for a new politics of water.

THE POWER OF RIVERS SPREADS

On the other side of the Pacific, American leaders of the Progressive Era embraced hydropower with a passion. With regulation after regulation, they promoted the development of American rivers for electricity production. The 1906 General Dam Act placed strict controls over privately owned installations. By 1907 a few states had introduced regulators for utilities. In a 1908 preliminary report to Congress by the Inland Waterways Commission, Theodore Roosevelt wrote that "our river systems are better adapted to the needs of the people than those of any other country. [. . .] Yet the rivers of no other civilized country are so poorly developed, so little used, or play so small a part in the industrial life of the nation as those of the United States."

Roosevelt believed that the development of rivers was going to be essential for the industrialization of an economic powerhouse. "The use of water power will measurably relieve the drain upon our diminishing supplies of coal." In the next decade, hydropower grew to become the mainstream technology of industrialization until the Federal Power Act of 1920 established a countrywide licensing regime. Industry had become the principal client for water infrastructure in America. Hydropower spread elsewhere too. After the first major plant at Niagara Falls, hydropower took off in similarly suitable locations, from Canada and Scandinavia to Switzerland.

Not all countries benefited immediately. Russia struggled, for example. Its first commercial utility, a thermal power plant, had been established in 1886, but Russia chose to stick to direct current rather than convert to alternating current supply, which kept power stations small and hydropower largely a theoretical option. The first hydropower station in Russia was eventually built in 1895 by the army, for a gunpowder factory. Then, industrial hydropower for mining was developed in the Caucasus, Siberia, and Georgia. But utility-scale hydropower to supply cities struggled to take off. By the time Russia entered the First World War, many of its cities still depended on imported coal, a considerably weaker position than anyone might have wanted.

Water management was also at the heart of the early industrialization of Italy, which, like Japan, was rapidly transformed by water-led industrialization. The country had limited-to-no raw energy sources that fit late nineteenth-century industrial models. Coal, imported from England, cost three times its price at origin. Italy's vast labor pool and competitive manufacturing made it an attractive market for intermediate goods and machinery. Growing railway connections through the Alps meant that large European markets were getting closer. But to fully industrialize, it had to solve its energy problem.

Italy's northern regions had steep rivers with enormous energy potential. The very first hydroelectric installation was a plant on the Gorzente River, near Genoa, on the northwest coast of Italy, built between 1880 and 1886. Reservoirs, mostly for water supply, were on a sufficiently steep incline so that the aqueducts leading to the city could also be equipped with hydropower capacity for about 750 kilowatts. Given the fragmentation of the industrial sector, however, the ramp-up of hydropower in Italy could not happen on the back of industry alone. It depended on cooperation between industrialists and the local public sector.

The company that took the lead in the development of hydropower was called Edison—the full name was the Committee for the Use of Electricity in Italy using the Edison System, and it had nothing to do with Thomas Edison directly—which, in 1883, had been the first in Europe to establish a thermal power plant to light a theater, the La Scala. The company had entered into an electricity contract for city lights in Milan, obtaining a first stable source of revenue. Power supply for electric lighting was not particularly profitable, however, as at that point it was not competitive with gas. Other electricity companies in Europe and America had dealt with this problem by diversifying their business into electromechanical equipment. But Edison's Italian lenders were not accustomed to diversified investments and did not allow the company to enter new businesses. It would have to make money generating electricity alone. This, as it turned out, proved to be a crucial incentive.

With no other option, the Edison company homed in on the huge hydropower potential of the Adda River near Milan. Because of the recession of the 1890s, the company needed a solid contract if it was going to find investors. The only real demand of sufficient size was the public transport system of Milan. The company proposed to electrify it. Local industrialists realized that, once developed, the plant could provide the infrastructure for the full electrification of manufacturing. They too got behind it.

The combination of lighting and tramway supply made the project on the Adda River solvent—investment capital came from Germany—and the plant became the largest in Europe. Equipped with ten megawatts, the most powerful generators on the continent, the Bertini hydroelectric plant became operational in 1898, and was only second to that at Niagara Falls at the time. By 1900, demand had already caught up, as hundreds of manufacturing firms still using steam engines decided to upgrade. And so the power sector in Italy grew. In 1898 there were fewer than a hundred million watts of installed capacity. Two decades later that number had grown to a billion watts.

Water and its power had helped Italy, as it had helped the United States, Japan, and many others, join the industrial twentieth century.

A NEW ROLE FOR THE STATE

The third major trend already evident at the start of the century was political enfranchisement. But its effects on the story of water are best seen through the lens of the state's role in investment. The enthusiasm for both agricultural and power infrastructure based on water resource development increased the need for investment. But in most cases, the financial crises of the end of the nineteenth century had dried up private capital. Besides, many of these investments had significant public-good attributes.

In the nineteenth century, as long as the majority of the population had no political agency, water investments on a significant scale would simply not have happened without a private arrangement. But now, as greater enfranchisement swept the West, and as a growing economy created a greater stock of savings, the problem of investing in water infrastructure shifted from a financial question to an economic and political one. This was the beginning of the modern hydraulic state: a powerful, public, economic entity, capable of investing substantial public resources to modify the environment in the service of the public good.

During the nineteenth century, the role of the state had been largely focused on military issues, policing, and clerical administration. Despite the early centralization efforts of Napoleonic France and Prussia, the state did not have a major direct role in delivering mass economic infrastructure. One of the indications of this was that government expenditure in industrialized countries like Great Britain or the Netherlands never exceeded 10 percent of national income. That is less than a quarter of today's average. Even countries like Italy or France, which were considered extravagantly profligate even then, spent only 12 to 18 percent of national income. But during the twentieth century, the complexity and ambition of the state increased everywhere, becoming the principal instrument of collective organization, and a dominant contributor to the national economy. This manifested itself in the growing involvement of the state in water resources.

The involvement of the state coupled with increased enfranchisement, in turn, increased the pressure on politicians and public administrators to allocate public resources in the service of justifiable policy objectives. From reclamation to contracts for power supply, decision makers faced a challenge: How should they choose which investments were in the public interest? The government needed a rule, some algorithm that could aid decisions and inform choices. It is at this point that cost-benefit analysis emerged as the principal basis for the legitimate allocation of public finance.

The practice of comparing the monetary costs and benefits of a particular piece of public infrastructure first developed in nineteenth-century France. Jules Dupuit was an engineer trained at the École Polytechnique, most remembered today for having designed the Paris sewer system. In 1844 he had tried to answer the seemingly simple question of what should the toll be for the use of a bridge or a canal. It was the first time an economist had analyzed the issue of "public utility." At that point, the dominant answer for any valuation was Ricardo's labor theory of value, which posited that the value of anything was proportional to the amount of labor that went into its production. That was clearly not a useful answer in this case, given that the benefits of public infrastructure seemed to be quite obviously independent of how much it cost to build. Dupuit turned economic thinking on its head. He realized that the value of public infrastructure ought to be measured in terms of its utility to beneficiaries.

If the value of infrastructure had to somehow be derived from the benefit accrued to its users, it begged the question of which benefits should be counted. In the case of transport infrastructure, one such benefit was the reduction in cost of delivering a good: faster delivery meant greater surplus for both consumers and producers. Dupuit also considered secondary impacts. Having a canal system might lead to new industries, which might not have existed had the canal not been in place. Those industries would add wealth to the economy, which could then in part be attributed to the canal.

Dupuit's theories were popularized by the economist Alfred Marshall, and eventually found their way to the other side of the Atlantic, where public administrators were looking for answers to that very question. In the United States, investments to harness a river of the size and force of the Mississippi, the Columbia, or the Colorado were well beyond the resources of any individual community living on their banks. They often needed financial support by the federal government, which then found itself having to choose among thousands of projects. Investing taxpayer dollars in the landscape required a formal mechanism to justify choices, lest the government end up playing favorites among its constituent states. The government needed some way of comparing the cost of infrastructure to the benefit that might accrue to society as a result of it being there.

A famous court case, *Willamette Iron Bridge Company v. Hatch,* made it all the way up to the Supreme Court in 1888 and was the trigger for the development of a solution to this problem, based on Dupuit's work. As its name suggests, the Willamette Iron Bridge Company had built a bridge to connect Portland and East Portland, across the Willamette River, a tributary of the Columbia. Hatch and Lownsdale were local entrepreneurs. They owned a wharf and warehouses, which large seagoing vessels used to be towed to, about seven hundred feet upstream from the site of the bridge. The latter would have impeded the passage of the former.

Hatch and Lownsdale lodged a complaint. When Oregon became a member of the United States in 1859, they argued, its navigable waterways were supposed to be free and open. The bridge infringed their access to the sea. The issue ended up in front of the Supreme Court, which ruled that, under common law, the federal government did not have authority to protect navigable waterways. If it wanted such authority, it could acquire it by introducing a new statute.

A statute was passed. Section 9 of the 1899 Rivers and Harbors Act declared unlawful the building of any structure on navigable rivers without the permission of the Army Corps of Engineers. This was a strong assertion of authority by the federal government. It converted the country's waterways into strategic national assets. But that political authority became operational when coupled with economic decision making. The 1902 River and Harbor Act established a board of engineers in the Office of the Chief of Engineers in the Army Corps, which would examine projects seeking public funding. It would make recommendations as to whether commercial and other benefits would warrant investment. Dupuit's valuation could provide the basis for the recommendations.

Over the subsequent two decades, first with the River and Harbor Act of 1920, and finally with the Flood Control Act of 1936, the government adopted what economists now call formal cost-benefit analysis as the basis for decision making. Through the oversight of the U.S. Army Corps of Engineers, the federal government had centralized decisions on much of the water resources of the country, establishing a crucial financial connection between local interests and national public policy. And through the increasingly important role the United States would play across the world on issues of water infrastructure, the approach to assessing the value of a public investment spread. When Dr. Sun described his plans for engineering the Yangtze River in meticulous detail, he was at pains to estimate how much each intervention would cost, and what commensurate economic value it would create. Writing in the 1920s, he knew China was going to have to resort to public underwriting—either its own or, more likely, that of wealthy Western countries.

The changing economic and financial argument for water infrastructure was a direct response to the changed political landscape. The twentieth century had opened in the throes of a profound socioeconomic transformation. Demographic pressures, the transition to electrification, and the increasing role of the public sector had all heralded the arrival of water-led national development. Had those trends continued to play out in the context of the nineteenth-century world order, the result might have been rather different. It was not to be. In the span of just a few years, the global architecture of powers and states that had ruled over the long century would be swept away, a victim of the first industrial-scale conflict in human history.

14

Crisis and Its Discontent

THE GREAT WAR

The cautious optimism of the first decade of the twentieth century was swept away, alongside its progressive politics, when the Great War broke out. Many believed it would be a quick affair. In an infamous 1914 editorial, H. G. Wells emphatically stated that "the defeat of Germany may open the way to disarmament and peace throughout the earth." It was the "war to end wars," his phrase, one that would be made famous three years later by Woodrow Wilson. They were both wrong. The world that emerged from it was not just transformed economically. It was upended politically. Water development, now a central public policy instrument, followed the world down this new path.

The Great War was a bloody affair. The destruction of Europe traveled along, and was fought across, its rivers. When the German offensive against French and British forces unexpectedly entrenched the conflict along the Western Front, shattering any hope of a rapid resolution, rivers hosted some of the worst battles of the war. The Marne River, an eastern tributary of the Seine, stopped the German advance to Paris between September 6 and 10, 1914. On October 19, 1914, a German offensive, which had aimed at cutting off British access to ports on the English Channel, managed to push through the Belgian defensive lines. The Belgian forces retreated to the Yser River. By October 25, the pressure from the German attacks was too strong to resist and the decision was made to let the river flood. It was the same tactic that William of Orange had adopted over two centuries earlier on the Meuse River, fighting for Dutch independence. The sluices near the coast at Nieuwpoort opened and water flooded through, covering an area a mile wide and ten miles long. The Germans withdrew and the front along the Yser held until 1918.

Fights along rivers continued throughout the Great War. Between July and November 1916, the fields of the Somme, a relatively small river in northeast France, were theater to ferocious fighting in one of the bloodiest battles in human history: over a million soldiers were either wounded or killed. The short Alpine Isonzo River, which flows through Slovenia in the Julian Alps between Italy and Austria, saw twelve bloody battles between 1915 and 1917. The Tagliamento River, between Venice and Trieste, played the role of backstop where the Italian troops retreated after the catastrophic Battle of Caporetto in 1917. The Piave River was the final setting for the battles that concluded fighting on the Italian front in November 1918. The list could go on.

In echoes of the Roman story, the conflict also exposed society's vulnerability to a fragile system of global trade. At the start of the war, Germany depended on trade for a third of its food and most of its fertilizers. The Germans had embraced a trade-based approach to water security, relying on other geographies to supply what they could not. But the British blockade disrupted their supplies. The Germans had not planned for the disruption, or for a long war. By the second year, food prices had shot up. Both the civilian population and the army suffered shortages. Severe restrictions ratcheted up, as the local agricultural system could not catch up with the gap in productivity left by the unwinding of trade.

Similarly, Britain's dependence on commodities from its distant domains was fragile. In 1914, the country imported around 60 percent of all its food and 80 percent of all its wheat. In the early stages of the war, as supplies from Europe collapsed, the British government sought to source them from other places, relying on the carrying capacity of the merchant fleet and the ability of the Royal Navy to protect it. But once German U-boats started targeting commercial transport, supply lines were severely damaged—in 1917, one in four merchant ships was sunk—and Britain discovered just what a gamble that approach to security had been.

The aftermaths of the war were equally disastrous. The victors were not in particularly good shape. To finance its military, the British government had had to borrow heavily, both domestically and internationally, particularly from America. By the twenties, interest payments commanded half of the government's budget, a situation it had not been in since the Napoleonic Wars. High levels of borrowing, a spike in taxation, millions of veterans to reabsorb into society, and a marked reduction in global trade had left the economy crippled. The recession of 1920 was the deepest Great Britain had ever experienced: unemployment reached double digits and stayed there for a decade.

The winner in all of this was the United States, but not without cost. The collapse of farming across Europe and the German blockade that isolated Russian wheat had led the United States to take over food production for its allies. This drove further expansion of the American agricultural sector. But to make enough grains available to war-torn Europe, the U.S. had to incentivize production while limiting domestic consumption. To achieve this objective, controls kept prices high at the farm. Food flowed towards Europe. Payments flowed back to the U.S. It was a massive transfer of hard currency towards the American economy. The boom inevitably attracted more farmers, who settled more western land, plowed it, pulling out grasses, growing wheat, and drawing more water from the ground. After the war and through most of the twenties the United States enjoyed high employment. But this transition of economic power across the Atlantic was also the prelude to a catastrophic collapse.

Many farmers that moved out west took out loans to invest in their farms. Most new settlers came from the east or from Europe and had no idea of the difficult environment they would find, particularly in the Great Plains. Many were lured with promises of plentiful water for crops, in the belief that "rain follows the plow." In truth, anyone with any knowledge of the long-term climate of the region would have known that rainfall in the Great Plains was unreliable at best. The landscape was too flat for long irrigation canals to source water from distant rivers, so the only viable option was to pump water from belowground. But with no rivers nearby and little to no industry, finding the power to lift water was a problem. The first solution had been windmills, which by the early twentieth century could be mail-ordered and constructed relatively easily on-site. They spread like wildfire, feeding the illusion of sustainable supply and pushing agriculture further out on a limb.

The Great War was an economic disaster. It vastly reduced most countries' capacity to mobilize finance for infrastructure projects both at home and in their territories across the world. Further, if rivers are the places of conflict during a war, they become places of diplomatic conflict once a settlement must be reached. In *The Economic Consequences of War,* John Maynard Keynes argued that the Versailles Treaty was too punitive towards Germany, pointing out that the use of rivers it depended on had been put under the control of other countries. In hindsight he was right: the conclusion of the First World War ended up setting the conditions for the Second.

But the Great War was not just an economic watershed. It was also a political fracture, which introduced a set of actors who tried to define an alternative to the liberal, imperial state that Britain had made so successful in the nineteenth century, and who, arguably, helped define how water resources developed for the rest of the century. Many also provided a counterpoint to the republican experience of the United States. And the first in line was, of course, Lenin.

COMRADE LENIN

In the middle of the First World War, in October 1917, Lenin's Bolsheviks had orchestrated a coup, and Russia descended into a long and bloody civil war. Vladimir Ilyich Ulyanov, Lenin, along with Leon Trotsky initiated a centrally planned social engineering project of unprecedented proportions. Most of it was built on water. Lenin was a radical, whose dreams more resembled those of Vera Pavlovna than those of Marx. He and Trotsky did not pursue an orthodox Marxist plan. Rather, they drew inspiration from the totalitarian strands increasingly common in Europe.

The German war effort, in particular, fascinated Lenin. It had shown how free-market capitalism could evolve into a capitalistic state monopoly to overcome bourgeois administration. Wartime Germany was the archetype of the socialized economy he had in mind, a command economy combined with extreme Taylorism, a theory of scientific management focused on economic efficiency. Lenin believed a break with history could only be delivered through centralized scientific and technical planning. Those who did not, or could not, understand should be educated to accept it. It was the birth of revolutionary vanguardism and of the single-party state.

Lenin hoped that the revolution in Russia would be quickly followed by a revolutionary movement across Europe, but by the end of the First World War it was clear that others would not follow Russia's lead. So, he concentrated on domestic industrialization and on consolidating the power of the Communist Party. In November 1920, he laid out his vision for industrialization. During the civil war that had followed the October Revolution, cities like Petrograd or Moscow, where the communist movement had its base, had been cut off from the grain-producing areas of Siberia, the Caucasus, and Ukraine. If the revolution was to succeed in transforming the economy, rural and urban areas had to be reconnected. Lenin believed electrification, particularly rural electrification, could reduce the wide gap that separated urban populations from the countryside. And electrification, in Russia, was about water.

Lenin was a technology enthusiast, a longtime believer in the importance of electrification. He wanted to prove that the dictatorship of the proletariat could transfer a vast number of peasants into higher-value industrial jobs, and for that he needed electricity. In his words: "Communism is Soviet power plus the electrification of the whole country." The scale of what the Soviet Union attempted is hard to fathom: a re-engineering of nature across a sixth of the world's emerged lands. The first electrification plan, GOELRO, was adopted in December 1921. Hydropower was at its heart. Lenin quickly moved to promote regional power stations: the first were green-lit on the Volkhov and Svir Rivers, both of which drained into Lake Ladoga, near Finland. Russia was finally going to catch up with the rest of the world.

Lenin's plan called for developing hydropower close to raw materials to support heavy industry and avoid their dependence on fuel imports. Aluminum production required the development of the massive river resources of the north and Siberia. New industries also required navigation and, in particular, irrigation, as agricultural exports were essential to generate the foreign currency to pay for parts and skilled labor needed for the construction of infrastructure.

In fact, exports were such a priority that the population was kept close to starvation in order to keep enough wheat to export.

GOELRO required investments and a huge injection of skills. Both were enthusiastically provided by American and German firms. Leon Trotsky, former commander in chief of the Red Army, had invited a number of foreign experts to participate in the planning of the Dnieprostroy hydro installation in Ukraine, approved in November 1926. At that point, America saw Russia as a potential partner in opposing British imperialism and invested in the relationship with gusto. Water engineering was its principal gift.

Colonel Hugh L. Cooper, who had been the chief engineer of the Army Corps of Engineers and had just finished building the Muscle Shoals Dam on the Tennessee River in 1925, supervised the construction of Dnieprostroy. American firms supplied technology and skills: Warner & Swasey, a lathe producer, and Foote-Burt, a producer of drills and boring machines, both from Cleveland, brought manufacturing to Stalingrad; McKee and Austin, construction companies, contributed to the development of steel mills and oil refineries; Albert Kahn, the "architect of Ford," assisted with the efforts to industrialize, while the Newport News Shipbuilding Company provided the Dnieprostroy Dam with the biggest turbines in the world.

At his death in 1924, Lenin was canonized as the hero of the new Soviet republic. He had succeeded in enshrining electrification and industrialization as the core objectives of the Soviet state. And both would be organized around the development of rivers.

FASCISM IS BORN

Russia had produced real communism through the electrification of its rivers. But the transformation of Russia into the Soviet Union was not the only water-led response to the consequences of the First World War.

The disruption of international trade due to the Great War had been economically disastrous for an export-oriented, largely agrarian country like Italy. America was its principal trading partner, but it had embraced protectionism. Italian-American communities in the United States, previously buyers of Italian products, had developed their own production. Pasta exports, as one example, dropped by a factor of thirty between 1913 and 1924. During the war, Italy's trade deficit grew. By the early 1920s Italy was importing twice as much food as it was exporting.

Trade was not its only problem. Before the war the United States had been a demographic valve for southern Europe. After the war, an increase in American nationalism led to immigration quotas. Only 10 percent of immigrants could come from southeast Europe, which eliminated a crucial outlet for rural, overpopulated Italian regions. To add to the demographic time bomb, more people faced fewer employment prospects. The war economy had pushed the emergent Italian manufacturing sector to unprecedented levels, but had also changed its structure, redirecting it towards higher productivity and away from mass employment.

The Italian population had turned politically more radical, as enthusiasm for the internationalism of Woodrow Wilson's League of Nations faded. Liberalization had struggled to provide the promised benefits, and socialism was spreading fast among the peasants of rural Italy. It was a combustible mix: international finance was applying pressure to increase profits by turning to a low-employment, high-productivity industrial sector, while a society under immense strain due to both chronic unemployment and increasing demands sought redistribution and state intervention. The unfortunate resolution of that tension ended up being fascism.

In October 1922, at a moment of particular difficulty for the Italian government, Benito Mussolini organized the infamous March on Rome. By then, he was the powerful leader of a movement that called itself the National Fascist Party. He threatened a siege of Rome unless he received a mandate to form a new government. The king, unwilling to test the military's resolve to protect the capital, and frankly more fearful of a socialist revolution than of the fascist threat, invited Mussolini to come to Rome to do just that.

Mussolini, named after Benito Juárez, the mid-nineteenth-century Mexican president, was by formation a radical. His political education had been shaped by unionism, nationalism, and futurism. At a young age Mussolini became a prominent voice of radical socialism, but was expelled from the socialist party because of his interventionist stance during the First World War. By then, however, his ideas had already begun to evolve towards a form of "national socialism." He had collected around him the disenfranchised and those returning from the bloody experience of the First World War. Violence was his principal credo.

In Mussolini's own definition, fascism was born without a real doctrine. Its ideological roots were shallow, focused on half-baked ideas like the celebration of youth, nation, and imperial power. Mussolini was a tactician, inclined to accommodate whatever policy was needed to support his climb to power. His shift from socialism and anticlericalism towards the right turned him and his violent Blackshirt followers into a supposedly useful instrument of a ruling class, in the throes of the fear that a Bolshevik revolution could overcome Italy. On January 3, 1925, Mussolini finally declared himself Duce, dictator. Parliamentary democracy in Italy had ended.

Fascism was a self-defined totalitarian regime: it demoted the individual to being entirely subservient to the state, which exercised its power not so much through the ownership of the means of production as in communist Russia, but as a violent, cultural hegemon. The waters of the country became an instrument of powerful propaganda as well as a fundamental platform for economic development and social control. Hydropower had proven an enormous opportunity for Italy. But to support its expansion and further industrialization, the regime needed to attract investment from abroad. On that basis, in 1927 Mussolini pegged Italy to the gold standard at an overvalued rate. He assumed Italy could resume exporting with the same labor-cost advantage it had had in the run-up to the world war, increasing profits for Italian industrialists while at the same time reducing exposure to foreign debt, as both the dollar and the pound sterling would have been relatively cheap.

For a while, it worked. American investors, who had been scared off by the rise of socialism and communism, returned to the country, as Mussolini appeared to be an antidote to radicalism. His capable propaganda machine portrayed him as a gentle authoritarian, who presided over social order at a time when America faced difficulties. Investment in hydropower companies soared. Between 1925 and 1928, of the over $300 million in bonds issued, half were government issue, the other were industrial, of which two-thirds were for hydroelectric development. Ironically, Mussolini became the darling of the liberal world he was intent on destroying, admired for his ability to use centralized government to steer the economy without threatening private property.

Eventually, though, his policies caught up with him. An overvalued currency deflated the economy, making wages expensive compared to the cost of goods. Unemployment increased. Domestic demand collapsed. Squaring that circle would have been difficult for any democratically elected government. To contain the consequences, the Fascist regime resorted to violence and tightened control over all the components of the national economy by creating vast monopolies, which were supposed to compete internationally. It called this practice "corporativism."

The brittle simplicity and feeble ideological basis of fascism did not prevent its mass appeal. No matter how intellectually inconsistent, the totalitarian mix of state intervention, militarism, and promise of prosperity took ahold of an entire generation, spreading a virulent contagion throughout Europe. And water development helped fuel its rise.

THE INTERNATIONAL DEVELOPMENT OF CHINA

The political situation after the Great War was unstable. The impact of this unexpectedly long, unfathomably destructive conflict was felt across the whole world. In China, Sun Yat-sen had been sorely disappointed by the failure of China's first republican government. His vision for a republican China, as described in his last work, the *Three Principles of the People,* was an enthusiastic blend of Confucian tradition and Western political and economic philosophy. But by then, he was writing in difficult circumstances.

The ideals with which he had mobilized the revolt that toppled the Qing dynasty in 1911 had proven far harder to implement in practice. The political situation in China had been deteriorating since 1916. As the world descended into the chaos of war and its aftermath, regional strongmen ran different pieces of the former empire as the center struggled to hold. The turmoil left China weakened on the international stage. At the Peace of Versailles, Japan was given the right to the German territories in China, which did nothing to soothe Chinese grievances, especially given Woodrow Wilson's much-advertised commitment to self-determination. By 1919, anger boiled over and students occupied Tiananmen Square, winning the sympathy of a large part of society. Dr. Sun feared China would descend into a Balkanized mess. The world was at risk of unwittingly lighting China like a gigantic powder keg.

In particular, Dr. Sun worried that, after the war, China would become the "dumping ground" of the production surplus of Europe and America. To avoid this fate, China desperately needed foreign investment to increase its own productivity and industrialize. It needed to harness its vast rivers. This is when Dr. Sun retreated to the French concession in Shanghai to write *International Development in China,* the blueprint that would guide China's approach to its resources thereafter. He specifically wrote it in English, in the hopes that it would reach Western leaders.

His ideas were radical. He proposed that a quarter of the money previously spent on the war effort by Britain and America should be invested in the industrialization of China. He imagined the creation of a single international institution that would represent the donor countries, and expected the productivity improvements to pay for the interest and principal. In many ways these were ideas that anticipated the Bretton Woods establishment of the International Bank for Reconstruction and Development, the World Bank. The only comparable program of postwar recovery was that outlined by Keynes in his 1919 *Economic Consequences of Peace.* Neither was taken up.

Dr. Sun was a progressive utopian in the nineteenth-century tradition. America and Europe were his models of industrialization. He believed that infrastructure on the Yangtze River—including the dam that would become Three Gorges Dam—would propel trade with China as far as Suez and Panama had pushed the world. A renewed, industrialized China would become a "New World," a gigantic market for the world's products, with effects as epoch-making as the discovery of America.

But he was also a twentieth-century visionary. He believed in international cooperation and the pursuit of development at a time when neither concept would have been understood in most diplomatic circles. Amidst the many strands of anti-imperialism, socialism, liberalism, and Chinese nationalism that flowed through his ideas, he focused on constitutional democracy as the anchor for his political project. It is worth remembering that an emphasis on democracy in the 1920s was in sharp contrast to the difficulties Western democracies were facing at the time. His own Chinese republic was failing, while the Soviet regime was seemingly heralding a new communist era.

At a time of nationalism, chauvinism, and extractive, zero-sum-game relationships between countries, Sun believed that development in China would happen in collaboration with the West. The American minister to Beijing, Charles R. Crane, thought Sun's ideas were "impractical and grandiose," but in truth Dr. Sun was prescient: Many of his plans were not that far from those put forward much later by Deng Xiaoping, the architect of China's opening up to the West. When Sun Yat-sen died in 1925, none of his dreams had been realized. But it is fair to say that in the mix of republicanism, socialism, and planned economy that came through his writing, he had captured something of the dominant wind that was sweeping the world.

A NEW ERA

The changes that followed the end of the First World War had a profound effect on the way water resources and political processes interacted. On November 6, 1924, a Thursday evening, John Maynard Keynes gave the Sidney Ball Memorial Lecture in the building of the Examination Schools at the University of Oxford. The title of his lecture announced "The End of Laissez-Faire." Keynes's opening salvo summarized the situation: "The disposition towards public affairs, which we conveniently sum up as individualism and laissez-faire, drew its sustenance from many different rivulets of thought and springs of feeling. [. . .] But a change is in the air."

The first two decades of the century had revolutionized, in some cases quite literally, the political landscape of the world. It was the end of the old imperial projects. China had lost an empire that had ruled more or less continuously for two thousand years. Russia had transitioned from the tsarist regime to a communist one. Europe was in ruins and captive to populist propaganda, while Britain, in the throes of an unprecedented financial crisis, was rapidly losing its grip on its empire. The sophisticated trade-based system that Britain had relied on was predicated on a financially strong global economy and on Britain's ability to maintain the system of trade relations that connected disparate water assets across the world. The war brought the latter to its knees, unwinding the former, leaving room for the United States to become the new hegemon.

Keynes saw change was coming and understood that the role of the state in the economy would have to evolve. In his early years, he had embraced the classical economic credo of Alfred Marshall, that social value was reflected in a market exchange. By the time of his speech at Oxford he had accepted that a self-governing economy, the system which Britain had long embraced and for which it had abandoned its long-standing eighteenth-century commitment to protectionism, was, in fact, an illusion.

The nineteenth-century trade system had gone hand in hand with a currency system based on the gold standard, which made commerce easier in a world of otherwise unconvertible currencies. In this system, the British pound sterling, which had been pegged to gold since the 1820s, had become the standard international currency. One of the limitations imposed by the gold standard had been that the supply of money was pegged to the stock of gold, constraining the amount of debt a country could sensibly issue and limiting the ability of the government to deficit-spend. Indeed, the twentieth century had started with a relatively low level of debt and government expenditure compared to national output.

But public expenditure was going to have to grow. The expectations for a welfare state and for a far more managed water landscape could no longer be avoided. After the better part of a generation had been sent to die in war, no one was going to accept the squalor of Victorian cities. No one was going to accept the abject poverty of rural communities. Where the state could act on behalf of the collective, it should, and the development of water resources was a natural candidate for state action.

The propaganda surrounding Leninist Russia and Mussolini's fascist regime put pressure on liberal states to demonstrate their ability to proactively intervene in society. The rise of Marxist socialism, which Keynes described as "little better than a dusty survival of a plan to meet the problems of fifty years ago, based on a misunderstanding of what someone said a hundred years ago," had given language to articulate strong opposition to the vast inequalities and injustices of society. The Anglo-Saxon liberalism that had conceived of the state as an immaterial economic actor could no longer withstand these social pressures. Wealth redistribution and social security, rather than latter-day Victorian-era objectives of individualism, private enterprise, and limited state intervention, were increasingly widely held political objectives. Having experienced the gains of the "economic Eldorado" of the prewar period, Keynes feared these gains would be jeopardized by politicians' inability to recognize the scale and depth of the opposition to its most extreme consequences.

The only hope of survival in the face of populist, fascist, and communist appeal was to embrace some form of a mixed-state system that could redistribute resources with a welfare system underpinned by state-owned enterprise, industrial policies based on subsidies and tariffs, and monetary policies aimed at mitigating business cycles. All of this ought to be underpinned by a mixed regime in which market-based instruments were accompanied by state intervention, redistribution, and high levels of taxation on income. The governments of the world would soon have to gear up to spend far more public resources than they had ever been able to, and a significant fraction of these resources would have to go into transforming the landscape in service of water security. By the 1930s the gold standard had been abandoned. That opened the door to a number of expansionary policies. By then, average government expenditure had already exceeded 20 percent in all richer countries. In the United Kingdom it approached 30 percent.

The colonial project of the nineteenth century, based on nineteenth-century principles, also began to unwind. One of the most famous building blocks of the colonial system was the freedom of navigation on the Congo River, which had prompted the Berlin Congress. The arrangement had seemed to make sense in the nineteenth-century world of economic laissez-faire governments, but had become highly problematic as planning and public investment turned into a common instrument of economic policy.

In 1934, a British national, Oscar Chinn, appealed to the Permanent Court of International Justice of the League of Nations against the Belgian government. Chinn ran a fleet of freight vessels on the Congo. The Belgian government was a majority equity holder in Unatra, the Union National des Transports Fluviaux, a commercial company that competed with Chinn. With the economic crisis of the 1930s, the Belgian government had imposed artificially low-cost transport on the river to support Belgian-Congolese exports, thus subsidizing Chinn's competitor. In court, Chinn argued that this state intervention constituted unfair competition.

The Permanent Court of International Justice of the League of Nations, in a contentious and split decision, ruled against the United Kingdom, representing Chinn. In the view of the majority of the court, the freedom of navigation principles had not been breached, because the economic conduct of the riparian countries was not the subject of the treaty. But five of the eleven judges dissented. They believed the Belgian government had violated the spirit of free navigation, which was predicated on commercial equality between the entities and nations using the river. The fact is that the treaty was based on assumptions from a different era. Sharing rivers when the sovereign governments do not operate as economic actors is a very different proposition from doing so when the fate of the sovereign itself is tied to the economic success of its policies.

Rivers had not changed in the twentieth century. But society had.

15

Industrializing Modernity

AMERICA EMERGES

Water infrastructure development accelerated across the world during the period between the two wars. It was a time of competition between economic and political systems that would define the century—America's liberal republic, the Soviet state, Italy's fascist regime. The transformation of the modern landscape was the physical manifestation of state architecture, the most visible, tangible evidence of a country's progress and one of its principal drivers.

In America, just as irrigation had supported the conquest of the western frontier, hydropower was the basis for the country's industrial development. Manufacturing was sensitive to the cost of electricity, and large, conventional hydropower had a clear cost advantage in the first half of the century. Once built, hydropower required limited maintenance costs, no fuel costs, and a very long amortization schedule. With advances in turbine technology and civil engineering, the limit to the size of installations quickly became the river itself, while the unit cost of electricity decreased with size. America had a number of powerful rivers whose potential could supply ample power to the country. The Columbia, for example, a steep, large river that crashes down on the Pacific side of the Rockies, ended up providing electricity for the whole Pacific Northwest. Because technology to transmit electricity over hundreds of kilometers was not yet available in the first decades of the century—high-voltage long-distance transmission would only appear in the 1950s—industrial manufacturing followed rivers. Water and its geography provided the master plan for industrial development.

America was also quick to capitalize on its advantages in expertise, projecting a water imperialism of sorts. In the 1921 *World Atlas of Commercial Geology,* the U.S. Geological Service assessed where in the world rivers could be harnessed to provide cheap power for mining. The Congo River alone was one-quarter of the world's estimated water power potential, which, some years later, led to a proposal for the largest dam ever to be conceived, Grand Inga.

What made American expertise distinctive was the complexity of the arid west. Rivers flowing through those environments had to be managed for multiple objectives: controlling floods and supporting reclamation, irrigation, and hydropower. The Colorado River was probably the most emblematic case.

For all its fame, the Colorado does not carry much water: only a tenth of the flow of the Columbia. The first driver for its development was agricultural. In the early part of the century, there had been a private effort to develop the "Imperial Valley" of California, an alluring name chosen to attract settlers to a large, arid valley in the south of the state. The valley was covered in a layer of baked topsoil, rich in organic material, which, once watered, promised to be extremely fertile. The California Development Company pursued the development. However, it quickly became clear that the project needed Colorado River waters to succeed. Private diversions followed. Immediately, they became a deadly threat. Prior to its engineering, the Colorado carried an enormous amount of silt, seven times that of the Nile or the Mississippi. Without expensive and continuous maintenance, floods could turn any attempt at diverting the river for agricultural purposes into a destructive stream of mud. The private conveyance channels turned into conduits for powerful uncontrolled floods. The solution to an irrigation problem had created a flooding one.

Those complexities propelled American engineers to the forefront of tackling difficult water challenges all over the world. The case of Harry Thomas Cory is emblematic. Cory had been the chief engineer of the California Development Company, responsible for the irrigation system of the Imperial Valley. The United States government appointed him to a commission called upon to rule on how to apportion water between Egypt and Anglo-Egyptian Sudan. Egypt had become independent in 1922.

At that point, the likelihood of an engineered integration of the Nile along the lines of what British engineers had imagined was practically zero. But because Egypt was so central to the security of the British Empire—the Suez Canal was the maritime gateway to India—the British were reluctant to relinquish control. The issue of who had authority over Sudan had been left unresolved at independence, so the British used its upstream position to control the river, creating a powerful point of leverage over the downstream, newly independent nation.

The international commission of three experts was appointed to rule on how water should be shared. The commission was skewed in British favor. The commission's chairman was an appointee of the Indian government, while a second commissioner came from the University of Cambridge. However, Cory's appointment by the United States was a sign that British hegemony on such issues was on the wane. Not surprisingly, the commission could not reach a verdict because Cory disagreed with his two colleagues. He thought the situation was similar to that of the Colorado River, an experience neither of his two colleagues would have had, operating as they were within the British colonial system.

The Colorado riparian states had been through the painful process of negotiating the apportionment of the river's waters in their Compact of 1922. It had been a legendary, drawn-out process, chaired by Herbert Hoover. The compromise stipulated how the four states of the upper basin and the three states of the lower basin would share water (notably, Mexico was not part of the deal). The complexity derived from the fact that, much like the Nile, upstream states provided the water, but most of the existing farmland was downstream. California, which contributed no water at all, had the most valuable agricultural production. The negotiation was about supporting downstream production without locking the upper riparian states into constraints that prevented their development. (What none of them could have known is that they were apportioning water they would later not have: the decade of the Compact happened to be the wettest period in the Colorado basin in five hundred years.)

Cory knew that the choices Sudan and Egypt made at that point would limit the degrees of freedom the two countries would face later on, and argued forcefully for considering the difference in needs and means of the two countries at different points in time and to apply a cost-benefit analysis to the system. The commission could not reach consensus. In the end Anglo-Egyptian Sudan and Egypt came to an agreement, quite independently of the commission, in 1929. However, Cory's ideas would become the basis of a subsequent agreement after the Second World War.

Back in America, the solution to the Imperial Valley problem led to what is likely the most iconic water project of the twentieth century. The most viable sites for engineering the Colorado River and protecting the Imperial Valley from floods were on the border of Nevada and Arizona. This meant that federal coordination was required. The Reclamation Service stepped in. At the heart of their intervention was a huge dam that could help manage the dangerous floods and regulate the flow. The choice for the site landed somewhere in Boulder or Black Canyon.

The huge dam was an immense financial commitment. Political resistance was high, also because once approved, the project would sell electricity to pay for its construction. At that time, private power utilities were still largely running as monopolies and saw this titanic project as a threat to their business. They spent vast amounts of money trying to turn public opinion against it. Then, nature unexpectedly intervened, breaking through the political deadlock.

Far from the Colorado, the Tennessee River flooded in December 1926, the Cumberland in January 1927. Rain poured on the entire Mississippi basin from January through April. The Missouri, the upper Mississippi, the Arkansas, and the Ohio swelled. On April 19, 1927, a mile-wide crevasse flooded about half a million hectares of land near New Madrid, Missouri. Then, on April 21, the river finally broke through at Mounds Landing. That crevasse was the biggest ever to have appeared on the Mississippi, unleashing a wall of water onto the lowland of the Delta. It covered over a million hectares in up to seven meters of water. The flood of 1927 was the worst in American history. In the end, it inundated seven million hectares of land, killed about five hundred people, and left seven hundred thousand homeless. Its damages were equivalent to a third of that year's U.S. federal budget.

The catastrophe on the Mississippi unleashed federal involvement in flood control, overwhelming any remaining political resistance. The Flood Control Act of 1928 provided the green light the Bureau of Reclamation needed to undertake the project on the Colorado, on the basis that it too was infrastructure for flood control. But to begin construction, the project still needed a contract for electric power against which finance could be released. William Mulholland, the Los Angeles chief of waterworks and supply, stepped in. He was always in search of water for his city and was prepared to pay for it by entering into a long-term contract to buy electricity. The hydropower Mulholland bought would pay for water storage for LA, which in turn would deliver flood protection to farming in the Imperial Valley.

The project represented a new paradigm in water resource management, one that American experts like Cory would disseminate across the world, changing the face of the planet: the multipurpose dam was going to be the instrument of water-led development. The Boulder Canyon Project Act authorized the dam in December 1928. At the dedication for the start of the works, Secretary of the Interior Ray Wilbur declared, to the surprise of most, that the dam would be named after the president of the United States. It was to be named Hoover Dam.

THE MAN OF STEEL

America was successful in mobilizing national resources in service of landscape development. The complexity of its republican architecture was evident in the multiple layers of governance and decision making that were involved in making something like Hoover Dam happen: from the city and state levels to the federal level, involving a complex mix of private enterprise, legislative action, and public underwriting.

America was not the only country to aggressively develop its water resources in the years between the two wars, but other political systems got there through very different institutional mechanisms. The Soviet Union is a case in point. Lenin had died in 1924. Joseph Stalin—then secretary general of the Communist Party—snatched power from the hands of Leon Trotsky and went on to rule for three decades. The "man of steel"—hence his acquired name, "Stalin"—was the archetypical authoritarian dictator of the twentieth century. His instruments of destruction were heavy industrialization and agricultural collectivization. The development of water resources was central to both.

Through a breathtaking acceleration imposed by Stalin's five-year plans, the USSR built a vast system to re-engineer the country's plumbing. Over the course of his regime, the Volga, the longest river in Europe, was turned into a canal thanks to thirty-four hydropower stations, part of the Great Volga scheme, and assorted canalization that engineered its basin and connected it to others. These were often massive installations. Kuybyshev Dam created one of the largest reservoirs in the world, flooding 300 towns and displacing 150,000 people.

But Stalin's real water legacy was in agriculture. Agricultural conditions in the Soviet Union were very different from those of the United States. Sixty-five percent of the agricultural land in the USSR was in an arid or semi-arid climate, especially the vast breadbasket of central Asia. Agriculture was irrigation in the USSR.

The Amu Darya and Syr Darya, the principal rivers in central Asia, come down the very steep Pamir Mountains and flow through the arid and desert steppes. During the Russian civil war, older irrigation infrastructure on those rivers had been destroyed, so the push to rebuild and modernize was an essential step in the inclusion of the region. Lenin had ordered three hundred thousand hectares put under irrigation. These were large-scale projects, powered by the modern tractor. The idea was to link industrialization and agriculture in the so-called agricultural-industrial complex, which should have accelerated growth in both. These vast complexes were to reach areas of twenty thousand hectares, coordinating industrial processing and agricultural activities in one unit.

This plan, progressive on paper, was soon distorted by Stalin's government. In Engels's and Marx's writing, collectivization had been a bottom-up process that was supposed to involve peasants. But in Stalin's execution it became an instrument of control and a top-down forced process. The political objective of controlling rural peasants superseded the economic objective of achieving a modern agricultural enterprise. The consequences were terrifying.

At the time of the first five-year plan in 1929, most agriculture was still based on small-scale farming. The push to collectivize committed the Soviet Union to low productivity and labor-intensive farming, destroying most of the incentives for productivity improvements needed to stimulate a modern agricultural sector. As state pressure and violence mounted during the thirties, the expropriation of agricultural produce coupled with lower crop yields and grain stockpiling led to famine. Estimates vary. Some historians put the death toll at around ten million people. Even the Duma recognized the death of at least seven million people. Up to three million of those were in Ukraine, the breadbasket of the country.

Central Asia bore the brunt of Stalin's policies. Between the end of the twenties and the early thirties a number of American experts visited central Asia in an attempt to help solve some of the challenges of irrigation. California was deemed to be the one place that could provide the Soviet state modern commercial experience with arid environment irrigation. But Stalin's Soviet Union was not open to advice. Those experts hoped to solve technical problems, but water infrastructure in Stalin's Soviet Union was above all an instrument of repression.

Most canals were built with hard, enforced labor and few tools in a feudal system of enslavement. Famously, in 1934, at the Seventeenth Party Congress, the Bolshevik grandee Vyacheslav Molotov—of Molotov-Ribbentrop Pact fame—railed against "gigantomania" in Soviet infrastructure, which was responsible in his mind for wasteful delays. The building spree was particularly intense towards the end of the thirties. In 1939 alone fifty-two canals were opened for over 1,300 kilometers of aggregated length.

Collectivization nearly broke the irrigation system. Canals were misused or neglected, and the efficacy of irrigation was heavily reduced as people had to constantly negotiate between illegally producing for their own survival and meeting centrally planned targets. The disaster that was Stalin's push to develop the water resources of the Soviet state was experimental and dogmatic. But from the outside, all that people could see was a state on a path of accelerated industrialization, an apparently successful planned economy able to harness the plentiful water resources of the country in a way not dissimilar to what the United States had done.

Stalin's efforts were eventually interrupted by the Second World War. Just a month before the Soviet Union entered the conflict against

Germany, Dnieprostroy Dam was demolished in an attempt to leave nothing useful behind for the invading German army. By the time a modern trade regime had been established after the war, the USSR had committed itself to self-sufficiency and economic isolation. Isolation would lead the Soviet experiment to spin into madness. Water would become the fighting ground for democracy all over the world.

MUSSOLINI'S BATTLE FOR WATER

Stalin certainly did not have the monopoly on water-led despotism between the two world wars. Mussolini was the inventor of the nomenclature "totalitarian state." He used water projects to great propaganda effect to maintain his grip on power. When the global financial world collapsed in 1929, Italy was far from ready to take the hit. The credit system collapsed, and the government had to create huge state-owned enterprises to nationalize the losses of failing private institutions. The inevitable loss of credit led to capital rationing and more expensive lending, in turn reducing consumer spending and production. The banking and industrial model that had supported Italy's rise during the early part of the century—and that had been instrumental in the early development of hydropower in the country—was finished.

Large-scale water projects are frequently thought of as despotic in nature. But confronted with the execution of large-scale interventions, real-world despots often end up with a poor track record. Mussolini's story is another case in point. In response to the economic crisis and in an attempt to stimulate import substitution and national production, Mussolini imposed tariffs and trade restrictions. Predictably, Italian exports—essential to the economy—dropped by over a third, a casualty of the inevitable retaliation by its trading partners. The regime then tried to spend its way out of the crisis. To try and keep up the pace of industrialization and keep the wealthy elite and international financiers on its side, the regime passed pro-cartel legislation, dividing the Italian economy into corporations, encouraging collusion among companies, and stifling competition. It enforced collective labor negotiation through a compulsory single union. Salaries were fixed in line with the deflation of the economy to ensure industrialists could maintain their profits.

Mussolini expected that a forced drop in wages would preserve employment, but because wholesale prices dropped more than retail prices, against which wages were indexed, and because most of the production in Italy was intermediate goods, the cost of labor actually grew compared to the cost of production. Unemployment exploded. Mussolini finally announced plans to move the country to autarky in the middle of the thirties.

That is when the country's water landscape stepped into the limelight of Fascist propaganda. Mussolini had to find employment for the vast impoverished population. If people had moved to cities in search of labor before industry had the capacity to absorb them, he would have faced significant civil unrest. The demand for housing, hospitals, schools, and transport would have strained the limited national resources, yet without them it would have been hard to maintain social order in densely populated areas. Besides, the concentration of industrial labor in cities would have made trade unions more powerful, even if there was a single, state-sponsored, Fascist one.

The regime tried to turn the secular tide of migration towards cities by actively encouraging an exodus back to the rural landscape. It was advertised as a ruralist, anti-urban return to the ideal of a simple life of the countryside. But for that it needed more land for low-productivity, high-employment agriculture. One answer was to add colonial land. That is why Italy ventured its second, ill-fated attempt at conquering Emperor Haile Selassie's Ethiopia in 1935. Another answer was domestic agriculture.

The problem was that much of Italian agriculture had been pushed towards high productivity to provide raw materials for the textile- and food-processing industries. The regime had set high prices for grains, and the well-capitalized farms of the north had responded with mechanized, fertilized monocultures. Tractors took the place of animals, which increased productivity and eliminated feed production, making more room for food. But all this vastly reduced the need for labor. So Mussolini turned to the impoverished south, where there was no organized labor, and he could rely on the landowners to control the rural population. Exploitation of farm laborers reached depths of inhumanity seldom witnessed in twentieth-century Europe. Still, he needed more land, and so Mussolini turned to reclamation.

Expenditures for infrastructure and land reclamation doubled over the decade leading up to the Second World War. On paper, the *bonifica integrale,* meaning total reclamation, went well beyond the hydraulic drainage of the first decade of the century. It included all water development, from canalization, to reforestation, to land conversion. It was a failure.

Reclamation was meant to create internal demand for industrial machinery and products that were struggling on the international markets. But the government did not have the finance required to actually implement the plans it had drawn up. The commissioner for internal colonization was charged with moving the excess labor in the north towards new expanded areas in the south. In reality, though, underlying population growth in those same regions made relocation difficult. Besides, there simply wasn't that much land left to reclaim.

Despite all the propaganda, most suitable land in Italy had been reclaimed long before the Fascists came to power. The grand scheme of the Fascist regime ended up focusing entirely on its flagship program in the notoriously malaria-ridden Pontine marshes, seventy thousand hectares of waterlogged lands, forests, and swamps south of Rome. The marshes became one of the symbols of Fascist modernization. But they were extraordinarily expensive to drain. The marshland was flat so gravity was of no help in moving water. Sixteen thousand kilometers of canals had to be built, aided by the continuous use of large pumping stations.

Whatever progress was achieved on the Pontine marshes was lost as the Italian government finally entered the Second World War. Towards the end of the conflict, German forces reflooded the Pontine marshes with the precise intent of inflicting a malarial epidemic on the civilian population, the only known instance of biological warfare in Italy. The grand reclamation plans of Mussolini had ended in disaster.

AMERICAN EXPERIENCE BEGINS TO TRAVEL

As the power of totalitarian states developed, leading the world to the destruction of the Second World War, America began establishing its own international weight by deploying the expertise it had developed on the landscape of its republic. American water experts showed up everywhere between the two wars. The distinction between how their expertise was deployed in the context of the republican institutions of America and how it was employed in service of far more authoritarian states is significant. Abroad, American water expertise—divorced from political processes—became an instrument to consolidate the power of local rulers and, on occasion, an instrument of a new brand of American imperialism. The most consequential instance of American water expertise deployment abroad was on the Arabian Peninsula.

After the First World War, Ibn Saud, ruler of the house of Saud, had defeated Ali, the son of the Hashemite king Hussein, to take over the Hejaz, home to Mecca and Medina, the holy places of Islam. He eventually consolidated his hold over a territory roughly two-thirds the size of India. It was a period of turmoil in the region, as the Ottoman Empire dissolved. Under the auspices of the League of Nations, much of the Ottoman territory was broken into British and French mandates in Syria, Lebanon, Palestine, Transjordan, and Iraq, ostensibly to guide them to independence. In 1923, Mustafa Kemal Atatürk turned Turkey into a secular republic. Atatürk's reforms in turn inspired Reza Khan, who a decade later became shah, founding the Pahlavi dynasty and renaming Persia "Iran." The region was turning towards independence and modernization.

As Ibn Saud established his power base in Arabia, he faced a significant problem. The country was desperately poor, with no income other than the stipend the British provided and what the pilgrims paid while visiting Mecca. Conquering a territory was far easier than maintaining a state. Achieving any level of security required settling Bedouin tribes. One instrument was faith: Ibn Saud recruited them into Wahhabism, a religious movement that from its inception had been closely associated with the House of Saud (and still is). The only other viable, long-term option to support settlement was to organize the population into agricultural cooperatives. But for agriculture to work, Saudi Arabia needed water.

Assistance came, as often seemed to be the case at this time, in the form of an enterprising American. After his stint as American minister to China, in 1919 Charles R. Crane was appointed by Woodrow Wilson to serve on the King-Crane commission, which was set up to determine what the Arab people wished for their future. The Wilsonian enthusiasm for national self-determination could not overcome the self-interest of European powers, which were prepared to fight

German imperialism to the death in order to protect their own version of it. The commission went nowhere. However, the experience stimulated Crane's interest in the Arab world.

A few years later, Crane visited Sanaa and met Imam Yahia, the ruler of Yemen. While regional production had improved after the war, the countries of Arabia remained hopelessly underdeveloped. Crane committed his philanthropy to the imam, to explore whether mining resources could help the country on the path to development. That is how a second enterprising American, Karl Twitchell, a mining engineer from Vermont, entered the picture. Crane hired Twitchell to be in Yemen from 1927 to 1932 to supervise the projects he intended to fund: prospecting for mineral resources, installing windmills and pumps, establishing experimental gardens and farms, and even developing infrastructure. This was not just a mining expedition. It was an economic development mission.

News of this extraordinary and unusual program traveled quickly, and in 1931 Ibn Saud invited Crane to visit him in Riyadh. If Yemen could get that sort of help, maybe his country could too. Ibn Saud was after water. Crane sent Twitchell to conduct a fifteen-hundred-mile survey across Saudi Arabia. He found no evidence of previously unknown sources of water. He did, however, have another idea. Mineral resource exploration might provide additional income. He continued to search for water and, now, minerals, heading towards the Persian Gulf.

Bahrain, an island nation in the Gulf, was exploring for oil right at that time. The British had been active in oil exploration on the other side of the Gulf. When Churchill became first lord of the Admiralty, he converted the whole imperial fleet from coal to oil, tying the interests of the British Empire to the Middle East, and to the fate of Iranian oil in particular. The Anglo-Iranian Oil Company—today known as BP—was the first and most productive. If they had struck oil in Bahrain, almost certainly—Twitchell thought—there would be oil on Saudi territory, given the similar geology. And strike they did, setting the stage for an energy revolution.

Saudi Arabia did not have any additional financial resources to devote to exploration. In 1933, the Standard Oil Company of California, which had been already engaged in Bahrain, called Twitchell. The company extended a loan to Saudi Arabia, and a joint company between Texas Oil and Standard Oil Company of California was formed, first called the California Arabian Standard Oil Company, and then the Arabian American Oil Company: Aramco. The age of oil had appeared on the horizon, the by-product of water exploration.

THE SUMMER OF 1931

China was an equally consequential recipient of American water expertise in the early part of the century. With Sun Yat-sen gone, China had turned far more ambivalent towards the republican tradition of America. By then, the Kuomintang had opened up to influences from the Soviet Union. Its leader, Chiang Kai-shek himself, had been trained there, and both the nationalists and the communists had learned how to organize a paramilitary one-party state thanks to Comintern. And yet, because the waters of North America had turned the United States into the breadbasket of the world, and because the First World War had opened up world markets to it, in time of need it was inevitable that China would once again turn to the other side of the Pacific. Such time of need began with the unusually cold winter of 1930.

The abundant snow that accumulated in western China melted in the spring of 1931, flowing into the Yangtze right when exceptional spring rains were adding water to the river. By July, seven sequential storms had hit the river valley, and as much water came down in one month as usually does in half a year. On August 19, 1931, the flood reached Hankou, part of modern-day Wuhan. The high-water mark measured 53 feet—6 feet above the bund, the embankment.

The flood control levees had given people a false sense of security. Far too many had moved to live in their shadows. When the water overtopped, single- and two-story houses went under. As the waters rose, thirty thousand people, who could not leave the city, moved to high ground, first finding refuge on a six-mile-long railway embankment that crossed the city. When that went under, those who could escaped by boat, while others scrambled on top of taller buildings and even on trees.

When the Huai River, a tributary of the Yangtze, flooded, the waters spread unimpeded because the river flowed through a flat alluvial plain. Higher ground was not an option and people scrambled to find anything that could serve as a boat. On the night of August 25, the water flowing through the Grand Canal reached critical pressure and broke through the dikes near Gaoyou Lake. Two hundred thousand people were caught in their sleep and drowned. By September 16, water had peaked in Nanjing.

In the end, the floods of 1931 inundated twenty million hectares of land, an area the size of New York, New Jersey, and Connecticut combined. Flooding extended as far south as Guangdong, just east of Hong Kong, and as far north as Manchuria in the northeast of the country. A vast freshwater ocean swallowed the landscape. Between twenty-five and thirty-five million people were affected, with hundreds of thousands presumed drowned. And what the flood didn't do, disease did. In Wuhan, the water system stopped working immediately. The mix of water, mud, industrial waste, and sewage that flooded the city turned it into a fetid, open sewer, making access to potable water impossible. No one knows exactly how many people died in this catastrophic event, one of the worst natural catastrophes in human history, but the estimates range from four hundred thousand to four million people. China discovered itself once again to be terrifyingly vulnerable to its waters.

The scale of the disaster had already become apparent to the government of Nanjing by the beginning of August. This was a weak government, keen to show leadership. Chiang Kai-shek, then leader of the Nationalist Party, decided that a centralized effort would be required to give an appropriate response. On August 14 he established the National Flood Relief Commission. T. V. Soong, the vice president of the Executive Yuan, the executive branch of the government—and Chiang Kai-shek's brother-in-law—chaired it. The commission operated under difficult conditions, trying to balance its work amidst the political conflict between communists and nationalists.

One of the immediate problems Soong faced was to procure grains for the inundated populations. The flood had destroyed much of the agricultural system, and, with winter coming, famine was a real risk. Soong had deep ties to America. The early republic had had a steady stream of advisors from the West, and he had maintained several commercial relationships in the country. He immediately began negotiating with the American Grain Stabilization Corporation for a shipment of wheat. It was perfect timing.

The United States, by then in the throes of the Depression, was looking for any opportunity to put people to work and sell its products. By September 25, China and the United States had reached a deal for 450,000 tons of western white wheat. Half of it would be delivered in the form of flour, so as to put American millers to work as part of the deal. American food assistance was quickly followed by technical assistance. But although up to then the expertise had been mostly economic, now was the time for water. Chinese experts traveled to America to visit its remarkable water achievements, often funded by American foundations like the Rockefeller Foundation. Over the subsequent decade, until the Second World War, China opened up to American technical assistance. Another building block in linking water development and geopolitics in the twentieth century had been laid.

Water resources had been the basis for the electrification and industrialization of the American republic. They had also ensured that the United States could assert its dominance in the world's agricultural commodities' markets. Being a technical leader in hydraulic engineering and, at the same time, the breadbasket of the world would position the United States for dominance as countries rushed to transform their own landscape to keep up with the growing number of citizens and their demands. By the mid-1930s this positioning was distilled into a modernist project, a synthesis of the American experience, which for a while would become the dominant model of development everywhere in the world.

16

FDR's Modernization Project

DEPRESSION

In the 1930s the United States produced a rare technical synthesis and political abstraction which, for a few decades, propelled the development of water resources at the very forefront of the state's toolkit. This synthesis started in the midst of a crisis.

When the bottom fell out of the stock market in 1929, it sent the United States into a deep economic contraction. The whole world followed. The infamous Smoot-Hawley tariff of 1930, which had begun life as an agricultural tariff, quickly spread to other goods. As other countries retaliated, it had a catastrophic impact on the lives of those it was designed to protect. The Great Depression had started. Given where the country had been in the 1920s, it is hard to fathom the scale of impact it had in the United States: in four years, gross domestic product dropped by about a third, and a fifth of the working population was unemployed. The very stability of society was at stake.

President Hoover tried to spend his way out of the problem, doubling public expenditure between 1929 and 1932. This was not deficit spending in a Keynesian sense, however, and was counterbalanced by an equally dramatic increase in excise taxes, the principal source of federal revenue at the time. To add to the pressure, a monetary straitjacket was put on the economy: between 1929 and 1933 the Federal Reserve failed to intervene to reduce the number of bank failures and limited the amount of money in circulation in order to maintain the gold standard. The Depression deepened.

Then the water environment played its own cruel trick. In the early thirties, a severe drought hit the Great Plains right as grain prices collapsed in the wake of the Depression. It was bad timing. The Dust Bowl had started. Overextended farmers went bankrupt. Land was abandoned at the same speed at which it had been developed, as John Steinbeck's Okies moved west. As farmers left their properties, the exposed topsoil baked and pulverized in the drought. The winds of the Great Plains then lifted up the dust into huge black blizzards, big enough to block the sun, worsening drought conditions further. When cold air from Canada and warm air from the Dakotas swirled over the plains igniting storms, the atmosphere became a huge planetary vacuum cleaner, sucking up into the sky hundreds of thousands of tons of dirt in squalls hundreds of miles wide and thousands of feet tall.

In 1933, as Franklin Delano Roosevelt opened the longest presidency in American history, the country was staring into an abyss of economic destruction. Roosevelt waded into the unfolding credit crisis with remarkable speed, stemming the run on banks. In his first hundred days, he set up programs, from the Works Progress Administration to Social Security, to inject cash into the economy (although admittedly preserving Hoover's fiscal discipline). Many of these programs modified the landscape in the process.

Electrification was a central concern, particularly for rural communities. In 1930, almost all cities in the U.S. had been electrified, but only 10 percent of farms had any access. When Roosevelt came into office, the power industry had a near monopoly in setting rates. For an unemployed population squeezed by an unaffordable cost of living and a manufacturing sector ready to modernize but starved for credit, cheap electricity was essential. Rivers were the answer. Roosevelt attempted a grand synthesis of democracy and technocratic planning on America's rivers, as part of his offer of a New Deal to the American people. The mighty Mississippi, the hardworking Columbia, and the powerful Colorado defined the American landscape. And they would be put to work in the service of an ambitious recovery agenda.

To represent the essence of this New Deal, Henry Luce had brought to *Life* magazine Margaret Bourke-White, a photographer whose work was in the realist tradition of other socially engaged artists of the time. But her beginnings were in industrial photography, so she was particularly well suited to capture the engineering that was unfolding, her aesthetics perfect for the magazine's first issue. To capture something of the grand scale of Roosevelt's New Deal, the choice could not but fall on the development of rivers. The initial subject was going to be Bonneville Dam, a big public works project on the Columbia River. But a massive dam on the Missouri seemed to better capture the essence of the period.

Fort Peck was planned to be the largest earth-filled dam in the world. The dam would create a reservoir that was second only to Hoover Dam's Lake Mead. Its construction had generated the town of New Deal, Montana, a boomtown that had sprung up to house the workers of the project. There, Margaret Bourke-White encountered a motley crowd of workers, prostitutes, charlatans: "Everything from fancy ladies to babies on the bar." The front page of the very first issue of *Life* was her striking photograph of the spillway of Fort Peck, before its filling. It looked like a massive defensive wall, complete with turrets and battlements, capturing the industrial scale of America's conquest of nature.

This was Roosevelt's America or, as Henry Luce described it, a "nation, conceived in adventure and dedicated to the progress of man." It was going to be the "first great American Century." A century, the foundations of which, Roosevelt planted firmly in the country's rich, life-giving waters.

LILIENTHAL'S DREAM

During the presidential campaign, Roosevelt had proposed that electricity prices should be dramatically reduced to break the monopoly of power companies. The way to do that was for the government to develop four major public power companies in each of the quadrants of the country: the St. Lawrence River in the northeast; the Columbia River in the northwest; the Colorado River in the southwest; and the Tennessee River in the southeast. Roosevelt hoped that having government-developed projects would provide a "yardstick" that would determine what the right electricity rates should be. But the projects were imagined as much more than power supply. They were economic development investments.

The Tennessee Valley, which, like the Colorado, was shared by seven states, was one of the poorest, most underdeveloped parts of the United States. The Tennessee was a tributary of the Ohio and, ultimately of the Mississippi, but the valley was far more isolated than might appear from a superficial view of geography. Because of its high variability, dangerous shoals, and steep gradients, the river was not navigable, and it flooded often, destroying whatever farmland there was. Over two million people lived there, three-quarters of whom were rural, a percentage far higher than in the rest of the country. Per capita income was also tragically low: under $700 per year, less than half the national average. A fifth of the population lived on less than $250 a year.

Electrification was practically absent in the Tennesse Valley. Only 4 percent of rural households had electricity, and only 3 percent had running water. Without adequate sanitation, typhoid, tuberculosis, and infant mortality were endemic. Access to education was also far below the national average, illiteracy correspondingly high. Steep and marginal lands had been cultivated, leading to further soil erosion. Forests were depleted, many had been clear-cut. Over 80 percent of the thirteen million acres of cultivated land suffered from visible or severe erosion and degradation. Mines were exhausted. The second industrial revolution had hardly reached the Tennessee Valley.

On April 10, 1933, thirty-seven days after his inauguration, Roosevelt submitted the Tennessee Valley Authority Act to Congress. The Tennessee Valley Authority, or TVA, was unprecedented in its scope and ambition. It was enormous, expected to use two and a half times as much concrete as had been used in the Panama Canal. But it did not just represent a project. It was an ideology, based on an organized use of natural and human resources in a framework that included both development and conservation. In Roosevelt's words, the use of the river "transcends mere power development: it enters the wide fields of flood control, soil erosion, afforestation, elimination from agricultural use of marginal lands, and distribution and diversification of industry. In short, this power development of war days leads logically to national planning for a complete river water shed involving many states and the future lives and welfare of millions."

The establishment of the TVA was a milestone of global significance. It contributed to changing the role of the state in society. It was to be "a corporation clothed with the power of government but possessed of the flexibility and initiative of a private enterprise." Roosevelt's act gave the new institution extraordinary powers to operate as a private corporation but also as a development agency. Without appropriate political safeguards, the risk of its degenerating into the authoritarianism inherent in the modernist ideal of scientific and technical order was significant.

The TVA was given the powers of several government departments. As a result, its board had to be able to interpret its political mandate by mediating between competing policy objectives to come to decisions: agricultural production against recreation, or power production against public health. This was an unusual condition to be in for a technical institution. The TVA industrialized public management. Unlike other public works, the authority did not contract out construction to private companies, but developed the capabilities and competences to build its own projects. Through the employment of local labor, the TVA also acted as an apprenticeship vehicle. It provided housing for workers. It acted as a relocation agency for the farmers who had to move because of the construction. There was a patronizing quality to the interventions of the TVA, which presupposed a belief in the incorruptibility of its institutions.

David E. Lilienthal knew that the success of the TVA was inherently dependent on its political framing. In 1933 he had been appointed as one of the three founding directors. He became its chairman in 1941. His book *TVA: Democracy on the March* was an extraordinary political argument, almost a treatise, in which he argued that "in tested principles of democracy we have ready at hand a philosophy and a set of working tools that, adapted to this machine age, can guide and sustain us in increasing opportunity for individual freedom and well-being." He took it as a matter of faith that the safeguards of representative democracy could be channeled through a regional authority to exercise a check on its extraordinary executive power.

Lilienthal's vision was utopian, with echoes of both Periclean rhetoric and Ciceronian paternalism. He termed it "grassroots democracy." If the TVA was an answer to the modernist dilemma of an economy that had industrialized to benefit the many, by disempowering the individual, then the enfranchisement of the population of the Tennessee Valley was going to have to be part of the answer. The empowerment of the beneficiaries of the project through bottom-up participation, Lilienthal reasoned, could deliver real efficiency.

A river has no politics, but the development of a river is an eminently political question. That much Lilienthal recognized. He believed in a balance of power between the political objectives of development and the technocratic experience that guided it, and saw the education and knowledge of the interested public as the only real safeguard against the distortion of interests by administrators. An informed public would be the principal driver of oversight.

The authority operated the river and its landscape as a single, functional system. It pushed the relationship between the different productive components of the river basin—the farm, the factory, even the household—to the ultimate optimization for collective welfare. The benefits could only be delivered if there was a way of enforcing that optimization. Through its regional focus and emphasis on grassroots participation, the TVA was a centralization of authority and a decentralization of execution. In this, Lilienthal had Tocqueville in mind.

He believed the TVA was a modern instrument of democracy, imbued with moral purpose. It was not just a regional authority, but an institution that operated according to specific principles with respect both to the systematic treatment of the resources and to the involvement of the population and other institutions.

A MODEL READY FOR THE WORLD

The Tennessee Valley Authority was an economic success. By the time work was completed, it was operating fifty-four dams in the Tennessee and adjacent Cumberland Rivers, making them fully navigable. TVA also managed fourteen more dams owned by others, from Alcoa, the aluminum producer, to the Army Corps of Engineers. During flood season the system could intercept about 10 percent of the rainfall of the basin, and had installed hydropower capacity of almost four gigawatts. During the war, power went to aluminum factories, to provide electricity to the Manhattan Project, and its electricity transmission lines, strung between buildings and along the roads, powered electric pumps, refrigerators, and factories, where once there were only cotton farms and tenant shacks. Tuberculosis and malaria were a thing of the past. Modern water management had fueled economic development.

The idea that an investment could drive economic growth and development was a relatively recent one. During the first decades of the twentieth century economists were not generally concerned with economic growth as a phenomenon. This was partly a legacy of the nineteenth century: while there was no expectation of mass social mobility and enfranchisement, the landed aristocracies were generally unconcerned with how to increase the size of the pie for all. After the First World War, the most pressing issue was how to use economic instruments to maintain employment through recessions and booms. Keynes had convincingly argued that the invisible hand of markets was not always a stabilizing force in labor markets, which opened the door to a far more interventionist state.

Then, two economists, the English Roy Harrod and the Polish-American Evsey Domar, independently proposed a theory for why income growth happened in the first place. Their basic insight was to use the tools of short-run economics introduced by Keynes—investment, aggregate savings, and demand—to deal with the long-run problem of growth. Harrod and Domar posited that the rate of growth of a country would depend on only two factors: the level of savings and the capital output ratio, that is, the amount of capital needed to grow the output of the economy by a unit. The implication was that there was a warranted growth rate that countries ought to get to, one in which all savings could be absorbed by investment.

The Harrod-Domar model had the narrative advantage of fitting recent experiences like the TVA. In truth, and seen with the benefit of hindsight, the impact of the TVA on the economy of the nation as a whole was more nuanced than those stories would suggest. Its effect on local manufacturing productivity was sufficiently large to make a measurable contribution at the national level. But whether or not it had been able to stimulate additional economic activity was less clear. It attracted industry, but that was at the expense of other places having less, so the net result was more or less neutral. But the anecdotal evidence on local economic development was so compelling that the TVA became the symbol of a concerted government response to manage the economy through its cycles.

From the point of view of this early theory of economic growth, the American experience might have appeared not that dissimilar to the Soviet one. After all, although their expectations for what would reward investments were vastly different—the United States trusted in the "animal spirits" of the private sector to repay investments through taxation, while the Soviet Union had largely relied on state-owned means of production—America and the Soviet Union had indeed pursued aggressive investments in the landscape, and indeed their economies had grown.

But unlike the Soviet state-directed approach, the TVA's exceptional executive authority made it vulnerable to strain, exactly like any other political institution in a democratic republic. Over time, the distance between increasing expectations and the limits of reality eroded its legitimacy. Political resistance to additional river basin authorities in the U.S. increased as people complained of federal overreach, and was ultimately too great to overcome. The TVA represented both the peak of state-directed water resource development in America, as well as its inevitable limit, evidence that in the American republican system of checks against the centralization of power, managerial efficiency could not, by itself, acquire political legitmacy.

But while limited by the political culture of the American republic, as a managerial intervention the TVA showed enormous promise on the international stage. Right when its fortunes were waning domestically, its experience grew to extraordinary significance worldwide. In 1944, Lilienthal reported that eleven million people from virtually all corners of the world had visited the TVA. Among them, Chinese, Indian, British, Australian, Brazilian, and Czech politicians and experts showed up to learn about it. The model was about to spread beyond the boundaries of the republic.

RETURN TO THREE GORGES

China was one of the first instances in which the experience of the TVA was exported in the form of technical assistance. When T. V. Soong struck a deal with the American Grain Stabilization Corporation in the aftermath of the 1931 flood, he had entered into a lopsided deal. The Chinese had made the mistake of accepting market prices, which pushed up the price on the commodity exchanges in America, making their subsequent purchases far more expensive. Despite this, the response to the floods cemented a close partnership between the Kuomintang and America.

That was the time when the Tennessee Valley Authority was being developed and, thanks to the intercession of the Rockefeller Foundation, Chinese experts were sent to study the TVA experience. It made a profound impression. Indeed, China was one of the countries Lilienthal believed could benefit from the TVA approach. The push to replicate the TVA happened as the context around the Chinese government changed.

While China was still reeling from its natural catastrophes, on September 18, 1931, Japan had attacked Mukden, in the northeast of the country. By the beginning of the Second World War, the Japanese controlled most of eastern China, where the country's primary economic infrastructure was. Britain had been unable to defend Hong Kong, Singapore, or Burma, so the Japanese controlled the south, while the Soviets had cut their support in the north to tend to their western front. The Chinese Nationalist government found itself isolated and landlocked in the west. T. V. Soong turned once again to the United States for assistance. Chiang Kai-shek was mostly hoping for military and financial assistance. The Americans sent John Lucian Savage instead.

Savage was chief designing engineer for the U.S. Bureau of Reclamation. He had worked on Hoover Dam, Grand Coulee Dam, the All-American Canal, and the first dams of the Tennessee Valley Authority. He was known as the first "billion-dollar" American engineer, a testament to the size of the projects he worked on. Savage's reputation was a powerful instrument of American influence.

In the second half of 1944, sponsored by the Cultural Cooperation Program of the State Department, John Savage made it to China. Sun Yat-sen's proposals from twenty years earlier were still lingering in the air, and Soong was a leading proponent of a project near the Yangtze's Three Gorges. Savage found five potential sites for the core of the river development, near Yichang, in the province of Hubei. His prescience was remarkable. He imagined a huge concrete gravity dam, over 220 meters high. The dam, as Savage conceived it, could produce over ten billion watts and irrigate around ten million hectares of land. Savage thought there should be one single lock that could lift ships up the entire height of the dam.

Savage believed such a project could drive industrial development: cheap energy, he thought, would drive industrialization. The U.S. Bureau of Reclamation agreed with Savage. The project would bring obvious advantages to China and would give the technical staff of the United States a huge opportunity to advance its skill and track record. By 1945 in China, the Office of War Information of the United States had distributed fifty thousand copies of David E. Lilienthal's book *TVA: Democracy on the March*. It worked. Apparently, Chinese experts of the Nationalist regime were quoting the "principles" of *Democracy on the March* to visiting American officials. This went hand in hand with Chiang Kai-shek's hopes for a "Yangtze Valley Authority." There was even talk of sending Lilienthal himself to China on a speaking tour.

At this time, the Yangtze was not the only river on which a TVA-like investment was being considered in the region. As the Japanese and the British fought over Burma, Indian officials had contemplated replicating the TVA model on the Damodar River, in the northeast of India, where the river cut across the supply lines that connected the British troops to Kolkata. Between July 14 and 16, 1943, an early monsoon had increased the flow of the Damodar in western Bengal. On July 17, the river had burst its northern bank and catastrophic floods destroyed villages and cut rail lines to northern India. It had been a military disaster: the disruption cut off both British and Indian armies fighting the Japanese. The event prompted the establishment of a commission to look into flood control and its implications for the infrastructure of Bengal and the health of its people. The Tennessee Valley Authority was seen as a successful "package," a solution that could be deployed in other parts of the world. The idea of a Damodar Valley Corporation was developed with the assistance of one of Lilienthal's engineers.

But when it came to China, not everyone believed in this vision. The State Department did not see how China could possibly absorb the amount of electricity that the project was predicted to produce. It seemed to them neither feasible nor economically sound. Besides, the political situation in China was unstable. As soon as the war ended and Japan surrendered in 1945, the conflict between the Communists and the Nationalists, which the United States had desperately tried to avoid, broke out in full force. Under these conditions, foreign policy experts saw little value in the U.S. engaging in such a massive project. Savage retired from the bureau and was immediately contracted to advise on the development of the Yangtze, but by 1947 Chiang Kai-shek had been defeated and had retreated to Taiwan. The communist era of China had begun, and the United States had lost the opportunity to bring the country into its sphere of influence.

Savage's engagement ended, but his legacy and that of the TVA, mixed with a good dose of Soviet hydraulic engineering, persisted. Sixty years later, China would finally succeed in building Three Gorges Dam, more or less following Savage's vision. It was part of a remarkable cascade of dams along the entire mid-stem of the Yangtze, which transformed the river into the industrial powerhouse of the country, as he had predicted.

AN AMERICAN BARGAIN

Roosevelt believed the world needed America's TVA experience. If China was not to be, maybe Saudi Arabia would provide the right testing ground. On his way back from Yalta, in February 1945, he had asked to fly at low altitude over the Arabian Peninsula. He had noticed the arid landscape and wondered about the lack of agriculture. He struck up a conversation with the army engineer who was traveling with him. Could irrigating the landscape solve the problem? Yes, it seemed, provided water could be pumped from the ground and used in a way consistent with the arid nature of the landscape. Roosevelt saw a universal answer in the American experience. And he believed it would bring wealth and development to those who adopted it, just like he believed it had in the Tennessee Valley. In a way, he was right. The pursuit of water did bring a form of development to the region. It also began a transition that would transform the world economy, eventually displacing hydropower as the star of the world's energy story.

Aramco had struck commercially viable oil in 1938, at Jabal Dhahran, eventually finding the largest oil deposit in the world. Oil was crucial to build wealth. But wealth was not the same as power. For that, Ibn Saud had to translate wealth into social control. The oil money had to be spent on projects that would improve livelihoods. Because most people depended on agriculture, water was still their principal rate-limiting factor. Ibn Saud still had to provide them with water. The royal family had a vision for what that looked like. In 1937, the king had started a project in Al Kharj, in eastern Nejd. Most of the land in the project belonged to the king, his brother, and the minister of finance. It was the first attempt at farm mechanization in Saudi Arabia, a precursor to large-scale farming. Their vision for agriculture was a modern, mechanized, commercial operation. But they needed help.

With the United States fully in the war, Roosevelt's interest in the region increased. His administration stepped in, providing additional resources and expertise from the U.S. to introduce modern farm practices and technology. The principal theme of those developments was always the same: extract water from the ground to irrigate and improve production. The U.S. Department of Agriculture sent a team of agricultural advisors, once again under the leadership of Twitchell, to provide the first comprehensive view of where water was, how much of it there was, and how much was usable. The recommendations of the mission were to engage in "reclamation" projects by pumping groundwater. It was the recipe of the west of the United States, applied to Saudi Arabia. By that point, government royalties from oil had grown enough to fund the development of agriculture. The technical expertise of Aramco's staff added infrastructure that would have been previously unaffordable. The technical and financial aspects of modernization had been resolved. But its politics were going to be another matter.

In February 1945, immediately after the Yalta conference, Franklin Delano Roosevelt hosted three regional leaders on the USS cruiser *Quincy*, on Egypt's Great Bitter Lake, near the Suez Canal. King Farouk from Egypt was the first to come on board. Then, the same day, it was Haile Selassie of Ethiopia. After him, the president hosted King Abdul-Aziz Ibn Saud. It was remarkable that Roosevelt would include Ibn Saud alongside Farouk and Selassie. Saudi Arabia was traditionally in Britain's sphere of influence, but Twitchell's oil had changed the calculus and it was already clear the king was a priority. The British were trying to convince Ibn Saud to stick with them, arguing that the Americans would retreat to the Western Hemisphere after the war, and that Saudi Arabia should stay within the pound sterling area, with the Royal Navy and the British Army as partners and protectors. Roosevelt, on the other hand, believed in a division of the Middle East: Persian oil should be the interest of the British, Kuwait and Iraq should be shared between the two, but Saudi Arabia was an American affair.

Ibn Saud, for his part, wanted America to continue to play a heavy role in his country to balance Britain's considerable influence in the region. And it was clear that Saudi Arabia's massive oil deposits made it strategically important for the United States no matter what. And so, King Abdul-Aziz Ibn Saud steamed across the Great Bitter Lake, sitting on a gilded throne on the deck of the USS destroyer *Murphy*, surrounded by his guards and a flock of sheep brought on board for the occasion.

Roosevelt and Ibn Saud spent five hours together. After a lengthy—and inconclusive—discussion about the issue of Palestine, Roosevelt turned to agriculture. For a new state trying to establish its own sovereignty, Roosevelt knew that the problem of supporting agriculture in such an arid context was going to be central to the leaders' concerns. Saudi Arabia could only really develop its agriculture if it committed to a development model, not just to a technical solution to get water. By this point, Roosevelt believed that the experience of the Tennessee Valley Authority was the pre-eminent water-led development model, and that Saudi Arabia should benefit from that intervention.

Whether or not Lilienthal's *Democracy on the March* could have directly contributed to the development of Saudi Arabia will never be known. In a prophetic moment, Roosevelt told his then secretary of labor, Frances Perkins, that once out of office he would devote himself to exporting the model of the Tennessee Valley Authority to the Middle East and to other countries, in the belief that such water-led development could bring progress. At Yalta, he suggested to Stalin that the TVA was the model to pursue for economic recovery in Europe. Even the Marshall Plan was described by some as a "European TVA." The TVA was so popular that it became a fundamental instrument of international relations, leading to the development of the Mekong as a regional project. As Lilienthal had said, there were "a thousand rivers" like the Tennessee that could learn from its lessons.

Roosevelt died in office two months after his historic meeting with Abdul-Aziz Ibn Saud, and did not live to see the idea of the TVA take hold in the rest of the world. Ibn Saud passed away in 1953 and was succeeded by his son Saud. But the impact of that conversation was long-lasting, as the global economy reorganized around the relationship between food and oil.

Roosevelt's belief in the power of the TVA's example was eventually fulfilled. At his second inaugural address, on January 20, 1949, President Harry Truman announced his Point Four strategy, committing the United States to using its scientific and industrial advances in the service of developing nations. The TVA—Roosevelt's central New Deal legacy—had become an American theory of modernization. It would dominate much of the second half of the twentieth century.

17

Cold War

COMPETING OVER GROWTH

After the Second World War, the United States had become the world's dominant economic and military power. The Bretton Woods agreement of 1944 had fixed exchange rates, turning the United States into the world's banker. Its capital could be deployed across the world as an instrument of development, often ending up in water infrastructure. With programs like the Marshall Plan, the United States hoped to provide financial support during reconstruction, not only to secure allegiance, but also to stimulate an expansion of markets for its goods and improve the productivity of its suppliers.

But in 1946 Churchill observed that "from Stettin in the Baltic to Trieste in the Adriatic, an iron curtain has descended across the Continent." The rise of the Soviet Union had created a profound geopolitical divide, fracturing Europe in half. In his election speech on February 9, 1946, Stalin celebrated the victory in the Second World War as testament of the viability of the Soviet state, claiming it had validated the system of five-year state planning. The Cold War was a battle between two economic and political systems that had emerged from the Second World War. That battle was to be fought also over the waters of the world, ultimately accelerating an unprecedented transformation of the landscape.

Countries that had already industrialized were swept up in the competition, often benefiting from the investments available in return for allegiance. Italy, for example, was a crucial ally to America, but given that it had the largest Communist Party outside the Eastern Bloc, it also posed a significant risk to the United States' containment strategy towards the Soviets in Europe. As a result, the Americans provided finance generously. Italy took American support, but instead of using it to stimulate demand for American products, it bet everything on exporting the products of its manufacturing industry, which in turn accelerated the development of water resources.

Most Italian industrial infrastructure, especially hydropower, had emerged largely intact from the war. Industrial installations had been mostly in the north, far from the Gothic Line that separated German-occupied Italy from the Allied invasion of the south. Most of Italy's proximate markets on the other hand—France, Germany—had been completely destroyed by carpet bombing and invading armies. Italy was ready to produce and the rest of Western Europe was ready to buy. Because the European coal market remained disrupted and supply from Great Britain was caught in union actions, to seize the market opportunity, Italy invested in hydropower development. Hydropower capacity doubled between the end of the war and 1960, in turn driving industrial production and economic performance to be among the best in the Western world. For most of the fifties and sixties the country's economy grew at over 5 percent per year, among the highest percentages in the world. Many described it as an "Italian economic miracle." Its success seemed to travel on its rivers.

Italy had already partly industrialized—the Cold War brought accelerated investment. But the main story of the period was that of the many countries across Asia and Africa that found independence out of the collapse of the old imperialist order. Over the course of three decades newly independent countries faced the problem of how to get on the path to industrialization in the first place. The economic orthodoxy of Harrod and Domar, backed by the example of countries like Italy, seemed to support a water-led investment approach as a way to stimulate industrial production, but it was of little use to poor countries, which could offer limited capital returns to begin with. Yet the answer to this problem, once again, seemed to lead to water. Economist Arthur Lewis proposed that the vast amounts of subsistence labor of poor countries could be mobilized from agriculture to increase returns in the capital-intensive industrial sector. The increased returns would then grow the savings rate, leading to a virtuous cycle of further investment. This theory was ultimately inadequate empirically, but still framed the approach to development for decades. Lewis wrote that "no country can fail to develop, if it has good government, adequate rainfall, and a reasonable system of secondary education." Many countries struggling to come out of poverty after the Second World War were in the tropics, where "adequate rainfall" actually meant vast quantities of water, unevenly distributed and erratically supplied. If the agricultural sector was going to contribute to growth and free up labor for industry, infrastructure to capture, convey, and manage water was going to be essential. This was going to be the story of the Indian subcontinent and of Africa.

What was at stake was more than economic growth and the shape of the landscape. In 1957 Karl Wittfogel, one of the most prominent Marxist sociologists of his generation, a founder of the Frankfurt School and a specialist of China, wrote a book titled *Oriental Despotism.* He interpreted the competition between the totalitarian, communist regimes of the East and the capitalist societies of the West as a battle driven by the landscape. Wittfogel believed that some societies were hydraulic, in that their economy depended on a managerial approach to "large-scale and government-directed water control." He updated and expanded Marx's ideas about an Asiatic mode of production.

Wittfogel was heavily influenced by sociologist Max Weber, who had also written about China—despite never actually visiting it—and who had focused on the country's transition from feudalism to centralized authority. Wittfogel adopted Weber's analysis of Chinese bureaucracy as an instrument of power and argued that the weakness of totalitarian communist regimes was in the stifling nature of their powerful bureaucracies, which were needed to operate large-scale hydraulic infrastructure. He believed that social development was ecologically determined: it was a "dialectic of geographical location."

As a work of sociology, *Oriental Despotism* did not stand the test of time. The book, however, was an unexpected success, provoking debates that have since generated rivulets of polemics. Crucially, despite the disagreements on its fundamentally deterministic argument, it captured an important point: that the architecture of the state, its structure and underlying political philosophy, could be seen in the landscape. It had a physical manifestation in the way in which a country chose to deal with its water resources. Both the Soviet communist system and the American capitalist one had made reclamation, irrigation, and hydropower pillars of their success. So, what was at stake in the race to develop the world's landscape was the supremacy of a political system.

The postwar world was the stage on which many transformations played out, from the creation of the largest democracy in the world to the descent into chaos of the Soviet regime, and from the development of the first, postwar international treaty on shared water resources to the fall of an old empire. In all cases, it was not just the economic value of water infrastructure that would be on display, but its political function.

A PAINFUL BREAKUP

The Cold War fractured the world. One of its most vulnerable fault lines was right in the heart of Asia, tied to the first and most consequential acts of independence of the postwar period, which marked the definitive end of the old world and the beginning of a new one.

The split actually came as a surprise. The two-country solution of Pakistan and India going their separate ways only really emerged at the eleventh hour, in the very last gasps of the British Raj. The increasingly isolationist Muslim League, led by Muhammad Ali Jinnah, and the Indian side, led by Jawaharlal Nehru, were unable to bridge the growing gap. When the Indian Independence Act of July 18, 1947, separated the two countries, everything happened far too quickly. The partition of the Indus, the most important river in the country, remained unresolved. In particular, the partition of the capital stock the British had left in the largest irrigation system on the planet was at stake.

The Raj appointed Sir Cyril Radcliffe to try and come up with a solution. Sir Cyril had never been to India, but he was a trusted bureaucrat, the former director-general of the Ministry of Information during the Second World War. The British must have thought he would have been impartial to the two sides, so he was made chairman of the Punjab Boundary Commission.

As he was drawing up the borders between the two countries, Sir

Cyril was keenly aware of the impact that partition would have on the irrigation system. He tried to convince Nehru and Jinnah to co-manage it. That was not to be: river unity conflicted with their ideas of national identity. Under the pressure of time, with imprecise maps that conveyed little of the ground realities, Radcliffe drew the border. He cut right through the upper reaches of the basin, dividing most of the tributaries. India got the headwaters and upstream reaches, while Pakistan got most of the floodplains. It was August 15, 1947.

The Indus is different from other rivers flowing down the Tibetan plateau. During winter, the westerly jet stream shifts southward from the plains of Russia, crashing into the plateau at the intersection of the Hindu Kush mountains, between Pakistan and Afghanistan, and the Himalayas proper. Its weather systems bring about a third of the annual precipitation of northern India and Pakistan, mostly as snow. Glacial and snowmelt contribute about one and a half times what rain does to the Indus. In fact, most of the basin is arid, receiving less than five hundred millimeters per year.

Any extensive agriculture in the Indus requires some form of irrigation. Flood irrigation was the most common form for several thousand years, from the early stages of the Harappan Civilization in the fourth millennium until the British arrived. By 1947 the Indus had the largest irrigated system in the world, with individual canals as large as the Potomac or the Thames. It was a marvel of hydraulic engineering but a nightmare of politics. At partition, India could claim only five of the twenty-six million acres of irrigated lands in the Indus canal system. Most of the areas of the basin that still needed to be developed were in India. Water allocation seemed similarly tilted: Over sixty-four million acre-feet of water went to Pakistan and just over eight to India. On a per-acre basis Pakistan had 50 percent more water than India, which had to spread its Indus water much more thinly.

The canal system was not as inseparable as one might have expected. A hundred and thirty-three canals could continue to operate in Pakistan as normal, as could twelve canals in India. So, 95 percent of the total irrigated area was seemingly unaffected by the partition. But the few canals that had to be divided were a massive problem. India inherited the Madhopur barrage and the upper reaches of the Sutlej Valley Project, specifically the Ferozepur barrage. From the former, the Upper Bari Doab Canal (UBDC) channeled water to the Baes River and ultimately to the Sutlej. From the west bank of the latter, the Dipalpur Canal extended towards Pakistan. This was a difficult setup. In principle, shutting down these two canals would dry up the Sutlej Valley Project, which was mostly contained in Pakistan.

Radcliffe's border arrangement did not specify anything about the use of water. He believed some agreement would be necessary, at least for the joint management of the headworks at Ferozepur, but it was not his mandate to come up with it. When the award of the Boundary Commission came, it became clear an agreement was missing. The chief engineers of East and West Punjab agreed to a one-year "standstill agreement" on the UBDC and at Ferozepur. Allocations at those two points on the river would be frozen until March 31, 1948. But on April 1, 1948, India stopped the water from flowing. The canals crossing the partition line went dry.

In retrospect, the decision appears to have been a unilateral one by the new East Punjab government. It is possible the act was more intended to force Delhi's hand than to antagonize Pakistan: East Punjab was dominated by Sikhs in a majority Hindu Congress. Be that as it may, the cutoff became a national act, which Nehru subsequently defended, knowing it would be an international matter of great significance.

On the Pakistani side, the 1948 water crisis fueled nationalistic rhetoric. The halting of water supply had suddenly put land out of irrigation. It was a national emergency. A hasty attempt to get the water supply restarted tried to circumvent India's headworks by building canals drawing water further up. But it was clear that India would have retaliated, moving its control even further upstream. There had to be a negotiated solution. Pakistan tried to argue that it had historical rights on existing flows, but India did not recognize them.

The problems of 1948 demolished trust between the two countries. It could only be solved through coordinated infrastructure replacement: East Punjab would have to slowly reduce water flowing to Pakistan while Pakistan built new canals.

For its part, India's first big step in controlling the Indus was the Bhakra-Nangal Dam, the first storage project to ever be developed on the Indus. Nehru called dams like Bhakra the "temples of new India," placing the national identity squarely in East Punjab and its water infrastructure. The dam was an assertion of national authority, but also a statement about the Indus. The project implied full control of the flow of the Sutlej and would have a material impact downstream on Sindh.

If prior to partition there had been any hope of joint management, the events of 1948 made it impossible. What no one had anticipated at the moment of partition was the lack of clarity on the border. The haste with which the process was conducted, the lack of ground control, would lead to panicked mass migration. Some fifteen million people—the biggest human migration of the century—took place in a matter of a few months. Building on the already tense communal violence that had plagued the end of the Raj, the migration left a long trail of victims. Estimates vary, but it is possible that one to two million people died in mass killings and ethnic cleansing. The idea of a joint agreement for the management of the river could not survive such a catastrophic event.

The partition of the system required engineers on both sides to implement separate national visions, and separate mechanisms for water security, based on extensive investments in water infrastructure in the Indus basin. The tension between the supremacy of the partition boundary and the unity of the water system would bedevil the two countries thereafter. It had created a fracture across the Himalayas. It was just about the most dangerous place for a crack to appear. Pakistan and India would have to manage their fragile negotiation while surrounded by Afghanistan, which had been the front line of conflict between Russian and British imperialism for over a century; the Soviet Union, China, and the United States, with its rising concern for a region that was in the growing shadow of Stalin's economic growth.

WATER MEGALOMANIA

Seen from the outside, the Soviet Bloc was indeed cause for concern. The idea that water infrastructure was a battleground for supremacy, as Wittfogel implied a few years later, seemed to be supported by Stalin's economic policies, which belied gigantic ambitions. In retrospect, water infrastructure was mostly a powerful instrument of propaganda, and Stalin made a point of using it for that purpose. But at the time it must have seemed like an unavoidable recipe for planning growth.

New hydropower was most certainly part of Stalin's plan. Part of his legacy was the construction of the Krasnoyarsk hydropower plant, which began on the Yenisei River with six gigawatts of installed capacity, or the development of the equally gigantic Bratsk Dam, which began on the Angara River. Some of the most pointless constructions in human history were built in Stalin's time. Solzhenitsyn's *Gulag Archipelago* vividly described the 227-kilometer-long canal between the Baltic and White Seas, which gulag prisoners dug through granite. It was useless, underengineered, and riddled with design mistakes. One above all: the canal depth was not enough to accommodate the draft of the seafaring ships it was designed to serve. But the intent was not usefulness as much as propaganda for the regime.

Cotton was of particular concern to Stalin. Cotton had been produced in central Asia since the tsarist regime. During the Second World War production had suffered, even though central Asia had been largely untouched by the conflict. Irrigation infrastructure had not been maintained, and productivity had decreased, both due to loss of labor and the lack of machinery. But unlike food, cotton was an industrial product. For the Soviet Union it was an essential ingredient for industrialization: anything from oils to clothing, lacquers, paper, and plastics, from tires to explosives, needed cotton. It was also the most obvious crop for the Soviets to invest in, because its cultivation could be mechanized and plugged into the industrial-military complex.

Stalin set the expectation that collectivized farming would achieve cotton self-sufficiency. In fact, he was singularly obsessed with meeting cotton yields from central Asia. Between 1946 and 1954 cotton production from its irrigated landscapes increased by 170 percent. It constituted a massive investment in both irrigation expansion and productivity improvements. The cultivation and the harvesting of cotton were mechanized, expanding both at the expense of food and feed grains. While agricultural production as a whole fell off towards 1952, cotton continued to grow. In fact, by 1954 it was double what it had been in 1940. It was a vast exercise in resettling forced labor, under the harshest conditions. Even a Ministry of Cotton was created to try and coordinate and push the production to new levels (despite all these efforts, the USSR never reached the productive capacity of the United States).

One of the most extraordinary attempts at landscape engineering came when Stalin issued a decree on October 20, 1948, to combat drought and increase fertility and production by reforesting close to six million hectares of land. The forests were mostly to be windbreaks along rivers and the perimeters of farms. This was supposed to protect the land from wind, thus retaining more moisture and dampening the climate.

In the first half of the twentieth century in Europe, afforestation was a favorite of autocrats. Germany's Hermann Göring was Reichforstmeister and had started a national afforestation program in 1934. Mussolini did the same with his Blackshirts in the National Forest Militia. Stalin directed afforestation to transform the landscape of the steppe of southern Ukraine and the south of European Russia.

The plan came to be known as the Great Stalin Plan for the Transformation of Nature. The plan was accompanied by similarly pharaonic investments in water infrastructure, including one of the largest hydropower installations in the world at Kuibyshev and Stalingrad on the Volga, and three massive canals: the Main Turkmen Canal, the Southern Ukrainian Canal, and the North Crimean Canal. The development encompassed some twenty-five-million hectares of land. All together they were known as "the Great Construction Works of Communism." Stalin had framed for the world a new ambition for landscape transformation. Newly independent countries, especially those led by revolutionaries schooled by the Comintern, looked at the Soviet model as the recipe for unmitigated industrial growth.

In reality, the plan largely collapsed upon Stalin's death. When Nikita Khrushchev took over power, he grew critical of the experience of collectivization. In 1953 he noted how the results had been unsatisfactory and that the efficiency of farms had only marginally improved. But Stalin's legacy was such that, despite this realization, the direction of travel was set. Under Khrushchev, and Brezhnev after him, virtually all land that could be put to use was plowed up, as farming transitioned from collectivization to the modern state-owned Agrarian Industrial Complex.

In the next thirty years, some thirty million hectares of land would be either reclaimed or put under irrigation. Farms were so inefficient that while the amount of water delivered increased, production did not. Prey to an unbreakable belief in the power of planning modernity, the Soviet leaders slowly drove the country towards implosion.

LILIENTHAL TRAVELS

The rise of a geographically intensive Soviet economic model in the heart of Asia was of great concern to American policy makers. In 1951, David Lilienthal went on a trip to India and Pakistan on behalf of *Collier's* magazine. He was operating in a private capacity but was not an ordinary tourist. After leaving the Tennessee Valley Authority, he had been the head of the Atomic Energy Commission, and, although he was now running his own consulting practice, he was still politically well connected. During his 1951 trip, Lilienthal was particularly concerned about the dispute over Jammu and Kashmir.

At partition, the princely state had been under a maharaja, Hari Singh. Singh was a Hindu, but his prime minister, Sheik Abdullah, and most of the population were Muslim. At the time of independence and partition, the various princely states of the subcontinent had to decide whether to join with one of the two countries or remain independent. Singh preferred independence, imagining himself governing a Switzerland of the Himalayas. But in the end, he joined India, fearing an uncontrollable revolt in Poonch, a majority-Muslim district west of Srinagar. When, in response, Muslim frontier tribesmen raided Kashmir, Singh asked for help from India's prime minister, Jawaharlal Nehru, who airlifted troops into the state. In turn, regular Pakistani army soldiers poured in. Fighting began. Eventually, a truce was called thanks to the intervention of the UN Security Council, and a cease-fire line was drawn between the two armies.

In 1951, the water frontier of regional politics had moved to the newly created control line between Pakistan and India. Lilienthal realized that the dispute was not just between the two countries. The problem also involved China, Tibet, and, most important, the Soviet Union. Lilienthal described Kashmir as "Communism's northern gateway to the great strategic materials and man power of the Indo-Pakistani subcontinent, and to the Indian Ocean." It was the frontier of the Cold War.

For the United States, India was a new opportunity to influence the region, one it had missed with China at the time of Savage. The conflict in Korea had just started and loomed large on the international stage. Just like China, India had a huge, impoverished population. The risk of an Indo-Pakistani conflict was grave. Had it been turned into a religious war, it would have brought the Muslim world to Pakistan's aid, pushing India to seek allies herself. The Soviet Union and China might take advantage by offering to help, in the hopes of spreading communism to the subcontinent. The opportunity to draw India into the U.S. sphere of influence, and out of that of the Soviet Union and of communism, seemed too important to miss.

Lilienthal realized that the sharing of water for irrigation was a dominant source of tension between the two countries. He wrote: "No army, with bombs and shellfire, could devastate a land as thoroughly as Pakistan could be devastated by the simple act of India's permanently shutting off the sources of water that keep the fields and the people of Pakistan alive." Kashmir was a very live issue at the UN, yet none of the debates seemed to reflect the fact that two-thirds of all the water flowing down the Indus originated there. In a meeting during Lilienthal's trip, Pakistani prime minister Liaquat Ali Khan told him that Kashmir was "like a cap on the head of Pakistan. If I allow India to have this cap on our head, then I am always at the mercy of India." That was the mindset that framed the conflict.

Lilienthal proposed a solution. He believed that only a fifth of the water of the Indus was being used for irrigation. The rest was "wasted" and flowed into the Arabian Sea. If this water could be put to use instead, most needs of both countries could be met. For Lilienthal, Pakistan's present use of water could be recognized in exchange for the country working with India on a joint use of the international river. The problem was how to store water that was now being lost at sea. He proposed an Indus Engineering Corporation, "jointly financed (perhaps with World Bank help)," which would develop dams, wherever optimal, in the service of both countries. The plan, "not political, but functional," could be financed because once the infrastructure was built, the land would have become productive, making it possible to repay any loan. It was the Tennessee Valley Authority all over again.

EMPIRE FALLING

While the complexity of the water landscape shared by Pakistan and India was setting up a conflict that would run hot and cold for decades, the pursuit of investments in water infrastructure accelerated everywhere. Even Britain engaged in it, in the hopes of saving its crumbling empire. Africa was its last hope.

Africa is a vast, ancient continent. Plateaus and high plains extend for thousands of kilometers before dropping precipitously close to the coast. Comparatively speaking, it has fewer mountain ranges than other continents: the Atlas Mountains at the very northwest, the Ethiopian Highlands, the Kenyan Highlands and the Albertine Rift Mountains, the Angolan Plateau. These are the "water towers" of Africa. Many long rivers start where rain and snow intercept those mountains, but then flow through vast arid or semi-arid environments.

Colonialism had broken the continent up into separate dominions, protectorates, and colonies, completely independent of the underlying geography. The nation-states of the post-colonial era inherited these political boundaries. This set of borders meant that an unusual number of landlocked countries, many dry or arid, had to share long rivers running through them. The challenge became even more evident as expectations of industrialization reached the continent.

After the Second World War, the British knew their time might be up. African nationalism, represented by the African National Congress on the one hand and the segregationist white Afrikaner settlers of South Africa on the other, were tearing the empire apart. Under increasing pressure from the United States and the Soviet Union, who were eager to extend their sphere of influence on the continent, the British embarked on a last-ditch attempt at revitalizing their empire. Investment in water infrastructure was at its heart.

In 1953, the British combined three colonies—Northern Rhodesia (Zambia), Southern Rhodesia (Zimbabwe), and Nyasaland (Malawi)—to form the Central African Federation. The federation ostensibly stood for "multiracialism" and partnership, but in truth it was a desperate and ultimately failed experiment to avoid losing valuable land and mineral resources. The plan was to give power and autonomy to a white minority that over time, it was alleged, would increase opportunities for Africans. For a while, the federation had some economic success: investments increased. Its focus was the Zambezi River.

Not quite as large as the Congo, the Zambezi was the biggest river in the southern part of Africa. It flowed right through the federation, so inevitably it was the main target for development. The river flows down three big steps. From the head to Victoria Falls on the border of modern-day Zambia and Zimbabwe, water flows through flat, spongy terrain, where it gently expands. This part of the river is dominated by the dambos, waterlogged anaerobic soils where no trees can grow. It is a source of perennial water covered in tall grass. At Victoria Falls, Mosi-a-tunya or "The Smoke That Thunders," the river plunges down rapidly towards the last gorge of Cahora Basa, where the river finally enters the Mozambique coastal plain and heads to sea.

The British realized that taking advantage of this huge power potential could supply the electricity-hungry northern copper belt of Zambia, whose only alternative up to that point had been to bring in coal by rail for thermal production. Long-distance transmission through 330-kilovolt lines, which had now become widely available, had unlocked the opportunity: large hydropower on the Zambezi could transmit energy over two thousand kilometers, reaching both the northern Copperbelt and Harare in the south. That is how Kariba Dam, between Victoria Falls and Cahora Basa, came to be.

Kariba was built between 1955 and 1959 as a double curvature concrete arch dam with installed capacity of just under one and a half billion watts. It was expected to triple the power capacity of the federation, the largest development project undertaken in the fifties anywhere in the world. It was a daring bet on state-led development, the engine that would provide electricity to industry both north and south, financed by private donors and the World Bank, and built by Italian and French experts.

The dam was inaugurated on May 17, 1960, ahead of time and under budget. It was seen as a great success, the height of the postwar modernist project. The architects of the federation may have fooled themselves into thinking they had managed to save the empire. But the dam was as big as the challenges it faced. The reservoir was gigantic. It is still the largest in the world, with storage of 180 billion cubic meters. It is right at the southern end of the African Rift, a tectonically active area, which meant that the huge weight of water could induce superficial earthquakes.

Kariba was originally planned to accommodate what the designers thought would be a one-in-ten-thousand-year flood event. But data on the river was scarce and unreliable. In 1957, during construction, the biggest flood on record, with a peak of over eight thousand cubic meters per second, shattered all expectations. The spillways had to be hastily increased. Another unexpected flood the following year led to further augmentation.

Lake Kariba transformed the arid landscape into an immense shoreline. The change appealed to the aesthetics of the white settlers, but displaced fifty-seven thousand Tonga people and completely destroyed all ecological processes in an area of 5,500 square kilometers. It was the fastest and most terminal destruction of ecology in one action in human history to that point. Kariba was an attempt to stitch together a vast landscape through an interconnected energy system, in the belief it would automatically power economic growth and provide stability. It did not. Calls for independence were brewing across the federation.

The story of water in the postwar world was one of investment. The development of large infrastructure was caught in the conflict of the Cold War. For all the fundamental competition between economic systems, their common nineteenth-century roots ensured that they were both committed to a water-led model of development. Both sought to harness the environment for the benefit of the nation. But the underlying competition between political systems meant that water resources development was not a neutral instrument in the hands of the state. Rather, through water infrastructure the state revealed its underlying architecture. Power resided where decisions about the landscape were made.

18

The Great Acceleration

FINDING SOLUTIONS

At the peak of the Cold War, the United States embraced its role as "hydrologist in chief" for the world. The U.S. Bureau of Reclamation had professionalized and expanded its international engagements. Already by 1953, it had over a hundred employees advising over twenty different countries. The United States' principal instrument were its technical agencies, whose experience had been forged on the complex hydrology of the American continent. The Indian subcontinent was the most prominent recipient of this assistance.

On his return from Pakistan and India, Lilienthal published an article in *Collier's* magazine. If anyone ever had any doubts that ideas can change the course of history, this one did. The head of the World Bank, Eugene Black, read it and decided to take up his proposal. The World Bank had found itself at an impasse due to partition. Pakistan and India both had a seat on its board. The bank had tried to provide loans to India to build the Bhakra Dam on the Sutlej, but Pakistan had objected. Then, India had objected to a loan to Pakistan for the Kotri barrage on the Indus. Anything that could be done to overcome the conflict would have been welcome. Black sent a letter to both prime ministers, offering the good offices of the World Bank, its technical advice, and its financing, directly referring to Lilienthal's proposal: that the problem could be solved on "an engineering basis."

The proposition came at the right time to break the deadlock. On those terms, both India and Pakistan accepted and the negotiation began. The bank's engineering advisor was General Raymond A. Wheeler, a retired lieutenant general of the Army Corps of Engineers. And in fact, the United States and its geography were the model solution. When describing the Indus, the negotiator for India, Niranjan Gulhati, described it as comparable to the Columbia. Wheeler's first attempt was to create an integrated plan along the lines of what Lilienthal had in mind. It failed. The negotiations had to diverge from the unitarian vision Lilienthal had proposed if a compromise was to be reached.

In 1954 the idea emerged of dividing the rivers into three tributaries each. A transitory period would accommodate the construction of canals to supply water to Pakistan from the western rivers, replacing the loss from the eastern ones. It was a partition of rivers rather than water, a call to make nature conform to political boundaries, rather than the other way around. The deal was agreed to in principle. A long negotiation followed to find a compromise on the details.

The issues of substance were three. First, Pakistan did not believe that the replacement of water conveyance could be done simply by replumbing the system, but that some amount of storage would be needed. This was a significant transformation of the system. India had reservations in part because it would have to shoulder some of the financing. Second, India also wanted the recognition that irrigation was going to be possible in Jammu and Kashmir, which it could do by diverting some of the western rivers. This was also agreed to in principle, although the amount of water involved was not specified. Finally, India wanted to be able to develop hydropower on all the rivers that went through its territory. However, Pakistan did not want to allow India to impound what it saw as its own rivers. A compromise was reached: only run-of-the-river hydroelectric plants—installations that did not store large amounts of water—could happen on the western rivers. No significant storage upstream of Pakistan's rivers would be permitted.

The proposal gave the Indus proper, the Jhelum, and the Chenab to Pakistan, while the Ravi, the Beas, and the Sutlej would go to India. Financing was secured to build the replacement works on the Indus and the Jhelum, about a billion dollars in 1960, to which India contributed about a fifth. The World Bank organized the Aid-to-Pakistan Consortium to bring together the donors and institutions that would contribute to the fund. The issue of additional irrigation and hydropower on the Indian side of the Pakistani tributaries remained tricky. For this reason, the agreement had also adopted a mechanism for dispute resolution. At first a panel of Indian and Pakistani expert commissioners would attempt to resolve a dispute. If that failed, the countries would have recourse to a neutral expert appointed by the World Bank or an International Court of Arbitration, depending on the nature of the disagreement.

In 1960, after protracted negotiations, the Indus Waters Treaty was signed by Field Marshal Ayub Khan and Prime Minister Jawaharlal Nehru. In many ways—as imperfect as it was—it was a monument to cooperation. Remarkably, the treaty withstood the test of time, surviving the conflict of 1965, the independence of Bangladesh in 1971, and the Kargil conflict of 1999.

If there was ever any doubt that international treaties could be anything more than words on paper, the Indus Waters Treaty proved it. Its physical impact was substantial. When the process started, the entire canal system, up until the development of the Bhakra Dam, could only rely on storage provided by snow and glaciers in the Himalayas. After the treaty, two dams—Mangla and Tarbela—nine inter-river-linking canals, and three additional barrages went into place, engineering the river to serve two independent countries.

The Indus irrigation system was and remains an astounding human achievement: it is still the largest contiguous irrigation system on the planet today. It commands an area of over fourteen million hectares, with storage of about twenty-two billion cubic meters, sixteen barrages, twelve inter-river-linking canals, two siphons, and forty-three main canals, for a total canal length of about fifty-seven thousand kilometers, a length that exceeds the circumference of the Earth at the equator. Infrastructure, indeed.

THE CONQUERING LION

Infrastructure development also continued apace in Africa and proved to be, once again, first and foremost a political instrument in service of

American foreign policy. Its first adopter was one of the few continuously independent countries on the continent: Ethiopia. Roosevelt had first developed a relationship with Haile Selassie during the latter's wartime exile. He offered American assistance to the development of the country, in part also because he was seeking the use of Addis Ababa as an air base in Africa for the expansion of American air transport businesses.

Ethiopia's relationship with America deepened under Truman, testament to the perceived strategic importance of the Horn of Africa. Consistent with his Point Four strategy, Truman offered the Ethiopian sovereign technical assistance to develop the country's agricultural sector. In 1954, Haile Selassie finally visited the United States. The tour included visiting some of the famous infrastructure projects that had been the hallmark of U.S. development. The Columbia River and the Grand Coulee Dam captured Haile Selassie's imagination, as did the evergreen Tennessee Valley Authority.

The tour was on the back of a 1953 investigation, conducted by Bureau of Reclamation experts, of the water resources of Ethiopia, which had focused particularly on the Blue Nile and on the Awash River. The Awash, one of the great rivers of Ethiopia, is a large drainage basin, covering parts of the states of Amhara, Oromia, and Somali. It starts in the high plateau near Addis Ababa and flows into Afar, ending into the salty Lake Abbe, near the border with Djibouti. The Awash had among the most productive lands in Ethiopia, and relatively easy access to market via Addis Ababa, making it the most obvious place to attempt a regional economic development intervention. The Italians had already begun hydropower development on the Akaki River, a tributary of the Awash, during their occupation of Ethiopia. After the war, they had been forced to pay reparations as part of the 1947 Paris peace treaties, and did so by contributing to the construction of the Koka hydroelectric dam on the Awash.

The development that the Italians were pursuing was distinctly different from that of the Americans. The Italians, and the Ethiopians with them, considered these as individual hydroelectric projects. The American Bureau of Reclamation staff, on the other hand, brought a macro-economic lens to their work: their focus was on the development of the entire basin. They could not but observe the substantial irrigation potential that went hand in hand with the hydroelectric one.

The purpose of the Awash development was not simply to produce power, but to allocate water resources to their most economically productive use.

In 1962, the Awash Valley Authority (AVA) was established on the model of the TVA. The idea was full regional development built around the water assets of the valley, including hydropower development, commercial irrigation schemes, and other ancillary industry. The presence of a railway line to Djibouti, where goods could go to market globally, made the choice of the Awash particularly attractive. But importing a model wholesale from America without its institutional context was a recipe for trouble.

The development of the Awash involved the settlement of the Afar people, herders who lived in the Red Sea territory between Eritrea, Ethiopia, and Djibouti. Their lands, which had been grazed for centuries, were now needed by the central government. The Afar were semi-nomadic people, so their way of life was essentially incompatible with the form of development piloted by the Awash efforts. Their nomadism conflicted with the concession of specific parts of their territory for agricultural production. Their grazing land was diminished, making them more vulnerable to drought. The distribution of benefits that were reaped from the concessions was also very unequal and, not surprisingly, the Afar were not the beneficiaries. As soon as the first drought hit, famines followed.

The problem with the authority was that it imposed a modernization model on an existing complex historical background of tensions and power structures that actively worked against its objectives of coherence and coordination. The experiment of the Awash Valley was an egregious example of how the experience of the United States could be misunderstood. For all its faults, the TVA was an institution that had as its ultimate target the residents of the Tennessee Valley. In contrast, the AVA was designed to deliver infrastructure and commercial agriculture to increase exports for the country, and its client was central government, not the local population of the Awash Valley. What had started as an imitation of an American experience quickly turned into a symbol of totalitarian rule.

WATER INDEPENDENCE

America reached the pinnacle of its expertise and experience with water infrastructure at the height of the Cold War in Africa. That is also when the newly independent nations of the African continent, finally freed from the shackles of colonialism, were trying to find models of development that would put them on a path to prosperity. Since the beginning of the century, nationalist movements had been growing across the African diaspora. Writers like W. E. B. DuBois and political leaders like Kwame Nkrumah, Jomo Kenyatta, and Julius Nyerere were getting ready to bring change. They would go on to shape the destiny of the continent after the Second World War.

By the time Kariba Dam was being built, the British were losing control of their far-flung territories. With India gone in 1947, the size of the empire was halved. After the war the domino pieces quickly fell one after the other: Ghana went first, in 1957; Nigeria in 1960; Sierra Leone and Tanganyika in 1961; Uganda in 1962; Kenya in 1963. Eventually, in 1963, the Central African Federation was dissolved. By 1964, both Zambia and Malawi were independent. On November 11, 1965, the white nationalists in Southern Rhodesia seized power in an act of unilateral independence. The British Empire in Africa was gone. And now a group of newly independent countries had to answer the question of how to deliver a better future to their people.

Now shared across borders, Kariba was a source of both tension and cooperation between Zambia and Rhodesia first, and Zimbabwe later. The British had planned the project's first power station on the south bank of the river, to favor the industrialization of Southern Rhodesia. When Rhodesia announced a unilateral declaration of independence, Zambia found itself in a difficult position. Its copper economy depended entirely on electricity from Kariba. Energy demand was growing in the postwar boom, so Zambia chose to develop another dam on the Kafue River, the largest tributary of the Zambezi, rather than go ahead with what was originally planned to be the northern powerhouse at Kariba (the latter would only be completed much later, in 1977).

The struggle to share the substantial resources of the Zambezi was at the heart of the relationship between the two countries. Power demand decreased for both Zambia and Rhodesia in the second half of the 1970s as copper prices fell, and Rhodesia entered the war that would culminate with the creation of Zimbabwe in 1980. By then Zambia was supplying energy to its southern neighbor, rather than the other way around, to the point that electricity was its second-largest export after minerals. The point is that the difficulty of the high modernist project based on water infrastructure was that its success was entirely dependent on a specific institutional context. The same technology with a different context could easily turn success into failure.

But while the dominant water planning approach in the world was that of America, most independence-minded, anti-imperialist movements in Africa were inspired by socialism. The pivot had happened at the seventh African Congress in Manchester, in October 1945, when the Pan-African movement, which up until the Second World War had been primarily a movement of elites and intellectuals, fully embraced socialism. The merging of nationalism and Marxist socialist ideology had introduced Marxism-Leninism across Africa. Many revolutionaries, like Ghana's Kwame Nkrumah, embraced the planned economy model with relish. Kariba Dam—the technology rather than the institutional arrangement—provoked a continent-wide switch to hydropower. On the back of the experience of both the United States and the Soviet Union, the independence movement committed to the technology.

In his 1961 book *I Speak of Freedom*, Nkrumah, the first prime minister and president of Ghana, wrote: "Although most Africans are poor, our continent is potentially extremely rich. [. . .] As for power, which is an important factor in any economic development, Africa contains over 40% of the potential water power of the world, as compared with about 10% in Europe and 13% in North America. Yet so far, less than 1% has been developed. This is one of the reasons why we have in Africa the paradox of poverty in the midst of plenty, and scarcity in the midst of abundance." Once again, the power of rivers was going to be the route to prosperity.

Nkrumah brought Arthur Lewis to the country as his pre-eminent economic advisor. As for many other leaders on the continent, the centerpiece of Nkrumah's development plans was a huge dam, the Akosombo on the Volta River. Lewis and Nkrumah fought and argued over the Volta project from the very early stages. While both believed the project had potential, Nkrumah saw the political value of a flagship project. Lewis on the other hand—in line with the economic development model he himself had been advocating—thought that cheap electricity without investments in import-substituting industries that could employ a lot of people would only lead to capital-intensive industrialization that favored foreign capital. In the end, as often happens, it was the narrow political calculation that won, and Lewis left the country.

The development of Inga Falls on the Congo, also under discussion at the time, could similarly have been the legacy of Patrice Lumumba, the first prime minister of independent Congo, had he survived Mobutu Sese Seko's coup. The Inga Falls had been identified at the beginning of the century in an American survey of commercial hydropower as a site of enormous potential. The falls are on a 30-kilometer-long descending curve, where the channel follows two sides of a triangle. The river narrows, falling about 270 meters over 350 kilometers and taking a sharp turn to the right as it drops the last 90 meters or so. The uniqueness of this site is that the two sides of the bend could be connected and exploited to produce an enormous amount of hydropower without requiring a huge reservoir, because the flow of the Congo is large and stable. The potential of this site was enormous: approximately twenty times that of Hoover Dam, the largest on the planet. It was concentrated close to the mouth of the river, in convenient proximity to manufacturing and a port.

Once again, an American speculator, Edgar Detwiler, managed to convince Lumumba to sign an agreement for the development of the "Grand Inga." But Lumumba backed out after pressure from other African countries and ultimately died in Mobutu's coup. Cahora Basa Dam in Mozambique followed. The mid-sixties and early seventies were the hydropower awakening of the continent. But it was happening late in the game. Rich countries had already soured on the technology, and a shift to oil was taking over the energy sector.

A WATERY DESERT

The United States' undisputed international leadership in water resource planning found its highest realization on the Pakistani Indus. In 1961, Pakistan's President Ayub Khan visited U.S. president John F.

Kennedy in Washington. The White House assumed that Ayub Khan wanted Kennedy to continue Eisenhower's policy of military assistance, predicated on the idea that Pakistan constituted a line of defense against the Soviet Union. But Kennedy had appointed John Kenneth Galbraith as U.S. ambassador to India, and Galbraith, who was close to Nehru, knew that military support to Pakistan would inevitably lead to India wanting the same. This would have fueled an arms race, escalating the conflict over Kashmir. It was a time bomb in the heart of communist Asia. It had to be avoided. If arms could not be given, what else could be provided?

For counsel, Kennedy called up Jerry Wiesner, the chair of his Science Advisory Committee. Wiesner was an MIT electrical engineer. As it happened, 1961 was the centennial of the founding of MIT and Abdus Salam, the Pakistani physicist, was due to give a talk at the event. Salam also happened to be the science advisor to Pakistan's government. Wiesner was chairing a discussion on science for developing countries and Salam was in attendance.

During one of the presentations, Salam debated vigorously with Patrick Blackett, an English physicist who had been Salam's boss at Imperial and was a close advisor to Jawaharlal Nehru. Blackett was making the case that all the technology required by developing countries was available in the "world supermarket of science." They just needed to find it and get it. Salam objected forcefully. He believed that one had to be able to read the labels on said "supermarket of science products." To make his point, he gave the example of the Indus in Pakistan. The basin had lost close to half a million hectares of land to salinization and waterlogging. This problem was not about technology alone. It required skill, understanding, and—most important—close study.

Wiesner saw his opportunity. He spoke to Salam immediately after the session. Maybe assistance could be offered to Ayub Khan and the government of Pakistan on this very topic. Salam agreed. That evening Wiesner called Kennedy back and suggested forming a team under the leadership of Roger Revelle, a scientist who had been involved in a number of government projects. The next day, Revelle, Wiesner, and Salam met in Washington to formulate a plan.

Revelle assembled a team of about twenty people, including Harold A. Thomas Jr. from the Division of Applied Science at Harvard.

Thomas was the director of Harvard's water program, which had recently applied systems analysis and linear programming to water resource problems. Computational methods would prove essential to solving the issue. The task force went to Pakistan several times, visiting Punjab and Sindh.

Revelle recounted flying over that landscape and seeing the devastation: "Salt lying like snow on the surface and villages of the canal colony grid, here and there derelict, like holes in a punched card." The area subject to severe waterlogging and salinity was probably around six and half million acres in West Punjab, increasing at a rate of fifty thousand to a hundred thousand acres per year. In upper Sindh, at least one million acres, out of the four million acres under irrigation command, were heavily salinized. In the lower Sindh, another million acres were too salty to be cultivated.

The problem was clear enough. Before irrigation was introduced, the water table reached the surface only near the rivers. The British introduction of large-scale irrigation in the plains of the Indus had caused seepage from the earthen irrigation channels. As canals spread out the water, some 40 percent of it would go back into the ground. Because the Indus was so flat, horizontal drainage was extremely slow, and so water accumulated belowground. Irrigation had become a new source of groundwater. A hundred years of it had filled the aquifer. The water was reaching the surface from below. Pakistan was turning into a watery desert.

The solution was also relatively simple. What was needed was vertical drainage: if water could not move laterally downhill, it needed to be pumped out and put back in the surface system. Tube wells for that purpose were already in use in the Punjab, but Thomas's computations showed that those tube wells drained too small an area. The speed at which the water flowed back in from the sides was a function of the perimeter around the area that was being pumped. On the other hand, the rate of evaporation, the speed at which the surface water was lost to the atmosphere, was a function of the area itself. As the ratio of the area to the perimeter increased with increasing area, evaporation would win out over lateral flow and the water table would lower. The team proposed dividing the landscape into million-acre parcels subjected to much greater pumping. It worked.

The program expanded in the sixties, sinking thousands of deep tube wells to lower the water table and manage the problem of waterlogging and salinization. Ultimately there were about fifteen thousand tube wells that went down anywhere between 40 and 120 meters in depth. In fact, the program worked too well. Eventually, the proliferation of state-owned tube wells was followed by the growth in private tube wells in all of these areas, especially in the Punjab. By the late sixties the number of private tube wells exceeded the state-owned ones. This inevitably increased the role of groundwater in irrigation. By 1996, Pakistan had three hundred thousand tube wells, most of which were in the Punjab, supplying about 40 percent of the irrigation water. By 2006 that number had grown to 60 percent.

The task force made another fundamental contribution. Thomas and Revelle realized that the waterlogging and salinity problems were a symptom of a much broader issue: low agricultural production. In stopping at a farm, Charlie Bower, a member of the team who worked for the U.S. Salinity Laboratory of the Department of Agriculture, a legacy of the U.S.'s expansion into the arid west, picked up a leaf from a stalk of maize. He studied it for several minutes. After he was done, he looked up and announced that the leaf showed signs of nitrogen deficiency. The problem was not just salt. It wasn't just a matter of efficiency. Farmers were not using water effectively. Water was applied poorly. Modern fertilization and irrigation practices were unknown. The problem was not technical but socioeconomic: Pakistan had to increase the productivity of the basin. In other words, the problem wasn't too much water. It was not enough water, and—more specifically—not enough agricultural production. Saving the Indus in Pakistan required using far more water to grow food. Indeed, the Green Revolution that followed transformed the Punjab.

ENGINEERS FOR A NEW EMPIRE

These examples show the extent to which American expertise shaped the postwar landscape across the world. In particular, the Bureau of Reclamation acted as the international technical agency of the U.S. State Department, assisting countries looking to develop their water resources. It assisted over a hundred countries. Some of its engagements were famous, or infamous depending on the particular point of view. Besides the Blue Nile, the bureau provided plans for the Helmand Valley in Afghanistan all through the sixties; the Han River basin in Korea in the second half of that decade; the São Francisco River basin in Brazil between 1964 and 1973; and the Mekong River basin and the Pa Mong Project in Thailand and Laos in the decade between 1964 and 1974. It was a remarkable range.

As the United States transitioned from opposing British imperialism to containing the Soviet Union's sphere of influence, it deployed its technical assistance as an instrument of diplomacy. In the absence of a single authority that could arbitrate or an international legal framework within which disputes could be settled, the competition on rivers was left to technical arguments and power relations. In the early stages of those engagements, the objective of Reclamation Bureau experts was to convince their counterparts that rivers were not just spaces for enterprise, as the imperialist British had conceived them, but national economic assets.

Hoover Dam and the Tennessee Valley Authority were the model development package: the first defined the multipurpose project; the second was the archetype of river basin development. Together they constituted a platform for development. It was a progressive, technocratic approach, the dream of any nineteenth-century utopian. It spread. Or, rather, it was spread. American expertise became an instrument of influence, a technical assistance strategy first pursued by Roosevelt, then formalized by Truman, finally pushed during the Kennedy and Johnson eras.

The technical nature of the knowledge deployed by the bureau at home and abroad obfuscated a crucial difference between the two. The federal government had led the first push to develop domestic water resources in the western United States, but it had done so in the context of a republican project, of a political system that was—imperfectly—attempting to deliver benefits to a commonwealth. In contrast, the international engagements of American experts were more geared towards creating opportunities for American contractors. The bureau experts might develop the blueprints for large river basins, but it was expected that on their heels private companies would develop the projects.

During the Eisenhower administration, the philosophy of development began to shift. Within the U.S. the mainstream had turned against public investment and New Deal–style interventions, veering more towards free enterprise. As this neo-imperialism developed, the moral authority that the American experience once commanded dissipated. By the mid-sixties the problem of water development on international waters transitioned to international institutions. This reflected the growing complexity—and the growing stakes—of dealing with sharing waters in an increasingly polarized world. This had already happened on the Indus with the intercession of the World Bank. But the influence of the United States, as the principal sponsor of the international order and of its institutions, was never far behind.

In the Mekong River, initially, the United States tried to bring the lower Mekong states—Thailand, and the newly independent Laos, Vietnam, and Cambodia—into its sphere of influence. The bureau engaged in its most intensive overseas effort in the Mekong. In particular, its experts developed plans for the Pa Mong dam, an enormous multipurpose installation on the main stem of the river, a few kilometers north of the capital of Laos. Eventually, the United Nations began getting involved through the Economic Committee for Asia and the Far East, which led to the intergovernmental Mekong Committee in 1957.

The long tail of American influence continued in this new era. The committee had been the result of recommendations by the retired General Wheeler, former chief of engineers of the Army Corps and consultant to the World Bank. American experts, led by the geographer Gilbert F. White, also developed a further set of recommendations in 1961, while on a mission funded by the Ford Foundation. Lyndon Johnson hoped the plan could bring peace to the region, and even sent David Lilienthal to consult. By 1965, the U.S. Army Corps of Engineers and the TVA had provided crucial technical consulting and had mapped the economic and social resources of the basin, in preparation for the implementation of a regional plan. But U.S. domestic opposition to involvement in the region stopped it in its tracks.

No sooner had the development of water resources infrastructure adopted the traits of an imperial project than its peak had passed.

19

The End of an Era

CONQUERING THE ENDLESS CHANGJIANG

The story of water in the twentieth century had been a political one: the tensions caused by a growing, increasingly enfranchised, consuming population had produced a transformation of the landscape. That transformation reflected attempts to develop political systems that would satisfy the needs of modernity through the instruments of territorial sovereignty. The differences in constitutional architecture determined how that sovereignty would be exercised.

Water-led development had been the hallmark of the United States' rise to becoming the model republic and a global hegemon. The sophisticated system of landscape development that had accompanied the ongoing negotiation between individual political power and collective action had been far from perfect. Alas, republics seldom have lived up to their promise. The full enfranchisement of their citizens is often the rate-limiting factor. In the case of the American project, the republic fell short of its ideal in its dealings with a number of constituencies on the landscape, from the Native American tribes it had displaced during its period of geographic expansion, to the urban and rural poor it had systematically excluded from the broad benefits of development. But despite its profound limitations, it had delivered security for many and a powerful platform for industrialization and wealth creation. The comparison with the performance of communist countries ended up confirming the intuitions of Wittfogel: in the end water infrastructure was a tangible realization of a political architecture. China proved to be one of its most important confirmations.

In May 1956, Mao Zedong, chairman of the Communist Party, famously swam across the Yangtze River, about four hundred kilometers downstream of where today Three Gorges Dam is. He chose the crossing from Wuchang to Hankou. He did it again in June the other way around, from Hankou to Wuchang. This became a tradition with profound symbolism for communist China. That June, Mao looked upstream. He wrote a poem titled "Swimming." He wrote that "A rocky dam shall stand athwart the western river, cutting asunder the mists and rains of Wushan, until the precipitous gorges yield a lake of tranquillity."

Water, and the infrastructure needed to manage it, was going to be the protagonist of China's evolution in the second half of the century. While Mao's swim across the river would become part of the mythology for the successful creation of the largest piece of river infrastructure on the planet, the Three Gorges Dam, it is easy to forget that water first tragically took center stage during one of the darkest periods of China's recent past.

Mao launched the Great Leap Forward in May 1958. Its origin lay in China's difficult relationship with its neighbor to the west. In 1956, Khrushchev had embarked on a process of revisionism and critique of the Stalinist experience. Mao profoundly disagreed with Khrushchev's direction and saw China as the true heir of Marxism-Leninism and of the Stalinist project. He determined to prove it.

At the celebrations for the fortieth anniversary of the October Revolution, Nikita Khrushchev announced that the Soviet Union would surpass America in fifteen years on economic indicators like steel or iron production. Mao knew full well that China was far behind the Soviet Union, let alone the Western economies, in terms of industrialization, but he nonetheless saw the opportunity for one-upmanship. He would show that China was also in a race with the West, and consequently in a race with the Soviet Union. After all, America was the ultimate enemy: the Korean War had pitted the U.S. against China, and America's support for Taiwan and the trade embargo against China were all evidence of its antagonistic approach. Mao could not put himself up against the United States, and so chose to define himself in competition with the second-largest country on the capitalist side: Great Britain. He announced that China would surpass Britain in fifteen years.

The Great Leap Forward, as this push became known, would restore China's greatness on the world stage, badly damaged since the opium wars the British had inflicted in the previous century, and the failures of the republican revolution and of the Nationalist governments that followed. And just as for the Soviet Union, the most important economic indicator in communist eyes was the production target. China should overtake Britain in steel production. Mao's plan to industrialize China followed the Soviet recipe: start with heavy industry. By that point, China was already industrializing. Its first five-year economic plan between 1953 and 1957 had already promoted heavy industry, aided by Soviet investments. But up until the decision to launch the Great Leap Forward, production targets for steel had remained relatively small. In 1957 China had produced just over five million tons. Mao called for a new production target of thirty million tons by 1962.

Mao faced a problem. Where would the money come from to support all the necessary investment in equipment, labor, industrial installations, and energy that would produce all this steel? The Soviets had provided some financial assistance, but they were in no position to finance the industrialization of all of China. After the 1956 nationalizations, there was no private sector in the country that could access capital markets. Foreign trade was also very limited. The only option was the one China had always turned to in times of need: agriculture.

If agricultural production could grow, its taxation could provide the source of income they needed. Of course, the problem was that at this point agriculture in China was mostly subsistence agriculture. It was hardly a sector able to produce the surplus required to support industrialization. Grain requisition targets, which were already high and tended to be set independently of what peasants needed to survive, were set at absurd numbers. The slogan of the Great Leap Forward was "more, faster, better, cheaper." The target was set for around four hundred million tons of grain per year. The tragic failures of the Great Leap Forward descended from this absurd expectation.

The American experience, alongside that of India and many other countries, had seemingly shown that the most important way of rapidly and decisively increasing the productivity of agriculture was to invest in irrigation infrastructure. Water infrastructure was the answer. Lots of it. The commitment to water projects was so substantial that it became the primary drain of labor from the farms. Between 1958 and 1959 an estimated hundred million peasants were assigned to dig canals and other irrigation projects. The drain of labor from the field meant that when harvest came there was no one to collect it and it remained to rot. By early 1958 one in six people were digging to transform the landscape of the nation. Six hundred million cubic meters of rocks and soil were moved during that year.

The human cost of these efforts was enormous. One estimate suggested that for every fifty thousand hectares under irrigation, a hundred lives were lost. Despite growing evidence that the Great Leap was not working, Mao did not relent. Criticism simply entrenched him further. The implementation through the provincial leadership varied, but the zeal of many of those wanting to please Mao, or scared of retribution, ensured that whatever negative impacts the projects caused, they would be absorbed by the local population while limiting the amount of bad news that would make it back to the center. Production dropped rather than increased: down 15 percent in 1959, and a similar amount in the following two years. For a system that was already so vulnerable, this was catastrophic. Famine and starvation followed.

The Great Leap Forward was an unmitigated disaster, resulting in the death of an estimated twenty to thirty-five million people. Not all of it was strictly Mao's doing. Because of the poor state of agricultural infrastructure and practices, weather—floods and droughts—could have a disproportionate effect on the stability of the economy. In 1960 and 1961 at least some of the drop in harvest could be attributed to weather, but this does not mean that human agency did not matter, far from it. It is unlikely that Mao fully realized the scope of the devastation as it was happening, and indeed once the scale of the disaster became clear, the Great Leap Forward finally came to an end. Yet a colossal weight rested on the leader's legacy.

Mao's failed efforts were simply the most recent evidence that infrastructure-led growth could not really live up to its promise, at least not without a commensurate effort in also addressing the underlying political architecture: water infrastructure was not a technology, it was an expression of power. Failure to understand this had significant consequences.

China's failure to catch the train of development ended up excluding it from participating in regional processes for cooperation, organized around the last gasp of the modernist vision for water. By the end of the sixties, conditions in Cambodia, Laos, and Vietnam had become impossible for U.S. personnel. When Richard Nixon was elected the U.S. president, he pushed for withdrawal. The idea of an integrated basin development receded on the Mekong, and leadership definitively passed to the World Bank and UN agencies. A Mekong River Commission exists to this day, but while it pursues technical cooperation among the lower riparian states, it no longer envisions the kind of integrated infrastructure-based plan that was at the heart of those first efforts. And now, of course, conditions have dramatically changed. For no one could have predicted at the time that the geopolitics of the region could change quite as dramatically as they did, when China finally hit its stride.

ETHIOPIA FAMINE

The initial enthusiasm for water-led development began to deteriorate also among its most recent adopters. Ethiopia had been the first African country to embrace water modernization. Even more than the Awash, the target of American development in Ethiopia had been the Blue Nile. Partly this was because it gave the United States leverage over downstream Egypt, which had proved a nexus of Cold War tensions. Egypt had begun contemplating the construction of the Aswan High Dam in the fifties and had failed to consult Ethiopia, which irritated Selassie's government to no end. After the Egyptian revolution, Colonel Nasser played American and Soviet interests against each other in the hopes of securing finance for the project. In turn, the Soviet Union saw the Horn of Africa and Egypt as important targets for influence, particularly for their proximity to the Suez Canal. The exploration of river potential on the Blue Nile was a veiled American threat to Egypt that the tap could be turned off.

The experts of the bureau saw this as another vast opportunity for water-led industrialization and development. The plan they provided in 1964 was extraordinary: a total of thirty-three projects, more or less equally divided between hydroelectric, irrigation, and multipurpose schemes, would have impounded a hundred and twenty billion cubic meters of water, supported seven gigawatts of power, and irrigated half a million hectares of land. But by then the geopolitics of the situation had changed.

The Ethiopian bureaucracy was more interested in visible individual projects to reinforce its patronage on specific constituencies than an integrated plan, mindful of the fact that the Soviets seemed to be successfully financing and building dams around the world without much concern for the broader economics. Technical assistance as a long-term strategy had fallen out of favor. In the end only one dam out of the scheme was financed by the World Bank, through the intercession of the Americans: the Finchaa Dam, completed in 1974.

By the seventies, Haile Selassie's regime was on its way out. As the politics of the region shifted with the rise of Mengistu's Derg regime, American influence in the Horn of Africa waned. Besides, its interests were rapidly shifting to the Middle East. According to Ryszard Kapuściński, the Polish writer, towards the end of Haile Selassie's reign he wanted to "leave an imposing and universally admired monument after himself." He imagined that the large dams he was planning to build would leave a legacy of development. He was wrong.

In 1973, Jonathan Dimbleby, a twenty-nine-year-old journalist working for the British channel ITV's current affairs program *This Week*, reported from the north of Ethiopia on "the unknown famine." His images shook the world. Scenes of starvation of this kind had not previously been seen on television and were broadcast around the world. The famine was long and deep; the country was mostly unprepared for it. The combination of drought and a disregard for the needs of the population created a combustible mix. Unrest grew.

In 1974, Haile Selassie was finally and dramatically deposed, but the next chapter for Ethiopia proved even worse. By the time Selassie fell, the Soviet Union was in a battle with the United States for control over the large continent. Most independence-minded, anti-imperialist movements in Africa were inspired by socialist movements. Many revolutionaries, like Ghana's Nkrumah, had embraced the planned economy model with relish.

The Derg regime—"Derg" means "council" in Ge'ez—a Soviet-inspired military dictatorship, took Selassie's place in Ethiopia. The Derg used Dimbleby's film for their own propaganda, broadcasting it in Ethiopia. In truth, the Derg had little to do with the pan-African movements. The regime had been born out of the grievances of the military against the landed elites of the imperial government during the famines. But its commitment to single-party communist rule was the symptom of the growing influence of the Soviet Union on the continent.

The Derg regime enacted land reforms. While initially hailed as the righting of a wrong, all the reforms did was collectivize land ownership in line with its Marxist-Leninist credo, but they did little to correct the cronyism and power relations on the land. By 1977, after a complex power struggle, Colonel Mengistu Haile Mariam remained as the sole autocratic dictator of the self-proclaimed Marxist-Leninist regime.

The regime followed a cycle of five-year planning. At the heart of those plans were attempts to improve the productivity of agriculture, and therefore the use of the country's water resources. They embraced large-scale mechanized state farming and cooperatives as the central pillar of its agricultural transformation. Cooperatives received preferential access to land, to subsidized inputs and labor. The problem was that the productivity of an increasingly degraded soil was lagging far behind the growth in population.

Increasing rebellions plagued the regime. They used resettlement and villagization—forcing peasants to live in easily controlled villages—to control them. These programs were so poorly run that all they did was increase land degradation and cause the state of the peasantry to deteriorate even further. If Selassie had taken his model of development from the United States, the Derg took theirs from the Soviet experience. They both proved equally incompetent at translating early twentieth-century industrialization plans into programs on the ground in Ethiopia.

The 1983–1985 El Niño–induced drought turned into a famine that killed over a million people. The drought mixed with volatile political and ethnic tensions. Eventually an insurgent group from Tigray, the Tigray People's Liberation Front (TPLF), rose to fight the Derg. By the mid-eighties the cover of Soviet power was no longer available, and the Derg regime collapsed. In 1987, Mengistu oversaw the transition to the Ethiopian Federal Democratic Republic and was finally chased away in 1991.

ONE NIGHT, IN OCTOBER 1963

The deterioration of the world's enthusiasm for water resource infrastructure was a long process, which ultimately played out through the seventies, when different economic theories displaced the older dogma of investment-led growth that had favored infrastructure development. But there were individual events that helped accelerate the demise of water infrastructure as the darling of development theory. Specifically, the death knell for modern hydropower development rang in the Italian Alps, in Longarone, on October 9, 1963, around ten-thirty in the evening, when a bright flash suddenly lit the sky.

Longarone was a small town of 4,500 people, nestled at the intersection of the Piave and Vajont river valleys in northeastern Italy. Hydropower developers had been eyeing the Vajont torrent, a tributary of the Piave, since 1900. The entire river system had been the target of extraordinary development during the first half of the twentieth century. The crown jewel of that system was the Vajont Dam, situated in a narrow, three-hundred-meter-deep gorge, which the river had cut since the Last Glacial Maximum.

When the project began execution in 1957, the height of the dam had been set at 261.6 meters. It was to be the tallest dam in the world. It was built quickly, in about two years, a marvel of modern civil engineering. Testing began in early 1960. Almost immediately it was apparent that there was a problem. The mountain on the left bank of the river was moving. As the dam filled up for the first time, terrain from the side of the reservoir began to move.

The citizens of Longarone complained that unnatural noises were coming from the mountain. Geologists realized that a number of deep, ancient fractures adjacent to the reservoir could lead to the catastrophic collapse of material into the reservoir. It was too late. The company, SADE, had no intention of stopping the commissioning process. Too much money was at stake. As the reservoir continued to go through testing, the first fill of water began to drain, and that is when things started to go very wrong.

On October 1, 1963, draining accelerated. The north side of Mount Toc began moving. By October 6, so many peaks and troughs had appeared along the road that ran along the left bank of the river that it was no longer accessible. Trees lining the road had begun to lean heavily. Some fell altogether. Further troubling noises were coming out of the deep. At this point, the project engineers raised the alarm with the company. But the decision was made not to alert the surrounding population.

By October 8, it had become clear that something was about to happen. The side of Mount Toc was going to collapse. The company finally decided to evacuate the people living around the reservoir. However, the assumption was that whatever happened, it would not affect those living downstream because the dam would be able to protect them.

And then it happened. That October 9 evening, around ten-thirty, some people were preparing to sleep in Longarone, while others were still in town. The European League football match between the Real Madrid and the Glasgow Rangers had promised to be a spectacle. Most families did not have a TV at home, so many had gathered in town to watch the broadcast. The match ended with a crushing 6–0 victory for the Spanish side. As people prepared to return home, a bright flash suddenly lit the sky. A booming, rumbling, unrecognizable thunder followed. The street and home lights went out. A strange wind picked up. Most wristwatches that were subsequently recovered from the disaster indicated the same time: 22:39.

At that moment, a two-kilometer-wide landslide broke off the mountain. Over two hundred and sixty million cubic meters of rock and terrain slid at a hundred kilometers per hour into the reservoir of the Vajont. Technicians had thought that an event of that magnitude would move extremely slowly. But the slide had such momentum that it filled part of the reservoir, crossed it, and kept going, climbing up a hundred meters on the other side of the gorge. Fifty million cubic meters of water moved out of the way in an instant. A huge wave reached over two hundred meters up the right bank of the reservoir, destroying transformers and producing a daylight flash, taking out power supply to the whole area.

It all came crashing back down. The wave headed towards the dam, overtopping its crest by two hundred meters. Like a bullet speeding through the barrel of a gun, it squeezed its way through the gorge pointing straight for the village of Longarone. As the wave traveled the 1,600 meters that separated it from its victims, it became a 70-meter-high wave, and 1,910 lives were washed away. A full third of the town, mostly young families with children, was gone.

The disaster of the Vajont was a defining moment for modern Italy's relationship with its water and for hydropower worldwide. It was not just the end of a technology strategy but the epilogue of decades of water-led development that had transformed a country, legitimized its institutions, and powered its economy. But the consequences of this fateful event were compounded by the geopolitics of energy, which were shifting fast towards oil. The disaster of the Vajont was not just an Italian disaster. It was an international one, contributing to seal the fate of the aggressive development of large dams driven by the West.

THE CRISIS OF THE 1970S

The seventies marked a watershed in the story of the hydraulic century. By then, most rich countries had developed most of their water resources. The United States had built about 70 percent of its viable hydropower potential, as had much of Europe. What was left was often uneconomic. Besides, the newest thermal power technology had displaced hydropower as the base choice for industrialization. In wealthy countries, at least, "peak water" for energy had passed, and so had their appetite for expensive infrastructure.

This transition occurred during a deep economic crisis. It is a well-known chapter of twentieth-century history. After the Second World War, America became the pre-eminent global consumer. This transition had been assisted by the Bretton Woods system in which the dollar was pegged to gold. The United States became a net importer of manufactured goods, in turn leading to its de-industrialization and to a consequent rise in unemployment.

According to economic orthodoxy, still informed by Keynes's policy prescriptions, unemployment called for monetary stimulus to sustain demand. So, the Federal Reserve pumped money into the economy in an attempt to stem the rise of unemployment. But the problem this time was not aggregate demand. American manufacturing was simply not profitable enough, its productivity too low and its products too expensive, to be able to increase supply. In an effort to shock the economy, in 1971 President Nixon released the dollar from gold, ending the Bretton Woods system and introducing the current floating currency system. Instead of reducing unemployment, inflation skyrocketed.

Meanwhile, on the other side of the iron curtain the Soviet economic system was rapidly collapsing. For much of the sixties, the Soviet Union had embraced food self-sufficiency. By the early seventies, domestic price controls had driven internal demand to outstrip the ability of the Soviet Union to supply. The USSR showed up on the global markets to buy grains right at the time when the United States and Canada, weighed down by their economic situation, had decided to remove the subsidies that had sustained exports since the Second World War, suddenly depleting their grain reserves.

Because of its years of isolation, the Soviet Union was a black box as far as traders were concerned. The fear that this new buyer would force prices up became a self-fulfilling prophecy. A speculative bubble and panic buying tripled wheat prices. At those prices, commercial trades overwhelmed and displaced food aid, leading to the famines and global food crisis of 1972–75. The crisis, coupled with the oil spike and consequent monetary adjustments of the Nixon era, left the Soviet state isolated, poorer and economically vulnerable to the market power of the United States.

ONE LAST FOLLY

The Soviets made one last major attempt to transform their landscape to produce more food. The Amu Darya and the Syr Darya Rivers emptied their waters in the Aral Sea, a huge saline terminal lake east of the Caspian Sea. Starting in the fifties and sixties the rivers had been so heavily diverted for irrigation that by the sixties the lake had started shrinking. The windswept, exposed, salt-covered bottom spread millions of tons of salt for hundreds of kilometers, destroying the ecology of the lake and of the surrounding region.

The Politburo decided that the only way to bring more water to the fertile but dry areas of central Asia and produce more food was to divert the northern rivers of Russia southward, away from the Arctic. The idea wasn't new. The company Gidroproekt had proposed a version of the plan in 1954, involving diverting the River Pechora in north Russia, to the Kama River and then into the Volga. The idea was revised during the tenth five-year plan of 1976–80. Two legs were considered: water would be diverted from the Arctic and Baltic drainage basins to the Volga and Dnieper rivers, while the Ob and Irtysh rivers in Siberia would be forced to flow down to central Asia. At a later stage, the Yenisei River could be added to the scheme. It was pure science fiction. Two hundred and seventy thousand hectares of arable land were to be flooded to make way for this replumbing, and a billion cubic meters of peat deposits would have been submerged. The matter was ultimately resolved when the Soviet Union disappeared a few years later, crushed under the weight of its economic failure.

The last quarter of the twentieth century appeared to have spelled the end of the era of water-led development. The economic stagnation, high inflation, and growing unemployment of the seventies destroyed faith in high-public-spending policies for a generation. America, which had been left the only hegemon after the Soviet collapse, shifted its economic thinking towards consumption and privatization. Water-led development and infrastructure had decidedly gone out of fashion.

In truth, what had emerged was a different idea of the state. Neoliberalism was on the rise. The notion that the state—not the government or the public bureaucracy, but the constitutional architecture that governs citizens' lives together—was no longer the organizing principle of society was gaining traction. Markets, not politics, would provide the glue for society operating as one. By the eighties, the so-called Washington Consensus had led most developing countries to embrace free markets as well, as the primary means of capital allocation, and to privatize national assets in exchange for financial assistance from international institutions. Development economists shifted their attention to structural adjustments and market reforms. Water infrastructure, which had been the backbone of early twentieth-century industrialization, was no longer a favorite in a world of neoclassical economists.

By the end of the twentieth century, water-led development appeared to have lost much of its appeal, much like central planning and industrial policy. It was a relic of an era that seemingly relied too much on public sector intervention, and supposedly stripping the individual of power. The failure of the Soviet Union seemed to further encourage the liberalizations of the eighties and nineties. Market valuation was all that mattered, and infrastructure, state-led development, trade protectionism all seemed to be expensive, bad ideas, as the world opened up for another major wave of globalization. It was supposed to be the "end of history."

It was a political transformation more than a technical transition. Individual agency, that of consumers more than that of citizens, focused attention on the life of individuals. Modern environmentalism in developed countries promoted increasing awareness of the complexity of ecological systems and their value for humanity, but did not see a republican constitution as a necessary infrastructure for governing the commons, preferring to moralize the individual rather than advocate for political governance. All too often, the state was seen as part of the problem, rather than a solution. If the engineers of the great projects had once been hailed as the heroes of the modern world, they were now seen as having left a legacy of environmental destruction. Large water projects and massive state-led investments were most definitely out of fashion.

But that is not where the story of water ended. The underlying millennial process of action and reaction, of institutional adaptation, was still very much active. The role of the territorial state as a vehicle for collective power was still the primary one. What had happened is that the negotiation between society and water had migrated to a different space, one mostly hidden from view.

PART IV

FINALE

20

A World of Scarcity

NASSER AND PROJECT ALPHA

By the 1970s the development of water infrastructure seemed to have peaked. Rich nations had developed most of what they could. Poor nations still could not afford it. The demise of hydropower as the principal power technology might have given the impression that the age of the hydraulic nation was finally over. It had accompanied the rise of the modern industrial state and supported the success of the Western republic. But the high modernist project that had succeeded for the wealthy was too expensive for everyone else.

In reality, the focus of society's struggle with water had shifted. If the first half of the twentieth century had been dominated by hydropower and multipurpose dams as the principal nexus of water and society, in the latter part of the century that focus moved to another energy-related terrain: that of oil. And if its principal political realm had been that of nation building, it now moved firmly in the territory of geopolitics driven by competition over regional and global markets. Over time, the heart of that competition also moved eastward. Reconstructing that shift and its implications requires taking a step back to the Middle East, and specifically to the Nile.

After the Second World War, discontent among the rural population of Egypt turned into a deep well of political energy. The obstinate poverty of the *fellahin,* the Egyptian peasants, and the absurd concentration of wealth in the hands of the pasha class, formerly the military and civilian Ottoman elite, fueled revolution. The disastrous Arab-Israeli conflict of 1948 accelerated radical change. King Farouk lost his grip on the country, and in July 1952 the "Free Officers" staged a coup. The king abdicated and a new military regime headed by General Naguib took over. But it was clear who the real leader was: Gamal Abdel Nasser, a thirty-four-year-old army colonel, who was about to show the world how far charisma could transform a region.

Egypt was an unusual country, at the intersection of the Muslim, Arab, and African regions. Nasser was its pre-eminent pan-Arab leader. He believed that the defeat of the 1948 war with Israel was due to Arab divisions. He was committed, above all, to resist British imperialism. He chose to play the U.S.A. and the USSR—locked in their Cold War—against each other in order to extract concessions and assistance.

Nasser had a structural problem, however. Given Egypt's population growth, the productivity of agriculture had to dramatically improve. During the Second World War, the cotton industry, the most important export of the country, had been disrupted when blockades interfered with fertilizer imports. The country needed to industrialize, to develop its own fertilizer production capacity if it was going to modernize agriculture. That, of course, meant electrification. In Egypt it meant water.

This wasn't exactly a new problem. Since the 1930s entrepreneurs had proposed harnessing the electric potential of the British Low Aswan Dam for chemical production. The idea had even entered the country's first five-year plan in 1943. But the scale of Nasser's ambition far exceeded what laissez-faire entrepreneurship could have delivered. Atatürk in Turkey had shown the Middle East what state-led industrialization could look like. Nasser knew how to take a cue, and grabbed an idea that had been circulating for some time.

In 1947, Adrien Daninos, a Greek-Egyptian engineer in the Ministry of Public Works, had proposed a daring project: instead of dealing with many small storage dams and barrages along the Nile, why not solve the regulation problem with one huge new dam built about four kilometers south of the old Aswan dam? The idea did not come out of thin air. Daninos had been one of those many foreigners who had visited Lilienthal's Tennessee Valley Authority and come away transformed. The High Dam at Aswan represented a shift in two ways. First, it was a decisive turn towards river nationalism: In contrast to the previous Equatorial Nile Project, the dam would give Egypt fully domestic storage for multiple years of water. Second, it was a shift towards economically driven water resource development. Classic TVA ideals.

Nasser knew right away this was going to be the heart of his development plan. To secure peace in the region, British and American diplomats wanted to offer the new leader something to entice him away from the USSR. It could not be military aid without a solution to the Arab-Israeli problem, but Nasser knew what to ask for: he wanted financing for Aswan. They called it Project Alpha.

FINANCING THE DAM

The story of Western financing of Aswan surely is one of the most egregious examples of miscalculation in modern diplomacy. It was plagued by misunderstandings and missteps. The primary concern of Nasser's government was the continued imperial aims of the British over the Nile, through their control of Suez and of the upper riparian states. To counter the British, Nasser wanted American support far more than he cared about aid from the USSR. It should have been easy for the United States to draw Nasser into its circle of influence.

But Eisenhower refused to intercede with the British on Egypt's behalf. Furthermore, in the U.S. there was significant domestic opposition to the project. The Americans worried the dam would alienate other Arab countries, especially those belonging to the U.S.-backed Baghdad Pact, an attempt at creating a regional defense pact similar to NATO or SEATO to contain the USSR in the region. The perception that the Americans were not in fully, in turn, pushed Nasser towards more radical positions. The mistrust came to a head when, in 1955, he both purchased arms from Czechoslovakia and recognized China.

Incensed by Nasser's behavior, Eisenhower's secretary of state, John Foster Dulles, prolonged the negotiation on the loan package for Aswan in order to extract the most from Egypt. At that point, Russia announced they were in fact ready to assist Egypt. The overtures of the Soviet Union were the last straw. Congressional support evaporated and the financing collapsed. As the U.S. withdrew its support, in a last, inexplicable moment of patronizing glibness, Dulles told his Egyptian counterparts that "Egypt should get along for the time being with projects less monumental than the Aswan Dam." Nasser instantly sealed the deal with the Soviet Union.

Soviet support still left Nasser with significant financial exposure and in need of additional revenue. Besides, Nasser had no interest in becoming a Soviet pawn in the Cold War chess game with the United States. Partly prompted by the Soviet invasion of Hungary, on July 19, 1956, Nasser, Yugoslavia's president Josip Broz Tito, and Jawaharlal Nehru of India met in Brioni, Yugoslavia, to sign the Declaration of Brioni. It founded the "Non-Aligned Movement" to establish sovereignty and independence against the dualism of the USSR and the U.S.A. Just a few days later, Nasser showed what he meant.

Surprising the world, at a speech on July 26, 1956, in Alexandria, Nasser announced the nationalization of the Suez Canal, taking advantage of the final withdrawal of British troops. The great nineteenth-century project that had heralded the era of infrastructure in the West, inspiring generations of visionaries to a planetary conquest of nature, was going to be owned by a once-occupied country. Besides, owning the canal would have given Egypt the resources to finish the dam.

The Suez Crisis followed, as a British-French-Israeli coalition invaded to take back control. The invasion had not been coordinated with the Americans, who were engaged in a presidential election and had little time for this sort of adventure. They pressured the invading countries to agree to a cease-fire. It was a humiliating withdrawal, which left the former colonial powers diminished, while cementing Nasser's reputation.

Naturally, the decision to build the dam created a problem with Sudan. By then, the Sudanese had exhausted the fixed amount of water that had been legally available under the 1929 agreement with Egypt. When a new Sudanese regime came in, it quickly decided it too wanted to build its own national piece of infrastructure, the Roseires Dam on the Blue Nile. But the World Bank financing it desperately needed could only be accessed if they settled the Nile issue with the Egyptians.

This situation pushed the two countries to land an agreement in 1959: fifty-five billion cubic meters went to Egypt, eighteen and a half to Sudan. A few details aside, the agreement reflected the proposal Harold Thomas Cory had made three decades earlier, when he was part of the Nile Commission. The enormous Aswan High Dam finally went ahead. It would take a decade to build. It created an impoundment five hundred kilometers long which could hold about two years of Nile flow and generate most of the electric power of the country. Egypt had also completed its conversion to perennial irrigation. At the time of commissioning the impoundment was called Lake Nubia. Now it is called Lake Nasser.

The development of the Aswan Dam coincided with a shift in American focus in the region, from infrastructure investments to commodities markets: grains and energy. The shift had been prompted by a domestic political problem that both America and Canada had faced in the fifties and sixties. The subsidies they had provided to support food production at home had created vast surpluses. Once European agriculture had returned to the market after the Second World War, overproduction risked a domestic collapse in prices. As a result, Canadian and American producers had to turn to developing nation markets, which up to then had been more or less self-sufficient.

The problem was that, while developing nations were unable to resist the competition of subsidized crops, most of them paid for imports in unconvertible currencies, which would have posed a challenge for the U.S. Treasury. Payments had to be in dollars. Where did dollars come from? The most important international source was the sale of oil from the Middle East, which constituted an extraordinary source of currency to pay for food. By the sixties, the American economy had become entirely dependent on cheap oil from the Middle East, ironically a product of the search for water that the Americans themselves had encouraged. The age of oil had arrived.

THE AGE OF OIL

The Jordan River runs for much of its length as a border between the Hashemite Kingdom of Jordan, Israel, and the Palestinian territories. It is a relatively short and limited river. Below its headwaters, the river flows into the Sea of Galilee, travels between Israel and Jordan, and ends up in the Dead Sea, an inland salty lake over four hundred meters below sea level. The lake is at the bottom of the rift that formed over twenty million years ago, when the Arabian peninsula pulled away from the African continent.

After the end of the British Mandate in 1948, the formation of the state of Israel, and the ensuing First Arab-Israeli War, interventions on the river accelerated. In 1951, the Kingdom of Jordan announced plans to divert the Yarmouk River, a tributary of the Jordan, via what was then known as the East Ghor Canal. Israel responded two years later by beginning construction of its National Water Carrier. It was a system of pipelines and canals that would draw large quantities of water from the Sea of Galilee to ensure Israel's own water security, especially in the arid south.

Sharing the Jordan River was going to be a problem. At the time of the Eisenhower administration in the early 1950s, special envoy Eric Johnston tried to persuade the riparian countries of the river to pursue an approach inspired by the Tennessee Valley Authority. The Jordan Valley Authority, which still exists today, is the legacy of that effort. The plan required developing the river so that benefits could be shared: just over half of the water would go to Jordan, a third to Israel, and just under 10 percent each to Lebanon and Syria. A critical contribution to the plan was the diversion of Nile River waters to the northern Sinai to resettle two million Palestinian refugees. Despite the efforts by the United States to broker a deal, the inherent conflict simmered to a boil, finally exploding in full force in the 1960s.

In 1964, the Israeli National Water Carrier began operations, drawing water from the Jordan River system. The surrounding Arab states saw this as a zero-sum game over the scarce resources. Nasser convened a summit—at which Yasser Arafat's Palestine Liberation Organization was formed—where the Arab states agreed to divert the headwaters of the Jordan, upstream of Israel, and channel them towards Lebanon, Syria, and Jordan through a series of impoundments. The diversions began in 1965. Israel declared this an infringement of its sovereign rights. Tensions rose.

In May 1967 Nasser mobilized the troops of Egypt, Syria, Lebanon, and Jordan to surround Israel. On June 5, 1967, and for six days after, Israel engaged in a rapid response, destroying the Syrian projects, taking over the Golan Heights, the West Bank, the Gaza Strip, and the Sinai. The country had secured for itself control of the Jordan River, including much of its headwaters and riparian access, as well as its possible uses downstream. The Israeli retaliation went all the way to Egypt, where two thermal power plants were destroyed, forcing the first two turbines of Aswan Dam to be brought online early to make up for the energy loss.

While these conflicts over water were playing out, the nexus between food and oil was also tightening in the region. Saudi Arabia had embarked on an aggressive expansion of its agricultural production. By this point, Aramco's technology had revealed deep reservoirs of fossil groundwater. The country's agriculture became addicted to them. The Saudis established a program of generous subsidies to wheat producers, encouraging them to draw as much water as they wanted, subsidizing mechanization, hybrid seeds, fertilizers, all the trappings of modern agriculture to boost production. By 1965, Saudi Arabia had three hundred thousand hectares under cultivation, and envisioned a regional system in which investments in Sudan would complement the domestic wheat program, creating a breadbasket for the Arab world. Oil money could have ensured the food security of the region. It was not to be.

The Gordian knot between oil, food, and water tightened when Nixon left the gold standard. At that point, most oil exporters, whose earnings were in dollars, suffered a big reduction in income from the depreciation of the American currency. This created an incentive for cartel behavior: producers would have welcomed nothing more than an artificial contraction in supply, to push up prices and restore their profits. That is exactly what they triggered during the second act of the Arab-Israeli conflict.

Anwar Sadat, who had come to power when Nasser died in 1970, waged war on Israel along with Arab allies in 1973, the Yom Kippur War. To try and isolate their opponent, the Organization of Arab Petroleum Exporting Countries (OAPEC), decided to enact an oil embargo on countries that supported Israel. The coordinated oil embargo only lasted a few months, but it sent the price of oil skyrocketing to four times what it had been in dollar terms.

The embargo sent shock waves through the supply chains of the world, sending streams of dollars towards the Middle East, while fueling inflation in the West. To try and stem the rise in hostilities, the United States offered food aid to Egypt, tied to the condition that

Sadat enter into peace talks with Israel. The conflict eventually ended with the Egyptian-Israeli Peace of Camp David in 1978. Those negotiations could have transformed the water geography of the Middle East. Sadat offered Nile waters to settle northern Sinai—along the lines of the thirty-year-old Johnston proposal—in a shared plan with Israel to develop the Negev desert. But Sadat was assassinated in 1981, presumably for his attempts at seeking a collaborative peace, and the plan came to nothing.

A second oil shock happened in 1979, when Iranian and Iraqi production collapsed in the wake of the Iranian revolution and the start of the Iraq-Iran conflict. The price of oil doubled again. In retaliation, the United States threatened a food embargo. This was just after the USSR had showed up on the markets after years of economic isolation, triggering a food crisis. The impact on food prices from this latest shock sealed the fate of the Saudi agricultural system and of its water reserves.

The Saudi leadership, frightened by the variability of the global food markets, abandoned dreams of regional integration and turned fully towards a policy of food self-sufficiency. The subsidy regime they adopted, coupled with a guaranteed purchasing price for grain, led to a severe distortion in wheat production. Farmers produced so much wheat they could not even store all of it, and grain rotted in the fields. Subsidies drove farmers to drill deeper and deeper wells. Water tables began collapsing. As supplies dwindled, social tensions followed. The eastern provinces of Saudi Arabia, which had originally inspired the water-led development of Ibn Saud, were majority Shi'a. When, at the end of 1979, the Iranian revolution gave the Shi'a an international voice, violent revolts exploded in Qatif, leading to violent repression by the Saudi government. Water-led development had become a security problem.

The oil shocks of the 1970s did not just change the course of water for the Middle East. Everything changed. The high oil prices that had emerged from the Middle Eastern conflicts had a dual effect. Exporting countries saw their earnings rise, which then pushed them to seek investment opportunities. Importing countries had to finance a trade deficit because of the higher oil price, which meant they had to borrow, growing their national debts. Then, America corrected course. The Federal Reserve restricted money supply and interest rates rose.

The rationale for this move was to try and stem inflation domestically, but the consequences were global. The higher American interest rates increased the cost of servicing debts for those countries that had raised finance in dollars. At the same time, the inevitable recession that followed contracted the American market for the products those same countries were trying to export. Many risked default.

The second half of the twentieth century had produced an intractable nexus between oil economics, food security, and water infrastructure investments. The net result of the crisis of the 1970s was that capital to invest in public infrastructure in developing countries dried up. The path that America, Europe, and the Soviet Union had followed to industrialization, a capital-intensive one focused on large financial flows directed at water infrastructure, had been closed.

THE DRAGON RISES

For a while, infrastructure development on the rivers of the world seemed to stop. In reality, while most observers' gaze was fixated on the merits or demerits of Western involvement in the replumbing of the world, the action had already begun moving east, where the neoliberal revolution had only partly changed the role of the state.

China's journey took a distinct turn when Deng Xiaoping rose to power at the end of the seventies. He recognized the need for a mixed approach, one in which—contrary to Mao's beliefs—markets could help allocate resources, while government maintained full control of infrastructure investments and strategic sectors to drive a structural adjustment from agriculture to industry and services. Taking advantage of Nixon's thawing of relations with the country, Deng also made the radical decision to decouple China's economy from the Soviet Union, betting instead on the unquenchable thirst for products of the American consumer market. "Made in China" became the most common three words of the last quarter of the twentieth century.

The catch-up of this aggressive export-led growth model was remarkable. In 1979, the Chinese government decollectivized the rural landscape, and agricultural productivity leapt by 60 percent. China was still exceedingly poor. In the 1960s, 60 percent of the population lived on less than one dollar a day, below the line of poverty. A growth led by exports and huge government investments in infrastructure took full advantage of the abundant cheap labor the country had to offer and the extraordinary rate at which the Chinese were finally saving money. China's economy grew. And grew.

China grew faster than any other country between 1970 and 2010. By 2001, China had joined the World Trade Organization. By 2010 the number of people living in poverty had fallen to below 8 percent. Between the end of the twentieth century and the first decade of the twenty-first, China became the second-largest economy in the world, second only to the United States. It was remarkable.

The consequences for the Chinese landscape were equally significant. Infrastructure continued to be an essential lever in government policy, and water in a land of floods and rivers could not but be at the heart of it. Over the course of forty years, hydropower installation grew twentyfold, reaching over 350 gigawatts of installed capacity, more than any other country in the world. China today accounts for more than a quarter of the entire world's installed hydropower capacity. Annual investment in water security in China in the first decade of the twenty-first century skyrocketed from around $5 billion in 2000 to seven times that after 2010. The one awe-inspiring symbol of this era of growth, the flagship project that more than any other proved to the world that China had succeeded, was the Three Gorges Dam, the dam that Sun Yat-sen had envisioned a hundred years earlier. When finally commissioned, it was the largest single piece of infrastructure in the world, holding back an immense six-hundred-kilometer-long reservoir.

Three Gorges was a regulating dam: it allowed engineers to smooth the peak of a flood by combining storage with timed releases. Other dams, like Aswan High Dam on the Nile, can hold multiple years of their river's flow. But the Yangtze River is far too large for that. Besides, the dam was meant not only to deliver storage and flood control but also to be an instrument of economic development. Flood control had to be balanced with other important uses, such as the twenty-two billion watts of installed hydropower capacity—ten times that of Hoover Dam—and transport locks letting a steady stream of cargo into its vast impoundment. It was a single development intervention—at once a means of protecting people, transforming the landscape, and creating economic value—as striking as what the U.S. had pulled off in the

West or in the Tennessee Valley a century earlier. It was a fitting legacy for Sun Yat-sen's dream, but one whose political implications are yet to be fully understood. For while the technical intervention had been successful, it had been enabled by a political architecture that was standing in an unstable balance between republican principles and authoritarian habits. After all, Three Gorges Dam had succeeded because the company responsible for its development had been given the power of a ministry of the central government.

And just as the United States had exported its approach to water security to the world with President Truman's Point Four strategy, so was China now ready to bring its expertise, technology, and finance all across the world. Three Gorges Dam had given the Chinese government enormous confidence in its ability to deliver on the most arduous project, and it was now time to deploy that confidence. Having survived the Asian financial crisis of the 1990s, China had accumulated savings. For a while, those savings were mostly invested in United States Treasury bonds, but as the performance of Western economies lagged early in the new millennium, China began looking elsewhere to deploy its money. It was ready to engage on the rivers of the world.

POWDER KEG ON THE ROOF OF THE WORLD

China invested everywhere. In part the reason was geopolitical. Unlike the United States, China has a number of significant neighbors with whom it shares a border, making its territorial security far more complicated. In the twenty-first century, the story of Chinese water development was not just an effort to project power abroad. It was also inextricably tied to its own domestic security.

Most of China's oil and gas supplies reach the mainland from the Middle East through its Pacific ports, via the Strait of Malacca. However, the strait is a narrow passage between Malaysia and the Indonesian island of Sumatra, and is, at least in principle, vulnerable. Were this seaway compromised, the commercial link between the Indian and Pacific Oceans would be severed from the Middle East, and China would be isolated, struggling to receive the supplies it needs to meet its demand for fuel.

The alternative is to transport supplies via an overland route through Pakistan, connecting the port of Gwadar, on the far western coast of Pakistan with Kashgar, in the northwest province of Xinjiang. That crucial security corridor became part of a water bargain that China struck with Pakistan. China would provide finance and technical assistance to develop infrastructure under the auspices of a China-Pakistan Economic Corridor—China had already financed the port of Gwadar. Hydropower was going to be an integral component of the investment, with Three Gorges Dam Company heavily engaged on the tributaries of the Indus like the Jhelum.

In Pakistan this was seen as a development program. Outside of Pakistan, though, and specifically in India and the United States, it was seen as a geopolitical play for a Chinese sphere of influence. And as a dangerous threat. The only way for this corridor to work was to pass through Jammu and Kashmir. David Lilienthal had remarked decades earlier on the dangerousness of this small state nestled in the Himalayas. Now, the former princely state was a keystone not just in the relationship between Pakistan and India, but also at the center of the interests of China. It should be no surprise that U.S. president Bill Clinton once referred to it as "the most dangerous place in the world."

The tension over economic development has played out on hydropower. The Neelum River, which flows in Jammu and Kashmir, is a tributary of the Jhelum, itself one of the tributaries of the Indus. The Neelum flows at a considerably higher altitude than the Jhelum, so that a barrage across the first could divert water down through a tunnel into the second, producing substantial hydropower. The line of control of 1949, which marked the cease-fire line at the end of the first war between Pakistan and India, happened to cut the two rivers: the Neelum ran inside India's Jammu and Kashmir, over the line of control, and down to meet the Jhelum in Pakistan's Azad Jammu and Kashmir. Either country could take advantage of this peculiar hydrological situation—but not both at the same time. Nature had set up a perfect zero-sum game competition to challenge the Indus Waters Treaty.

Pakistan was first. It began contemplating a Neelum-Jhelum scheme in the late eighties. The project—a massive billion-watt installation—was supposed to begin construction in 2002 and be finished by 2008. However, it was beset by problems and delays, not least the Kashmir earthquake of 2005, which forced a redesign. Around 2008, China stepped in with financial and technical assistance to support development. By 2011, China had established CSAIL, a holding company of Three Gorges Dam Company, with its branch office in Islamabad, with the purpose to acquire, develop, build, own, and operate power generation like hydropower.

Meanwhile, India was looking to build many more dams upstream of Pakistan. One such dam was the Kishanganga, a 330-megawatt plant operating under the exact same principle as the Neelum-Jhelum scheme, except upstream of the Pakistani installation. The Indians were faster.

When Kishanganga was inaugurated, the Pakistani Foreign Office stated that "the inauguration of the project without the resolution of the dispute is tantamount to violation of the Indus Waters Treaty." On the basis of a provision of the treaty, designed to ensure that India could not build schemes that would interfere with "existing uses" in Pakistan, the government of Islamabad initiated a legal escalation, attempting to bring India to a Court of Arbitration. Eventually modifications were made to Kishanganga, but both projects went ahead, with interactions between the two as yet unclear. Through its state-owned enterprises, China found itself at the heart of a dispute between its belligerent neighbors, one with existential implications for both countries.

China's water geography makes it difficult to separate security and economic development in river investments. The Mekong River, once the focus of American interests, is a case in point. The river starts in China, where it is called the Lancang. China had been excluded from the Mekong Commission at a time when no one predicted that it would become the biggest regional power by far. China now planned twenty-three large dams on the river, starting in its upper reaches in Tibet and through Yunnan, before it flows down into Myanmar and Laos. The cascade of infrastructure would turn the Lancang into one of the great hydropower installations of the world, with more than thirty gigawatts of installed capacity from the large dams, and several more from the eight hundred smaller dams planned on its various tributaries.

Some of these dams have enormous reservoirs. Xiaowan and Nuozhadu Dams together have greater storage capacity than Lake Mead. These types of installations are enough to regulate the flow of the river from one season to the next and, at least in principle, interfere with the downstream seasonal flow that sustains the largest freshwater fishery in the world. By building this infrastructure, China in theory owns the tap of the Mekong.

The Chinese argue that the development of the river is purely a domestic development strategy. That it is about a vision of industrialization for Yunnan's provincial economy. They point to the renewable energy that would be produced as an indication that these investments are nothing more than the symbols of a peaceful rise to power and sustainable prosperity. From the Chinese perspective, any international concern for its development of the Lancang is the hypocritical stance of societies that have done exactly the same in decades past, and who simply want to prevent China from doing the same.

From the point of view of those living downstream, Chinese development risks giving an upper riparian hegemon absolute power over the lifeline of four countries. It was true for the British and Egypt on the Nile, for the United States and Mexico on the Colorado. Why wouldn't it be true on the Mekong?

At the start of the twenty-first century, the idea that the twentieth century had been unusual, that the landscape development that accompanied the rise of America had been an unrepeatable anomaly, appears misguided. The rise of China has restarted the process of landscape development in the twenty-first century that may well dwarf what has happened so far. It may still be that if the twentieth was the American Century, the twenty-first will be the Chinese one.

21

A Planetary Experiment

A REAL-WORLD MODEL

Without the benefit of hindsight, it is possible to see only a faint outline of the political undercurrents that are shaping the contemporary world's relationship to water, yet what is visible reveals both new routes and the groove of old mechanisms.

During the twentieth century, inspired by the success of the model republic of the modern age, most rich societies replumbed the planet to insulate their citizens from the impact of the planet's climate and give their economies a comparative advantage. To do so, they harnessed the power of water while allowing everyone to live their lives at the sole beat of industrialization. For all intents and purposes, in wealthy countries at least, the climate system had mostly disappeared from people's lives. Never before had water always been available, when and where needed, and always of a quality fit for its purpose. Never before had people been able to move around the landscape unimpeded, going about their technology-laden day, streams paved over, rivers contained, and all floods avoided. But while technology has changed people's relationship to climate, the thousands of years of layered institutions, which have defined the relationship between society and water over time, continue to play the dominant role in shaping the outcome.

The climate system is changing. There is a very good chance that it may change far beyond anything in recent experience, thanks to modernity's impact on the chemistry of the atmosphere. The injection of carbon dioxide from burning fossil fuels and the transformation of landscape on an unprecedented scale have both had a measurable impact on the energy balance of the planet. It may turn out to be the biggest since the Last Glacial Maximum. As the climate system responds to this perturbation, the hydrology of the planet is responding along with it, forcing countries to develop new solutions to old problems. The question of whether the current institutional arrangements are going to be resilient to those changes is of paramount importance. The twentieth century has not delivered the same answer to all. It has left behind countries with vastly different endowments of infrastructure and institutions. Time and time again, the story of water has shown that society is disrupted at its point of weakest security.

It is still too early to see the full extent of the challenge posed by a changing climate. However, occasionally, the climate system offers a natural experiment that makes it possible to have a glimpse of what might be in store. The year 2010—the year of Three Gorges Dam's first test, the episode that opened this book—was a year of a particularly strong ENSO. ENSO, the "El Niño Southern Oscillation," is a quasi-periodic phenomenon in which the warm waters of the equatorial Pacific, to a depth of a few hundred meters, slosh east to west from Peru to Australia and back, like a gigantic seesaw, over three to five years. This phenomenon manifests itself most at the sea surface: during a cold state, called La Niña, colder, deeper waters reach the surface in the eastern Pacific as warm water piles up in the west; during the opposite, warm state of El Niño, warm waters spread back into the east. ENSO is one of the most significant, measurable, natural signals in the climate system. It also happens to be a natural experiment in what happens when the climate system changes over multiple years.

During its cold phase, La Niña establishes a strong temperature gradient at the sea surface from Peru to Australia. It should not be surprising that an anomalous patch of cold temperature straddling the equator, in the biggest ocean on the planet, stimulates a response in the atmosphere. It also has profound impacts on water distribution across the planet. During a typical La Niña, rainfall increases over South Asia and parts of the United States. Those shifts, in turn, interact with other, short-term weather phenomena, producing a myriad of anomalous, localized events that test the resilience of communities everywhere. Some of those were responsible for the bursts of rain that fed the Yangtze in July 2010. In the first months of 2010, La Niña happened to be particularly strong. It was the perfect experiment to reveal how society would respond.

A TALE OF TWO COUNTRIES

On April 27, 2010, the U.S. National Weather Service had predicted that something was going to happen over the Mississippi basin. Moisture does not move from the tropics to the midlatitudes uniformly. Microwave satellite images show that it often travels in long filaments of air a few hundred kilometers wide, wound around weather systems and extending for a few thousand kilometers. These atmospheric rivers are big. When added up from top to bottom, one of these streams in the sky can carry as much water as the Amazon, or ten times that of the Mississippi. In late spring 2010, meteorologists saw one of these atmospheric rivers finding its way into the Cumberland basin in Tennessee, a tributary of the Ohio River. It was a very coherent plume, extending all the way from Central America, taking water from both the eastern Pacific and the Gulf of Mexico. It was as if nature had decided to plant a hose in the tropical oceans to suck water out and use it to soak the Mississippi valley. It brought a lot of moisture. With moisture came instability. With instability came the storms.

Over the course of two days, the weekend of May 1 and 2, 2010, two large storms positioned themselves over central and western Tennessee and parts of Kentucky, dumping an unprecedented amount of water on the landscape. The Cumberland River was a highly engineered system, which the Army Corps of Engineers managed through ten dams at different points. Four of these constituted the system of flood management. Wolf Creek Dam, the biggest, was on the Cumberland itself, while the other three were on tributaries. Just over half of the Cumberland basin sat above this flood control structure, while 44 percent was uncontrolled, meaning that between the river and the population were only navigation and hydropower projects, with no significant storage capacity. Unfortunately, most of the rain fell in the uncontrolled part of the basin, so that while storage capacity sat upstream unused, rain pummeled the lower part of the basin.

The first wave of storms on May 1 filled rivers and streams, saturating soils. At 7 p.m. the Cheatham Lock and Dam just below Nashville was overwhelmed and overtopped. The second wave of rain on May 2 did the rest. By May 3, the only one of the four flood storage projects to be usefully placed—J. Percie Priest Dam on the Stones River, a tributary of the Cumberland—had filled up. The dam overtopped and water had to be released to avoid a catastrophic failure. Little could be done at that point. The water levels of both the Cumberland and the Tennessee Rivers broke all records. Twenty-six people lost their lives, mostly from flash floods on torrents and streams in the river basins below. As the river failed to accommodate the additional water that the sky threw down, the concert halls, sports facilities, businesses, and homes of Nashville were covered in water and mud.

The events left over two billion dollars' worth of damages behind. Not even a century of experience, built into the very cement and fabric of institutions intended to protect people, was enough for those living in the Cumberland Basin to ignore a river the size of the Amazon flowing above their heads. Water moves. No stationary solution is going to work under all conditions. Yet, the events of the Cumberland, as tragic as they were, were small compared with what happened only a few months later in South Asia.

In the shift to a strong La Niña, by late spring surface temperatures in the Indian Ocean were above average. So was the amount of water in the atmosphere. Blocking cyclones—atmospheric stationary waves that sustain high and low pressures for several days—hovered over western and central Russia. Those cyclones acted as obstacles to propagating weather systems, which crashed into their flanks, stimulating storms. One such flank was stationed over northern Pakistan.

Pakistan had been dry since July 13. As sun shone through the cloudless sky, the ground warmed, to the point that the atmosphere became unstable: just like a boiling pot, upward air plumes began to punch through from its lower layers. On July 18, low-level winds dragged in moisture from the Arabian Sea, delivering it right over the hot terrain. The fuel for the storms had arrived. Satellite images from those days show a first storm formed at the foothills of the Himalayas. On July 19 it started to rain. Temperatures dropped. Rain increased the moisture, feeding convection further.

The first warnings from Pakistan's Meteorological Department came as early as July 20. A weather system flanked the blocking anti-cyclone over Russia, reaching northern Pakistan on July 21. The bubbling atmosphere turned into a proper large-scale storm. Lahore flooded. Several people died across all provinces except Sindh. The army entered into action on July 22. By then, tens of thousands of people needed to be evacuated.

Meanwhile, the Indian monsoon formed a low pressure over the Bay of Bengal and shifted inland. On July 24 it made its way to north Pakistan, bringing with it additional moisture. It was a perfect storm. The stationary high pressure that had been occupying central Russia had deflected powerful waves towards Pakistan, while the monsoon was dragging to the same area water from both east and west. On July 27, all of that energy fell on the heads of the Pakistani population sitting below. Another flood warning was issued.

It was catastrophic. Two dams—Tarbela and Mangla—provide all the available flood storage on the principal tributaries of the Indus. The rest was uncontrolled. Floods washed through the Indus basin, taking with them thousands of lives. The Kabul and Swat Rivers burst their banks. Hundreds died as floods crashed through. Landslides accelerated in the northwest of the country, killing hundreds more. A fifth of Pakistan went underwater. By July 29 the situation was unprecedented. By July 31 the UN declared them the worst floods in living memory. Through the first part of August, as waters cleared, the scale of the disaster revealed itself. Almost two thousand people had died and twenty million needed shelter, food, and care, more than during the Indian tsunami, the Kashmir and Haiti earthquakes, and Cyclone Nargis combined.

The events of 2010 tested the resilience of the United States and Pakistan. The extent to which they succeeded or failed was measured against the expectations of water security that had been set in the twentieth century. Up until then, the vulnerability of most people in the world was far closer to that of modern Pakistan than that of modern America. Both the Cumberland and Indus events were disastrous. But it is unquestionable that the United States fared better because of its wealth. But the difference in experience was not just due to wealth.

People in Tennessee had agency not just through their local infrastructure, but also through a state that is, on the whole, responsive to their needs and resourced to meet them. And it was able to do so because of how it took advantage of those same water resources over a hundred years. The same week that saw Pakistan overwhelmed by the floods, Three Gorges Dam, on the other side of the Himalayas in China, had passed its first test, containing an equivalent flood on the Yangtze.

This, in the end, is the fundamental legacy of a century that drove industrialization along the rivers of the planet: countries that are rich can manage water better; but it is often the case that countries are rich because they have found a better way of managing water, one in which water on the landscape reflects a civic contract that was able to accommodate a growing, enfranchised, economically productive citizenship.

THE DEEP ROOTS OF INSTITUTIONS

But the undercurrents of water's long relationship with human society flow deeper still. The foundations of modern trade were set by the emergence of property rights, the rule of law, and the rise of the nation-state. By 2010, those institutions had formed a planetary network through which climate signals could propagate.

In the summer of 2010, the unusually low pressure that drew water over Pakistan during the Indian monsoons was part of a planetary wave that also kept western Asia under a persistent high pressure. The landscape warmed under the summer sun. Between July 25 and August 8, the worst heat wave since at least 1880 hit western Russia, with temperatures up to eighteen degrees above average in Moscow. Over the first two weeks of August, scorching temperatures burned up a million hectares of farmland. It was also unusually dry. Fires are not unusual in Russia, but that year they were particularly bad, with heavy impacts on the urban populations. The extraordinary heat and air pollution killed an estimated fifty-five thousand people.

But the biggest impact of the drought was on food. By 2010, Russia had regained a pre-eminent position in global wheat markets. Long gone were the 1980s, when Soviet agriculture was desperately backward and the USSR was a permanent net importer of wheat. In 2010, Russia had completed a very long journey back to the top of the food trade hierarchy and was earning as much from selling wheat as it was from the sale of weapons.

The heat wave of 2010 led to a drought that largely overlapped with the grain-producing areas of Russia. Ultimately 17 percent of the crop area was destroyed. In the Volga region, the biggest producer in Russia, the harvest dropped by 70 percent. Overall, the harvest decreased by a third. When it became clear that production was going to be severely affected, the price of grain rose. In fact, prices began rising before production dropped, as speculators held back their supply while buyers flooded the market in a panic to buy up what was available, in anticipation of a rise in prices. This created a self-fulfilling prophecy.

The Russian government intervened by mid-August, releasing some reserves to try and contain the impact. At the same time, the government established an export ban on grains. The aim was to secure domestic supply, although because those sellers were now free to sell their product at whatever price they could, prices rose even further. The expectations of buyers and sellers, and the distortions introduced by government trying to intervene in their domestic market, are powerful amplifiers—so much so that a one-third drop in Russian grain production, which represents about a fifth of the world exports, ended up leading to a doubling in price.

By autumn of 2010, China joined the picture. Between the summer of 2009 and the spring of 2010, rain had been so unusually low in north China that sixteen million people and eleven million head of cattle had gone short on drinking water. The drought was severe. Four million hectares of crop land were affected, and at least a quarter produced no yield at all. Some rivers lost between 30 and 80 percent of their water. Others went completely dry. Despite some localized floods, lakes emptied, leaving fish behind to desiccate, and the landscape turned to desert. Early summer of 2010 did not relent. A once-in-a-century drought hit, and by October, the drought that had afflicted Russia had reached northeast China.

Fearing imminent catastrophe would threaten its food supplies, the Chinese government also began buying large amounts of wheat on the global markets. The shopping spree had consequences, compounding the effects of the export bans from Russia. The price of wheat in July was about $180 per ton. By August it had risen to $250 per ton. It would eventually reach $350 per ton by February 2011.

Grains, and wheat in particular, are still the most important part of the human supply of food. Over 20 percent of the world's food depends on the harvesting of wheat. The legacy of postwar economies and of Western subsidies to agriculture has left most developing countries as wheat importers. They consume more than three-quarters of all exported wheat production. Most of those governments have to then subsidize prices for consumers, limiting food insecurity at great cost for public finances. Those countries are highly vulnerable to changes in prices on the global markets.

All of this made Russia's market share in wheat and the impact of the 2010 heat wave a consequential policy issue for much of the world, and for the Middle East in particular.

ECHOES OF ANTIQUITY

By autumn, the doubling of crop prices due to the Russian heat wave and the Chinese drought had traveled halfway across the world. The Middle East and North Africa, where consumer prices were often sensitive to wholesale prices of grain, were particularly hit. Pressure grew on public finances, which attempted to shield consumers through subsidies and household incomes. Then, on December 19, 2010, a fuse was lit and the entire region blew up.

The city was Sidi Bouzid, in Tunisia. Twenty-six-year-old Mohamed Bouazizi, a street vendor, had refused to pay a bribe. In retaliation, the police had confiscated the scales from his vegetable cart, his family's only means of subsistence. Bouazizi set himself on fire in front of the provincial governor's office. It was a desperate act. Tunisia was quickly engulfed by protests. By January 14, 2011, ten days after Bouazizi died of his burns, President Zine al-Abidine Ben Ali fled to Saudi Arabia, ending twenty-three years of continuous autocratic rule. The Jasmine Revolution of Tunisia had kicked off the Arab Spring.

Whether food prices were the proximate cause of the revolts is less relevant than the fact that the food system, one of the important instruments of social control in the hands of the state, had proven vulnerable. With the fall of Ben Ali, revolts quickly spread westward to Algeria. Then eastward to Lebanon. Then Yemen. By January 25, 2011, the situation had gotten out of hand in Egypt.

Egypt was the largest wheat importer in the world and spent about 10 percent of its revenues from exports and remittances on food imports: over a third of per capita income was spent on food. The combined effect of food prices and democratic aspirations ignited the country. Protesters flooded the streets. Egyptian president Hosni Mubarak first attempted to appease the crowd. When that failed, he tried to crush them. His choice of fighting protesters in Tahrir Square was a fateful one. On February 11, 2011, after thirty years of absolute reign, the Egyptian dictator fell.

After Egypt came Libya. In Benghazi, the Libyan security forces reacted violently to protesters. An armed revolt exploded against Colonel Muammar Gaddafi, who was soon eliminated. The fall of Gaddafi's regime had sudden and unexpected consequences.

Below Libya stood the great Sahara desert and the Sahel. This region of West Africa between the Sahara and the south coast included Niger, parts of Mali, southern Algeria, Burkina Faso, Chad, and northern Nigeria. It was poor, with limited infrastructure and weak institutions, and highly vulnerable to changes in rainfall. Many who lived off agriculture in precarious conditions frequently moved to cities in search of employment, or to other countries in search of a better life.

During the eighties and nineties, Gaddafi had wanted Libya to be the center for pan-African migration and a destination for labor. He encouraged the development of substantial migration infrastructure—desert routes, refueling points—to help people of the Sahel cross the Sahara and come to Libya. When his regime fell, that infrastructure became a springboard for migrants to cross the Mediterranean towards Europe. And so, a population escaping drought, famine, and war walked through those floodgates towards the shores of Europe. Alongside the Syrian refugees escaping war, grew the migratory fluxes that provided fuel for populist anger across Europe.

The impact of water events is often revealed through the most vulnerable. It does not mean that every migration is violent, nor that its destinations cannot be welcoming of those displaced. But it is the weakest that ultimately reveal everybody's vulnerability.

THE DANCE AHEAD

The water conditions experienced by society are not just physical. They result from the interplay between the climate system and the institutions that society has developed over time to deal with it. Droughts alone do not cause food price spikes, nor are shortages the single driver of social unrest. Failures of policy are required for both. The economic integration of the new millennium created a complicated global stage in which the dialectic with water played out at an unprecedented scale.

The events of 2010 provided a perturbation—small compared to what is likely to come—that revealed some of the modern world's fault lines. Those events shared basic traits with what had been happening for centuries and millennia.

There are some indications that countries are already adjusting to climate change, in preparation for what is to come. As the center of economic gravity has shifted east, the emphasis on water security has shifted to reflect the concerns of its emerging economies, first above all, China. China's recent primacy in water infrastructure and hydropower has proven to be a strategic choice, positioning the country as the dominant model of development in the early new millennium. Like the United States before it, China is likely going to use it as an instrument of international relations. But unlike the United States, China has so far focused on technology, leaving explicit political propaganda to one side. That may well change, as its global weight increases. The first indications of this transition are already beginning to show.

Africa is the frontier of water security in the twenty-first century and the canary in the mine for the geopolitics that are to come. China's vast financial resources and state-owned enterprises have been eager to export their domestic experience of infrastructure-led development to Africa. Between 2000 and 2017, China's loans to Africa added up to $143 billion, disbursed over a steep growth: in 2000 China's loans were just $130 million, while between 2010 and 2017, loans to the continent were about $15 billion a year, a hundredfold increase. This was mostly about financing extraction, its power supply, and transport to move commodities to ports headed for a rapidly growing China, hungry for raw materials. Not coincidentally, though, the principal recipients were the two biggest water towers of Africa: Angola and Ethiopia.

On February 11, 2011, in the wake of the Arab Spring and after thirty years of absolute rule, Hosni Mubarak was deposed. Egypt was in the throes of chaos. That week, a stunning piece of news appeared on the front page of the *Addis Fortune*, the main English-language newspaper in Addis Ababa, Ethiopia: teams had just broken ground on an enormous hydroelectric project on a tributary of the Blue Nile, the Abay River. The timing of the announcement could not have been more symbolic. Egypt had always threatened military retaliation should any upper riparian country—and particularly Ethiopia, the source of the Blue Nile—decide to develop storage infrastructure on the river. But with Mubarak gone, a newly assertive Ethiopia could not wait to execute its development plan.

The Renaissance Dam, a six-billion-watt installation, was going to be the biggest dam on the continent. The investment exceeded $5 billion, a quarter of the GDP of Ethiopia at the time of the announcement. China, once again, stepped in. China was an extraordinary model of development for Ethiopia, as much as the United States had been for China a century earlier. But whereas early twentieth-century China was in the throes of a political process informed by nineteenth-century republicanism, Ethiopia at the beginning of the twenty-first century was a socialist republic, which embraced a planned economy regime.

The influence of Chinese experience on Ethiopia is often framed as the experience of one developing country helping another. Indeed, Deng Xiaoping's China, with its mix of state-directed infrastructure and export-led entrepreneurship, has been the most successful case of poverty alleviation in human history. To many African leaders, the Chinese experience seemed far more relevant than that of countries that had industrialized in the nineteenth century. So, it should not be surprising that Meles Zenawi, the prime minster of Ethiopia at the time, was eager to accept Chinese financing to build his vast project on the Nile. Ethiopia was one of the great water towers of Africa, and if the Chinese experience was any indication, the development of its rivers was going to be crucial for economic development.

This story may represent the first of several salvos in a transition in traditional spheres of influence. Climate change may well accelerate that transition. If it ends up being another momentous shift in the geopolitics of water, then one thing is clear: the nature of the political institutions that seek to decide how water should be used will be far more important than the technology of choice to manage it.

Coda

The story of water does not end here. It will continue to evolve, driven by its deepest tension: that of a sedentary society trying to live together while negotiating a world of moving water. Ever since the first communities wrestled with water as it streamed down from retreating glaciers—ever since people began telling stories about it—water has been the dominant agent in people's relationship with the environment. Stories of great floods, of landscapes transformed by water, of rivers whose force was the expression of deities have reached modernity. They captured a deep sense of vulnerability that has been a common trait of human cultures since.

In response to that vulnerability, the vast majority of water management, even today, consists of modifications of the landscape that have been technically available for centuries and, in some cases, millennia. The Grand Canal in China, still the longest in the world, was dug between the seventh and the thirteenth centuries CE. The Cloaca Maxima, part of the current sewer system of Rome, dates from the sixth century BCE. Lagash's canal between the Tigris and the Euphrates is full of water after over four thousand years. The story of water simply isn't mostly a story of technological progress. That is not because technology is missing or unaffordable. It is because the story of water is principally a story of political institutions.

The birth of the state, the development of trade, cultural adaptation to difficult conditions, political institutions that emerged or were inherited from the deep past, the accumulation of economic resources that multiplied the power of society: all of this happened as society learned how to confront floods, storms, and droughts, deciding where to settle and how to live, and how to adjust its particular organization in the process. Institutions and behaviors were shaped by them. No other material condition impacting society today is so inextricably tied to the nature and development of its political institutions, no other intervention so well suited to both take advantage of, and reinforce, the power of the state.

This book started with the story of Dr. Sun Yat-sen. He was a revolutionary. His were dreams of a better future for his country. The blueprint he left behind informed subsequent governments, providing historical justification for China's own high modernist dreams. But Dr. Sun did not just describe a landscape. He imagined republican institutions. He dreamed of a constitutional democracy, one in which the balance of powers protected its citizens from the tyranny he had experienced in the latter days of Qing China. He imagined a state focused on the res publica, committed to the welfare of its citizens, and a nation empowered by its sovereignty.

The model republic that Dr. Sun imagined was the product of centuries of negotiations—at times violent—between individual and collective agency. Its most prominent real-world realization in his time was the United States, which had developed a complex system of local, state, and federal governments subject to a complex balance of powers. The evolution of that system had been a response to the unusual water landscape of the country. It was born out of the relationship between the individual farmer and the overwhelming nature of the western frontier, between industrialization and the country's rivers, between its irrigated agriculture and its trade surplus. It was on its landscape and its strange geography of water that America trained its social contract. In this particular sense, Dr. Sun's model republic, just like the United States, was a hydraulic state. In fact, it was *the* hydraulic state.

There is a continuous thread that runs through the traces left by Plato and Aristotle, through Livy and Cicero, to Machiavelli and Montesquieu that defines a complex constitutional architecture whose primary objective is the commonwealth, the res publica. That blend of institutional authority and citizen empowerment, of individual rights and collective responsibility, is tested in the sophisticated transfer of individual agency to the state that modern water management requires. The bargain that the modern republic has entered into is this: making the climate system disappear in exchange for the authority to exercise ultimate sovereignty on the landscape, and for the financial resources to underwrite the development of infrastructure. The state, the primary instrument of political organization in society and its principal economic actor, has become the single protagonist of water management. Its domestic and international conduct has defined society's response to the power of water on the planet.

When, a hundred years after Dr. Sun's dream, parts of it came to pass with the inauguration of Three Gorges Dam, the casual observer might conclude that the great construction was the necessary technical response to the contemporary material problem of flood management and hydropower production. In that view, Dr. Sun might well have been prescient in identifying it as a need, but he did nothing more than inspire successive leaders to contemplate a technical intervention, which only came to pass when economic and technical conditions were met. But that is not the whole story.

The relationship between water and society over thousands of years has shown that society's response to the power of water consists of an inseparable blend of technical and institutional measures. Society's agency is reflected mostly in the adaptation of its social contract, not in the engineering it applies to the landscape. It was the whole of Dr. Sun's dream that represented a response to China's landscape. The pursuit of water security is never separable from the fate of the institutions that drive it.

This brings the story to modern China. Over the first two decades of the twenty-first century China has poured more concrete over its rivers than any other nation in history. It has relentlessly pursued an engineered water landscape in its quest for economic development. At face value, that modernization is simply the reflection of a state investing in the commonwealth of its citizens. But one of the great illusions of modernity, at least since markets have become the main organizing principle of social activity, is that contemporary life is possible because the profound relationship that has tied water to society has finally been severed. It is an illusion that argues that the gift of an engineered landscape is the emancipation of society from the impacts of a variable climate. It is an illusion that serves a purpose: to justify the delegation of all environmental agency to state bureaucracy, far from the heart of public debate. The question, once again, is what will happen when—not if—that illusion is shattered.

At no other time have societies across the world valued so highly the ideals of individual rights in citizens and consumers, while at the same time delegating so much power away from them on a matter of such existential consequence. They live lives in the service of a productive economy, uninterrupted by the variability of nature. A sustainable bargain between individual security and the delegation of collective power to the state is a supremely fragile one. It is susceptible to corruption and to mismatched expectations, to distortions and capture.

This is clearest when republics exceed the boundaries of their sovereignty and become empires. The United States employed the same technology it had adopted at home in its dealings with the rest of the world, but these engineered solutions faced different institutions in the recipient countries. Water management was no longer a means of entering into a fair bargain with society, a component of a collective agreement in which landscape control was delegated in exchange for a daily life unfettered by the interference of water. Rather, it became an economic exchange between states, an act of patronage for extraction or allegiance.

Today, China is pursuing a similar approach. Its state-owned enterprises are offering Chinese technology, built by Chinese labor and experts, and financed by Chinese savings, much in the same way. What is less clear is what international order those interventions pursue. What is less clear is what political principles China is going to disseminate.

This crucial question will soon become urgent. As the climate system evolves in response to changes in the chemistry of the atmosphere, it will challenge engineered and institutional solutions in unprecedented ways. The fractures that may appear will test the robustness of society's adaptation. Catastrophes are a powerful source of political energy, which can easily lend themselves as much to the next step in nation-building as they can to oppression. The ideal republican state served society by intermediating collective resources in the service, broadly speaking, of the common good. But it has always been unstable. The authoritarian state that may take its place—or the imperialist state abroad—used society's struggle with nature as an instrument of control.

When the struggles with water break through the surface, wherever the walls of institutions and infrastructure are thinnest—the destruction of communities, the forced migration of a people fleeing drought, the displacement of others to make room for infrastructure—all can destroy that carefully constructed illusion that holds up the authority of the state on the landscape. The impacts of climate change on water will travel not via the rivers and floodplains of the world, but through the institutions of human society. Today more than ever society is bound by its expectations of water security. What will happen when a society anywhere in the world cannot or will not meet those expectations will shape everyone's future on the planet.

Acknowledgments

My interest in water and institutions was transformed by countless conversations over several years with John Briscoe. The breadth and depth of his experience on water issues was second to none. I had hoped there would be many more such discussions, but it was not to be. This book exists, at least in part, as an extension of those conversations. The memory of his good humor, combative style, and intellectual integrity fueled my writing.

Several people were travel companions on the journey that led me to water: among them Lee Addams, Dave Allen, Tony Ballance, Peter Brabeck-Letmathe, Nick Brown, Mark Burget, Jean Michel Devernay, Andrea Erickson, Adam Freed, Carl Ganter, Christopher Gasson, Bill Ginn, David Harrison, Jon Higgins, Mike Kerlin, Rachel Kyte, Brian Leith, Brian McPeek, Usha Rao Monari, Alex Mung, Jeff Opperman, Stuart Orr, William Rex, Brian Richter, Dan Schemie, Martin Stuchtey, Alex Tate, Richard Taylor, Mark Tercek, Steven van Helden, Kari Vigerstol, Andy Wales, Catherine Watling, and Dominic Waughray. I learned from all of them. The World Economic Forum, its Young Global Leaders and the Global Agenda Council, hosted many stimulating conversations.

My agent, Max Brockman, helped me turn an idea into reality. I was able to discipline a sprawling topic into a sensible structure thanks to my editor at Pantheon, Erroll McDonald. My colleagues at TNC were infinitely patient as I finished the manuscript. Stella Cha, Andrea

Erickson, Joe Quiroz, Lynn Scarlett, and Bianca Shead were generous with their time and kind with their feedback. Amanda Dickins lent her usual, astounding wisdom to the entire enterprise, as she has done for over twenty years. Years ago, Antonio Navarra and George Philander taught me how to harness the most important skill a scientist can have: curiosity. Jeremy Oppenheim and Martin Stuchtey turned that training into real-world purpose.

My parents, Susan Williams and Paolo Boccaletti, always present, even when oceans and continents separate us, made it all possible. Andrea Mattiello spent countless hours listening to my soliloquies on water, helped me find my bearings amidst the historical literature, and was the best thought partner anyone could wish for. Whether we were trying to find a hidden aqueduct in Rome, looking for cisterns in Matera, or examining a fresco in Arezzo, life together is an amazing quest for knowledge and meaning that makes it all worth it.

Notes

Prologue

x Modern water infrastructure: Wisser et al., "Beyond peak reservoir storage?"
xiii His point of reference was America: Sun, *The International Development of China*, v.
xiii His political philosophy: Wells, *The Political Thought of Sun Yat-sen*, 93.
xiii Then, he imagined a dam: Sun, *The International Development of China*, 46.

1 Standing Still in a World of Moving Water

3 So, whatever water is found: Hallis et al., "Evidence for Primordial Water."
4 The amount of water vapor in the atmosphere: Schneider et al., "Water Vapor."
4 Ice accreted enough water: Clark et al., "The Last Glacial Maximum."
6 During the Younger Dryas: Wanner et al., "Holocene Climate Variability."
6 at its peak around 12,000 BCE: Lambeck et al., "Sea Level and Global Ice Volumes."
6 Be that as it may, hunter-gatherers: V. Smith, "The Primitive Hunter Culture."
7 Natufian communities in the Levant: Valla et al., "What Happened in the Final Natufian?"
7 The first step in that transition: Snir et al., "The Origin of Cultivation."
7 In north China, for example, millet: Jia et al., "The Development of Agriculture."
8 The journey of modern man had begun: Lev-Yadun et al., "The Cradle of Agriculture."
8 The sedentary population grew: Bocquet-Appel, "The Agricultural Demographic Transition."
9 Eventually, larger, walled centers: Wilkinson et al., "Settlement Archaeology in Mesopotamia."
9 Transport of cereals was limited: Widell et al., "Land Use of the Model Communities."
9 Productive wetlands or estuaries could not form: Kennett and Kennett, "Early State Formation in Southern Mesopotamia."
9 Estuarine fisheries alone could yield: Day et al., "Emergence of Complex Societies."
10 They offer a myriad of food sources: Pournelle, "Marshland of Cities," 133.
10 If the energy used in the global economy: Global primary energy use in 2019

was roughly 600 exajoules. A hurricane releases about 50 exajoules of latent heat of condensation every day. The energy released by the Asian Monsoon and by global precipitation are easily calculated from estimates of water volume released multiplied by the latent heat of condensation of water. Emanuel, "The Power of a Hurricane."

11 The cycle repeats: Lau et al., "Seasonal and Intraseasonal Climatology."

12 Neolithic communities began settling in the mid-basin: Zhang et al., "The Historical Evolution and Anthropogenic Influences on the Yellow River."

12 The plateau is covered in a few hundred meters: Sun, "Provenance of Loess Material."

12 During much of the fourth millennium: Dong et al., "Mid-Holocene Climate Change."

12 The resulting drought: An et al., "Climate Change and Cultural Response."

13 That affected the rate of breaching: Chen et al., "Socio-economic Impacts on Flooding."

13 which pushed the population: Dong et al., "Mid-Holocene Climate Change."

14 After they disobeyed him, he entrapped them: Birrell, "The Four Flood Myth Traditions."

14 The Lenape, the original inhabitants of Manhattan: Bierhorst, *Mythology of the Lenape*, 43.

14 The Navajos believed they originated: Capinera, "Insects in Art and Religion," 227.

14 When in the sixteenth century, Cristóbal de Molina: Bauer, "Introduction," xxv.

14 Similarly, Maya populations left accounts: García, "The Maya Flood Myth."

14 In Scandinavia, the Old Norse frost giant Bergelmir: Anzelark, *Water and Fire*, 11.

14 Even the aboriginal societies of Australia: Nunn and Reid, "Aboriginal Memories of Inundation."

15 It told an episode of an Akkadian epic: George, *The Epic of Gilgamesh*, 1.

15 In a moment of lyrical abandon: Cregan-Reid. *Discovering Gilgamesh*, 37.

2 The Rise of the Hydraulic State

17 However, the narrow view that attributes: Goldstone and Haldon, "Ancient States, Empires, and Exploitation."

17 But explaining *why* things happen: Carneiro, "A Theory of the Origin of the State."

17 The name first appeared in a second-century CE history: J. Finkelstein, "Mesopotamia."

18 In the Babylonian poem *Enuma Elish*: Foster, *Before the Muses*, 379.

19 As a result, winter rainfall is distributed: Hoskins and Hodges, "New Perspectives on the Northern Hemisphere Winter Storm Tracks."

19 Local weather turns dry as the Indian monsoon: Rodwell and Hoskins, "Monsoons and the Dynamics of Deserts."

19 Their peak flow did not correspond: Wilkinson, "Hydraulic Landscapes and Irrigation Systems of Sumer," 33.

20 Below the rainfed north, the Tigris and Euphrates: Adams, *Heartland of Cities*, 7.

20 They became unstable, meandering: Adams, *Heartland of Cities*, 130.

20 Over time, the rivers began traveling: Cole and Gasche, "Second- and First-Millennium BC Rivers in Northern Babylonia."

20 People took advantage of this configuration: Rothman, "Studying the Development of Complex Society."

20 Over time, sediment from the floods: Wilkinson, "Hydraulic Landscapes and Irrigation Systems of Sumer," 36.

20 The space between the canals: Widell et al., "Land Use of the Model Communities."

20 The effect on productivity was dramatic: Liverani, *Uruk, the First City*, 15.

21 Soils needed care: Wilkinson, "Hydraulic Landscapes and Irrigation Systems of Sumer," 42.

21 As the first Sumerian city-states appeared: Eisenstadt, *The Political Systems of Empires*, 33.

21 When the monsoon shifted north or south: Chen et al., "Socio-economic Impacts on Flooding."

22 Yu the Great: Birrell, "The Four Flood Myth Traditions of Classical China."

22 By the fourth millennium BCE: Rothman, "Studying the Development of Complex Society."

22 In early Uruk, however, that wealth: Liverani, *Uruk, the First City*, 19.

23 Even that intercession was related: Jacobsen, "The Historian and the Sumerian Gods."

23 In *Atrahasis*, one of the oldest Akkadian epics: Dalley, *Myths from Mesopotamia*, 9.

23 In other words, those who wrote *Atrahasis*: Kilmer, "The Mesopotamian Concept of Overpopulation."

23 In the levee system, cultivable land was limited: Wilkinson, "Hydraulic Landscapes and Irrigation Systems of Sumer," 42.

24 The canal system became a transport network: Yoffee, "Political Economy in Early Mesopotamian States."

24 At the point of exchange, evidence: Liverani, *Uruk, the First City*, 40.

24 That was the setting for the oldest: Rey, *For the Gods of Girsu*, 76.

24 Lagash was a state: Cooper, "The Lagash-Umma Border Conflict," 8.

24 It was a "beloved field" of the god Ningirsu: Adams, *Heartland of Cities*, 134.

25 This was probably the Shatt al-Gharraf in Iraq: Cooper, *Reconstructing History from Ancient Inscriptions*, 19.

25 By the eighteenth century BCE: Altaweel, "Simulating the Effects of Salinization on Irrigation Agriculture."

25 Sumerian control over southern Mesopotamia collapsed: Jacobsen and Adams, "Salt and Silt in Ancient Mesopotamian Agriculture."

26 These changes left abundant: H. Weiss, "Megadrought, Collapse, and Causality," 1.

26 When a northern state arose: Liverani, *The Ancient Near East*, 115.

26 Yet, between the twenty-seventh and twenty-sixth century BCE: H. Weiss, "Megadrought and the Akkadian Collapse," 93.

27 This was not the beginning of Moses's story in Exodus: Westenholz, *Legends of the Kings of Akkade*, 38.

27 He then united southern Mesopotamia: A. Westenholz, "The Old Akkadian Period."

27 The grain trade was centralized: Brumfield, "Imperial Methods," 3.

27 about a hundred years after Sargon: Cullen et al., "Climate Change and the Collapse of the Akkadian Empire."

28 The angry god brought famine: Black et al., "The Electronic Text Corpus of Sumerian Literature."

28 They descended from the mountains: Weiss et al., "The Genesis and Collapse of Third Millennium North Mesopotamian Civilization."

3 Bronze Age Globalization

29 Mesopotamia may have sourced its tin: Radivojević et al., "The Provenance, Use, and Circulation of Metals in the European Bronze Age."

30 It had been carrying copper and tin: Kristiansen and Suchowska-Ducke, "Connected Histories: The Dynamics of Bronze Age Interaction and Trade."

30 As a result, the Near East ended up: Kristiansen and Larsson, *The Rise of Bronze Age Society,* 32.

31 Lake Victoria, the source of the White Nile: Yin and Nicholson, "The Water Balance of Lake Victoria."

31 As June approaches, the rain band: Conway, "The Climate and Hydrology of the Upper Blue Nile River."

32 By the time the Ethiopian waters flooded the Nile Valley: Hassan, "The Dynamics of a Riverine Civilization."

32 But basin irrigation, as it is called: Butzer, *Early Hydraulic Civilization in Egypt,* 18.

32 Five million people in the third millennium BCE: Kremer, "Population Growth and Technological Change."

33 Emmer was robust, able to tolerate: Chernoff and Paley, "Dynamics of Cereal Production at Tell el Ifshar."

33 During low waters, it might take two months to travel: Hassan, "The Dynamics of a Riverine Civilization."

33 Water transport was far more energy efficient: Hughes, "Sustainable Agriculture in Ancient Egypt."

34 Cities would produce artifacts: Hassan, "The Dynamics of a Riverine Civilization."

34 While in the Levant or Mesopotamia: Hughes, "Sustainable Agriculture in Ancient Egypt."

34 The integration of the landscape was so strong: Kemp, *Ancient Egypt: Anatomy of a Civilization,* 19.

34 He longed for Egypt: Parkinson, *The Tale of Sinuhe,* 21.

34 "Now there was a famine in the Land": Genesis 12:10, Hebrew Bible.

36 Over time, Syro-Palestinian people had moved: Booth, *The Hyksos Period in Egypt,* 9.

36 The expulsion of the Hyksos: Finkelstein and Silberman, *The Bible Unearthed,* 52.

37 The idea that a climate shift: Weiss, "The Decline of Late Bronze Age Civilization."

37 These were not short-lived conditions: Kaniewski et al., "Drought and Societal Collapse 3200 Years Ago in the Eastern Mediterranean."

37 During Merneptah's reign: Kaniewski et al., "Late Second–Early First Millennium BC Abrupt Climate Changes in Coastal Syria."

38 The fact that, at the end of the thirteenth century BCE: Singer, "New Evidence on the End of the Hittite Empire."

38 Their identity is disputed: Kaniewski et al., "Environmental Roots of the Late Bronze Age Crisis."

38 It was the beginning of the Greek Dark Ages: Finné et al., "Late Bronze Age Climate Change."

38 From there, part of the invasion turned seaward: Kaniewski et al., "The Sea Peoples."

38 "The enemy advances against us": Astour, "New Evidence on the Last Days of Ugarit."

39 "My father, behold, the enemy's ships came": Astour, "New Evidence on the Last Days of Ugarit."

40 The crisis marked the end of the Bronze Age world: Liverani, *Israel's History and the History of Israel*, 33.

4 An Article of Faith

42 In response, the philosopher argued: Needham, *Science and Civilization in China*, 223.

42 Destructive floods, which in the Yellow River basin: Dodgen, *Controlling the Dragon*, 1.

42 He suggested that a bureaucracy: Needham, *Science and Civilization in China*, 232.

44 There, in 256 BCE, Li Bing constructed: Dodgen, *Controlling the Dragon*, 35.

44 A similar story involves the spectactular Zhengguo Canal: Sima Qian, *The First Emperor*, 102.

44 The productivity of these lands: Zhuang, "State and Irrigation."

45 Levantine farmers were incorporated: Brody, "From the Hills of Adonis Through the Pillars of Hercules."

45 Then, the total collapse of the Hittite Empire: Liverani, *Israel's History and the History of Israel*, 41.

45 Between the ninth and seventh centuries BCE: Joffe, "The Rise of Secondary States in the Iron Age Levant."

45 Not a lot—these were small numbers: Broshi and Finkelstein, "The Population of Palestine in Iron Age II."

46 It was rich enough to support a monumental culture: I. Finkelstein, *The Forgotten Kingdom*, 87.

46 The city was in the grips of a terrible famine: 2 Kings 6:24, Hebrew Bible.

46 A population exceeding the carrying capacity of the land: Finkelstein, *The Forgotten Kingdom*, 109.

47 In the end, North Israel proved to be an inevitable target: Finkelstein and Silberman, *The Bible Unearthed*, 200.

47 The western border was trickier to define: Kletter, "Pots and Polities."

47 In the eighth century the population of Judah: Finkelstein and Silberman, *The Bible Unearthed*, 208.

47 The Amarna Letters mention it: Dever, "Histories and Non-Histories of Ancient Israel."

47 When the kingdom of North Israel collapsed: Finkelstein and Silberman, *The Bible Unearthed*, 2.

48 Grain was a superior choice: Faust and Weiss, "Judah, Philistia, and the Mediterranean World."

49 Judah's King Josiah attempted: Jeremiah 44:1 and 46:14, Hebrew Bible.
49 But when Assyrian rule was replaced by the Neo-Babylonians: Luckenbill, *The Annals of Sennacherib*, 17.
49 He had to move to the Jordan Valley: Genesis 13:5–10, Hebrew Bible.
49 The names Isaac gave those wells: Genesis 26:19, Hebrew Bible.
50 The well was closed by a large stone: Genesis 29, Hebrew Bible.
50 Later still, in Numbers, Moses asked the state of Moab: Laster et al., "Water in the Jewish Legal Tradition."
50 "For the land that you are about to enter and possess": Deuteronomy 11:10–12, Hebrew Bible.
51 "Beware, lest your heart be seduced": Deuteronomy 11:13–21, Hebrew Bible.
51 William James believed religion played a moralizing function: Snarey, "The Natural Environment's Impact upon Religious Ethics."
51 harsh environments tend to promote: Mathew and Perreault, "Behavioural Variation in 172 Small-Scale Societies."
51 religious beliefs are a cultural manifestation: Richerson and Boyd, "Cultural Inheritance and Evolutionary Ecology."
51 moralizing gods: Purzycki et al., "Moralistic Gods, Supernatural Punishment and the Expansion of Human Sociality."
52 monotheism, as the philosopher: Kaufmann, "The Bible and Mythological Polytheism."
52 When the Jewish diaspora began: Botero et al., "The Ecology of Religious Beliefs."

5 The Politics of Water

53 Dorian tribes swept in to replace Achaean ones: Drake, "The Influence of Climatic Change on the Late Bronze Age Collapse."
53 By the twelfth century BCE, the Mycenaean civilization: Chadwick, *The Mycenaean World*, 188.
53 Agriculture reverted to subsistence: Morris, "Economic Growth in Ancient Greece."
53 In *Work and Days*: Hansen, *The Other Greeks*, 27.
53 These were all symptoms: Gallant, "Agricultural Systems, Land Tenure, and the Reforms of Solon."
53 In Homer's *Odyssey*: Homer, *Odyssey*, 543.
55 It was the age of the polis: Crouch, *Geology and Settlement: Greco-Roman Patterns*, 380.
55 In the seventh century BCE, people fled: Herodotus, *The Histories*, 291.
55 In fact, the combined effect: J. D. Hughes, *Environmental Problems of the Greeks and the Romans*, 111.
55 Karst was not unusual: Horden and Purcell, *The Corrupting Sea*, 11.
55 Over a relatively brief period, the Greeks spread: Crouch, *Geology and Settlement: Greco-Roman Patterns*, 9.
55 Because colonies were dispersed: Krasilnikoff, "Irrigation as Innovation in Ancient Greek Agriculture."
55 The Greek world was successful: Morris, "Economic Growth in Ancient Greece."
56 Its moral boundaries were as important: Manville, *The Origins of Citizenship in Ancient Athens*, 45.

56 These were infantrymen: Forrest, *The Emergence of the Greek Democracy*, 88.
56 After all, what made the phalanx effective: Aristotle, *The Politics*, 267.
56 The hoplite agrarians: Hansen, *The Other Greeks*, 566.
57 So, to keep the peace: Koutsoyiannis et al., "Urban Water Management in Ancient Greece."
58 The pressure on the majority of farmers: Foxhall, *Olive Cultivation in Ancient Greece: Seeking the Ancient Economy*.
58 Much of the real power resided in the *gerousia*: Forrest, *The Emergence of the Greek Democracy*, 131.
58 The constitution called for nine thousand Spartiates: Herodotus, *The Histories*, 497.
61 But given that much of what could be stored: Bresson, *The Making of the Ancient Greek Economy*, 158.
61 He defined property rights on land: French, "The Economic Background to Solon's Reforms."
61 In his reforms, the individual Athenian citizen: Manville, *The Origins of Citizenship in Ancient Athens*, 133.
61 His fundamental innovation: Hansen, *The Other Greeks*, 109.
62 Agricultural production, in turn: Pnevmatikos and Katsoulis, "The Changing Rainfall Regime in Greece and Its Impact on Climatological Means."
62 He organized the citizenry: Manville, *The Origins of Citizenship in Ancient Athens*, 187.
63 The silver mines of Laurion: J. D. Hughes, *Environmental Problems of the Greeks and the Romans*, 136.
63 By the fourth century BCE, houses: Morris, "Economic Growth in Ancient Greece."
63 In the fourth-century deme of Kephissia: Morison, "An Honorary Deme Decree."
63 Around 320 BCE, Chairephernes: Krasilnikoff, "Irrigation as Innovation in Ancient Greek Agriculture."
63 Chairephernes was essentially made a partner: Bresson, *The Making of the Ancient Greek Economy*, 164.
64 Deforestation on a grand scale followed: Hughes, *Environmental Problems of the Greeks and the Romans*, 69.
64 In his dialogue *Critias*: Hughes, *Environmental Problems of the Greeks and the Romans*, 136.
64 Even though Plato had argued: Plato, *The Republic*, 176.
64 In the dialogue between an Athenian stranger and Clinias: Plato, *The Laws*, 299.
64 Under this legal regime, people: Demosthenes, *Orations*, 55.1.
65 Buried in the folds of his political philosophy: Aristotle, *The Politics*, 114.
65 The polis was at once: Manville, *The Origins of Citizenship in Ancient Athens*, 45.

6 Res Publica

66 From the third century BCE: Büntgen et al., "2500 Years of European Climate Variability."
66 Some regions received more: Lionello et al., "Introduction: Mediterranean Climate—Background Information."

66 In many cases, yields: van Bath, *Agrarian History of Western Europe*, 238.
66 But while a good year could yield eight: Potter, *The Roman Empire at Bay*, 16.
66 Rome alone housed a million people: Kessler and Temin, "The Organization of the Grain Trade in the Early Roman Empire."
67 The capital needed roughly: Erdkamp, *The Grain Market in the Roman Empire*, 2.
67 Production in the region of Lazio: Kessler and Temin, "The Organization of the Grain Trade in the Early Roman Empire."
67 Where geology allowed, they used: Thomas and Wilson, "Water Supply for Roman Farms."
67 In some cases, they shared water: S. P. Scott, *The Civil Law*, 8.3.2.
67 Field drains, whether open ditches: Quilici Gigli, "Su alcuni segni dell'antico paesaggio agrario presso Roma."
67 And despite the Tiber's frequent floods: Thomas and Wilson, "Water Supply for Roman Farms."
67 To deal with this tension: Scott, *The Civil Law*, 39.3.
67 Only what could be shared without impact: Cicero, *On Obligations*, 19.
68 Navigation was particularly: MacGrady, "The Navigability Concept in the Civil and Common Law."
68 Critically, though, public control: Scott, *The Civil Law*, 43.12.1.7.
68 But this was an unacceptable option: Cicero, *On Obligations*, 9.
68 For Sallust, the Roman Republic: Sallust, *Catiline's War, The Jugurthine War, Histories*, 6.
69 Crucial to freedom in Rome: Wirszubski, *Libertas as a Political Idea at Rome*, 1.
69 Cicero defined it as: Cicero, *The Republic and the Laws*, 19.
69 He used the word *populi*: Polybius, *The Histories*, 371.
69 The Centuriate Assembly, for example: Lintott, *The Constitution of the Roman Republic*, 55.
69 Centuries of conflicts followed: Lintott, *The Constitution of the Roman Republic*, 34.
71 The distribution of rainfall: Ulbrich et al., "Climate of the Mediterranean: Synoptic Patterns."
71 Even today, while the amount of water: Dünkeloh and Jacobeit, "Circulation Dynamics of Mediterranean Precipitation Variability."
71 To meet Rome's food demand: Temin, "The Economy of the Early Roman Empire."
72 There is even evidence: Kessler and Temin, "The Organization of the Grain Trade in the Early Roman Empire."
73 When it came to agriculture: Keay, "The Port System of Imperial Rome," 34.
73 The Mediterranean grain market: Duncan-Jones, "Giant Cargo-Ships in Antiquity."
73 The system was linked by two canals: Keay, "The Port System of Imperial Rome," 35.
74 If Augustus was the "architect": Reinhold, *Marcus Agrippa*, vii.
74 At the time of Augustus: Temin, "The Economy of the Early Roman Empire."
74 At the beginning of the imperial age: Evans, "Agrippa's Water Plan."
75 Rural communities used them: Wilson, "Machines, Power and the Ancient Economy."
75 The complex could produce: Hodge, "A Roman Factory."
75 Pliny described how water: Wilson, "Machines, Power and the Ancient Economy."
76 Paying an army its wages: Kay, *Rome's Economic Revolution*, 87.

76 An expanded empire coupled with more limited resources: S. Williams, *Diocletian and the Roman Recovery*, 264.
76 Ditches silted up, sites were abandoned: Cheyette, "The Disappearance of the Ancient Landscape."
76 The unraveling of the Western Roman Empire: Alexander, *Der Fall Roms.*
76 Once fortified, it created: P. S. Wells, "Creating an Imperial Frontier."
76 In the second part of the third century: Pitts, "Relations Between Rome and the German 'Kings' on the Middle Danube."
76 His reforms included the tetrarchy: Williams, *Diocletian and the Roman Recovery*, 264.
77 If Augustus commanded twenty-five legions: Stanković, "Diocletian's Military Reforms."
77 It depended on production and type of crops: A. H. M. Jones, "Capitatio and Iugatio."
77 Control of the treasury: Wickham, *Framing the Early Middle Ages*, 58.
77 This locked the state's ability: Wickham, *The Inheritance of Rome*, 100.
77 Changes in climate: Columella, *De Re Rustica*, 29.
78 Development in the landscape reversed: Cheyette, "The Disappearance of the Ancient Landscape."
78 The Huns were dangerous fighters: Heather, *The Fall of the Roman Empire*, 592.
78 By the first century BCE: Chen et al., "Socio-economic Impacts on Flooding."
78 As nomadic tribes followed the steppes: Lattimore, "Origins of the Great Wall of China."
78 Pressured by a growing sedentary population: Di Cosimo, "China-Steppe Relations in Historical Perspective."
79 Pelagius, who was living in Rome at the time: Salzman, "Jerome and the Fall of Rome in 410."
79 Saint Jerome said: Jerome, "Letter 127, to Principia."

7 Fragments of the Past

84 While the loss of food sources: A. H. M. Jones, *The Later Roman Empire*, 824.
84 The shipment was to travel: Hodgkin, *The Letters of Cassiodorus*, 515.
84 Traffic on the great river: Hallenbeck, *Pavia and Rome.*
84 Grass returned, land converted to pasture: Cheyette, "The Disappearance of the Ancient Landscape."
84 Rivers ran clearer and more regularly: Hoffmann, "Economic Development and Aquatic Ecosystems in Medieval Europe."
85 Smaller houses replaced larger constructions: Lewit, "Vanishing Villas."
85 Society turned to rivers, swamps, and marshes: Squatriti, *Water and Society in Early Medieval Italy*, 66.
85 During the sixth century: Lamb, *Climate, History, and the Modern World*, 149.
85 In 538 CE, Cassiodorus wrote: Hodgkin, *The Letters of Cassiodorus*, 518.
85 The effects of these eruptions: Newfield, "The Climate Downturn of 536–50," 447.
85 Whether that was the specific cause or not: Haldon, "Some Thoughts on Climate Change, Local Environment, and Grain Production in Byzantine Northern Anatolia."
85 Then, the Justinian plague of 541 CE: Procopius, *History of the Wars*, 451.

85 During the first of many outbreaks: Newfield, "Mysterious and Mortiferous Clouds," 89.
85 The last quarter of the sixth century: McCormick et al., "Climate Change During and After the Roman Empire."
85 Gregory, the bishop of Tours, described: Gregory of Tours, *The History of the Franks,* 295.
85 Open fields became rare: Verhulst, *The Carolingian Economy,* 11.
85 Climate conditions improved: McCormick et al., "Volcanoes and the Climate Forcing of Carolingian Europe."
85 One possibility is that the temporary regrowth: Goosse et al., "The Origin of the European 'Medieval Warm Period.'"
86 Whatever the cause, the milder climate: Büntgen et al., "2500 Years of European Climate Variability."
86 Rather, it administered the landscape indirectly: Mann, *The Sources of Social Power,* 376.
86 When Pachomius, the fourth-century ascetic: Brooks Hedstrom, *The Monastic Landscape of Late Antique Egypt,* 114.
87 The only city Christians ought: Brown, *Augustine of Hippo: A Biography,* 285.
87 Ethics and theology trumped practical politics: Weithman, "Augustine's Political Philosophy."
87 Both Pliny and Tacitus had described it: Di Matteo, *Villa di Nerone a Subiaco,* 125.
87 It was held back by a dam: Ashby, *The Aqueducts of Ancient Rome,* 252.
88 Managing water in the early days of Saint Benedict: Wickham, *Framing the Early Middle Ages,* 558.
88 Rivers and waterways, whose flow: Gardoni, "Uomini e acque nel territorio Mantovano."
88 Boatmen were once again allowed: Lambertenghi, "Codex Diplomaticus Longobardiae," 17.
88 Charlemagne's political consolidation was extensive: Verhulst, *The Carolingian Economy,* 58.
88 Absent a strong administrative state: Mann, *The Sources of Social Power,* 390.
88 During this time, monasteries: Arnold, "Engineering Miracles."
89 Emperors gave monasteries and bishops: Dameron, "The Church as Lord," 457.
89 Therefore, the Church and the empire: Rinaldi, "Il Fiume Mobile."
89 Canossa chose Mantua as his capital: Wickham, *The Inheritance of Rome,* 513.
90 Churches and monasteries were often involved: Malara and Coscarella, *Milano & navigli,* 11.
90 The Cistercian Abbey of Chiaravalle: Boucheron, "Water and Power in Milan."
90 While monasteries like Chiaravalle: Grillo, "Cistercensi e società cittadina in età comunale."
91 The most important system of rules: Christensen, "Introduction," xii.
91 It was concerned with providing a framework: Rodes, "The Canon Law as a Legal System."
91 Irnerius's teachings restored: Muller, "The Recovery of Justinian's Digest in the Middle Ages."
91 For example, in December 1125: Torelli, *Regesta Chartarum Italiae,* 138.
91 Four jurors were called in: Gardoni, "Élites cittadine fra XI e XII secolo."
92 The celebrated English Magna Carta: Helmholz, "Continental Law and Common Law."

92 It was full of language and references to the *ius commune*: McSweeney, "Magna Carta, Civil Law, and Canon Law."
92 Between the two lines of poles: "The 1215 Magna Carta: Clause 33, Academic commentary," the Magna Carta Project, trans. H. Summerson et al. http://magnacarta.cmp.uea.ac.uk/read/magna_carta_1215/Clause_33; accessed June 17, 2019.
92 But clause 33 was not really about fishing: Helmholz, "Magna Carta and the Law of Nature."
93 In fact, the glossator Vacarius: Helmholz, "Magna Carta and the Ius Commune."
93 During the twelfth century: Wickham, *Sleepwalking into a New World*, 1.
94 The settlement of the Peace of Constance in 1183: Skinner, *The Foundations of Modern Political Thought*, 3.
94 Countless legal conflicts emerged: Boucheron, "Water and Power in Milan."
94 The fact that cities: Skinner, *The Foundations of Modern Political Thought*, 9.
95 supposedly inspired by a walk: Cavallar, "River of Law."

8 The Republic Returns

96 These texts had resurfaced: Skinner, "Machiavelli's Discorsi and the Prehumanist Origins of Republican Ideas."
97 as a matter of course, farmers: Temin, "The Economy of the Early Roman Empire."
98 Because of this, most scholastic theologians: Epstein, *An Economic and Social History of Later Medieval Europe*, 138.
98 The modern banking sector was born: Cipolla, *Storia Economica dell'Europa Pre-Industriale*, 227.
98 The Aposa was too small: Bartolomei and Ippolito, "The Silk Mill 'alla Bolognese,'" 31.
99 By the fifteenth century the city: Racine, "Poteri medievali e percorsi fluviali nell'Italia padana."
99 This expropriation, an act that would be replicated: Pini, "Classe Politica e Progettualità Urbana a Bologna nel XII e XIII secolo."
100 Grasses were fuel for horses: Putnam et al., "Little Ice Age Wetting of Interior Asian Deserts and the Rise of the Mongol Empire."
100 Much like the Huns: Pederson et al., "Pluvials, Droughts, the Mongol Empire, and Modern Mongolia."
100 By late 1242 the Mongols had withdrawn from Hungary: Büntgen and Cosmo, "Climatic and Environmental Aspects of the Mongol Withdrawal from Hungary in 1242 CE."
101 Work started more or less at the same time: Haw, *Marco Polo's China*, 68.
101 When the Ming dynasty inherited it: Huang, *Taxation and Governmental Finance in Sixteenth-Century Ming China*, 316.
101 For centuries, Western visitors: Hanyan, "China and the Erie Canal."
102 In the summer of 1315, the archbishop of Canterbury: Kershaw, "The Great Famine and Agrarian Crisis in England, 1315–1322."
102 The change marked the shift: Nesje and Dahl, "The 'Little Ice Age'—Only Temperature?"
102 The climate shift also took its toll: Newfield, "A Cattle Panzootic in Early Fourteenth-Century Europe."

102 Malnutrition and cramped living quarters: Jordan, "The Great Famine: 1315–1322 Revisited."

102 By the end of its first cycle: Epstein, *An Economic and Social History of Later Medieval Europe*, 171.

103 Standardization of thread sizes: Blanshai, *Politics and Justice in Late Medieval Bologna*, 500.

103 Successful drainage only made bogs sink further: De Vries, *The Dutch Rural Economy in the Golden Age*, 28.

104 Eventually, the windmill made its first appearance: Kaijser, "System Building from Below."

105 With the 1493 papal bull *Inter Caetera*: W. H. Scott, "Demythologizing the Papal Bull 'Inter Caetera.'"

105 More's *Utopia* relied on an extraordinarily benign river: More, *Utopia*, 57.

105 More's writing was an indication: Davies, "The First Discovery and Exploration of the Amazon in 1498–99."

106 For example, Friar Gaspar de Carvajal: Medina, *Relación que escribió Fr. Gaspar de Carvajal*, 1.

106 Unlike arid or semi-arid climates: Wang and Dickinson, "A Review of Global Terrestrial Evapotranspiration."

106 They had the far more complicated: Erickson, "The Domesticated Landscape of the Bolivian Amazon."

106 People used artificial mounds: Erickson, "An Artificial Landscape-Scale Fishery in the Bolivian Amazon."

107 Several studies suggest a pre-Columbian population: Meggers, "Environmental Limitation on the Development of Culture."

107 Because farmers cultivated tracts of land: Willis et al., "How 'Virgin' Is the Virgin Rainforest?"

107 The regrowth in forest was so significant: Koch et al., "Earth System Impacts of the European Arrival."

108 It too eventually dissipated: Skinner, "Machiavelli's Discorsi and the Pre-humanist Origins of Republican Ideas."

108 The mind behind that revolution: Masters, *Machiavelli, Leonardo, and the Science of Power*, 260.

108 He tripled them again the following year: Mann, *The Sources of Social Power*, 451.

108 Machiavelli noted that financial resources: Barthas, "Machiavelli, Public Debt, and the Origins of Political Economy."

108 He believed "the republic should keep the public rich and its citizens poor": Machiavelli, *Discorsi sopra la prima Deca di Tito Livio*, 94.

108 That had been the fundamental tension: Nelson, "Republican Visions," 193.

108 In his day, admiration for the Roman world: Warner and Scott, "Sin City."

109 There are some passing comments: Colish, "Machiavelli's Art of War."

109 Eventually, Machiavelli abandoned the project: Masters, *Machiavelli, Leonardo, and the Science of Power*, 240.

9 Water Sovereignty

111 Migration increased, leading to exchanges: Parker, *Global Interactions in the Early Modern Age*, 3.

111 In the High Middle Ages: David, "The Scheldt Trade and the 'Ghent War' of 1379–1385."

112 By the second half of the fifteenth century: Bindoff, *The Scheldt Question to 1839*, 32.

112 The former sought free passage: Cornelisse, "The Economy of Peat and Its Environmental Consequences," 95.

112 In practice, though, by that point: Wijffels, "Flanders and the Scheldt Question."

113 By 1581 the northern provinces: Bindoff, *The Scheldt Question to 1839*, 82.

114 No wonder: the first half of the seventeenth century: De Vries, "The Economic Crisis of the Seventeenth Century After Fifty Years."

114 When Charles V deployed Spain: Ogilvie, "Germany and the Seventeenth-Century Crisis."

114 Similarly, in Italy population declined: Parker, "The Global Crisis of the Seventeenth Century Reconsidered."

114 In the middle of the century, China: De Vries, *The Dutch Rural Economy in the Golden Age*, 1.

115 One possible culprit: Gray et al., "Solar Influences on Climate."

115 Volcanic activity may have also played a role: MacDonald and McCallum, "The Evidence for Early Seventeenth-Century Climate from Scottish Ecclesiastical Records."

115 The number of famine years per decade: Zhang et al., "The Causality Analysis of Climate Change and Large-Scale Human Crisis."

115 "Men had need to pray for Faire Weather": Bacon, *The Essayes or Counsels Civill and Morall*, 45.

115 Court depositions of Protestants: Zhang et al., "The Causality Analysis of Climate Change and Large-Scale Human Crisis."

116 During the wars of the seventeenth century: Cyberski et al., "History of Floods on the River Vistula."

116 Dutch-Flemish emigrants brought: Ciriacono, *Building on Water*, 194.

117 While his direct engagement: Israel, *The Dutch Republic: Its Rise, Greatness, and Fall*, 272.

117 The undertaker was the Bedford Level Corporation: Roberts, "The Earl of Bedford and the Coming of the English Revolution."

117 The project faced significant technical challenges: Knittl, "The Design for the Initial Drainage of the Great Level of the Fens."

118 The consequent expropriation of property: Bowring, "Between the Corporation and Captain Flood," 235.

118 Prominent cartographers: Degroot, *The Frigid Golden Age*, 57.

119 It became an attractive mode of transport: De Vries, *Barges and Capitalism*, 26.

119 The treaty was, in effect, about sovereignty: Baena, "Negotiating Sovereignty."

119 Article 14 in particular: Rowen, *The Low Countries in Early Modern Times*, 179.

119 The closing of the Scheldt: Wijffels, "Flanders and the Scheldt Question."

119 Spain was forced to recognize: Bindoff, *The Scheldt Question to 1839*, 102.

119 By the end of the 1660s: De Vries, *Barges and Capitalism*, 34.

120 Evidence from logbooks suggests: Degroot, *The Frigid Golden Age*, 97.

121 In continental Europe, the Peace of Münster: Gross, "The Peace of Westphalia, 1648–1948."

122 after 1688 in England, river improvement proposals: Bogart, "Did the Glorious Revolution Contribute to the Transport Revolution?"

122 Return on invested capital: North and Weingast, "Constitutions and Commitments."

10 American River Republic

123 "those who would give up": Franklin, "Pennsylvania Assembly: Reply to the Governor, 11 November 1755," Founders Online, National Archives, accessed September 29, 2019, https://founders.archives.gov/.

124 They were the source: Bailey, *Description of the Ecoregions of the United States*, 126.

124 He knew that without a cost-effective way: Doyle, *The Source*, 17.

124 After independence, he decided: Littlefield, "The Potomac Company."

124 Large landowners and wealthy traders invested: Littlefield, "The Potomac Company," n12.

126 The Mount Vernon Compact, as it became known: Rowland, "The Mount Vernon Convention."

126 On January 21, 1786, the Virginia legislature: Madison, *The Papers of James Madison*, 470.

126 If the Articles of Confederation of 1776: Kramnick, "Editor's Introduction," 16.

126 The founders of America: Richard, *The Founders and the Classics*, 1.

126 He invited all of the states: Fiske, *The Critical Period of American History*, 212.

127 "the power of regulating trade": Hamilton, *The Papers of Alexander Hamilton*, 686.

127 It ultimately led to a draft constitution: Littlefield, "The Potomac Company."

127 James Madison brought up the example: Farrand, *Records of the Federal Convention of 1787*, vol. 1, 330.

127 Luther Martin, attorney general of Maryland: Farrand, *Records of the Federal Convention of 1787*, vol. 1, 439.

127 Gouverneur Morris of Pennsylvania: Farrand, *Records of the Federal Convention of 1787*, vol. 2, 586.

127 In his impassioned case for independence: Paine, *Common Sense*, 27.

127 These were typically publicly recognized private corporations: Trew, *Infrastructure Finance and Industrial Takeoff in England*.

128 When the panic of 1837 hit: Doyle, *The Source*, 28.

129 Most large cities, such as Boston: C. Smith, *City Water, City Life*, 57.

129 The business competition for New York: Murphy, "'A Very Convenient Instrument': The Manhattan Company."

130 It first merged with Chase: J. Salzman, *Drinking Water: A History*, 61.

130 As far as water is concerned: Cutler and Miller, "Water Everywhere."

130 Peak floods can carry: Niebling et al., "Challenge and Response in the Mississippi River Basin."

131 For one, the Louisiana Purchase was fateful: Hobsbawm, *The Age of Revolution*, 299.

131 Over the course of the nineteenth century: Portmann et al., "Spatial and Seasonal Patterns in Climate Change, Temperature, and Precipitation Across the United States."

131 By the end of the century: Lipsey, "U.S. Foreign Trade and the Balance of Payments."
132 There were echoes of Greece's democratic project: Forrest, *The Emergence of the Greek Democracy*, 35.
132 Without a comparable federal policy: Rogers, *America's Water*, 48.
133 The arid region of the southwest: Powell, *Report on the Lands of the Arid Region of the United States*, vii.
133 The Ingénieurs des Ponts et Chaussées: Belhoste, "Les Origines de L'École Polytechnique."
134 As the country expanded: Doyle, *The Source*, 36.
134 The channel had to be kept deep: Twain, *Life on the Mississippi*, 400–3.
134 "bonds that unite us as one People": Everett, *Address of Hon. Edward Everett*, 82.
134 He drew generously: Thucydides, *The Peloponnesian Wars*, 91.
135 Niccolò Machiavelli had warned: Machiavelli, *Discorsi sopra la prima Deca di Tito Livio*, 25.
135 Its exceptional size, geography, demographic complexity: De Tocqueville, *Democracy in America*, 655.

11 Global Water Empire

137 In 1858, the philosopher John Stuart Mill: Harris, "John Stuart Mill: Servant of the East India Company."
137 Mill largely ignored pre-existing water infrastructure: Mill, *Memorandum of the Improvements in the Administration of India*, 52.
137 These projects supported settler colonization: Bell, "John Stuart Mill on Colonies."
137 The project was eventually transferred: Atchi Reddy, "Travails of an Irrigation Canal Company in South India."
138 That was the opening shot: Gilmartin, *Blood and Water*, 27.
138 British farmers were unprepared: J. Brown, *Agriculture in England*, 1.
139 Water resource development of India accelerated: Brezis, "Foreign Capital Flows in the Century of Britain's Industrial Revolution."
139 Canals and other water projects: Cain, "British Free Trade, 1850–1914: Economics and Policy."
139 In the 1890s, canal building in the Indus basin: Gilmartin, *Blood and Water*, 147.
139 They represented the mix of social philosophy: Gilmour, "The Ends of Empire."
140 Opium from India: Richards, "The Indian Empire and Peasant Production of Opium."
141 By the 1870s, when the revenue of the Raj: J. B. Brown, "Politics of the Poppy."
141 Of course, all of this trade went: Richards, "The Indian Empire and Peasant Production of Opium."
141 The plant was not indigenous: Asthana, *The Cultivation of the Opium Poppy in India*.
141 He was convinced this would be its demise: Karl Marx, "Revolution in China and in Europe." *New York Daily Tribune*, June 14, 1853.
142 Addiction is hard to estimate within China: Brown, "Politics of the Poppy."
142 In the spring of 1876: Hao et al., "1876–1878 Severe Drought in North China."

142 During the summer of that year: Cook et al., "Asian Monsoon Failure and Megadrought During the Last Millennium."
142 The imperial state, already struggling: Janku, "Drought and Famine in Northwest China: A Late Victorian Tragedy?"
143 Chinese philanthropists started a fund-raising campaign: Edgerton-Tarpley, *Tears from Iron.*
144 British prospectors kicked off mineral explorations: Reader, *Africa,* 16.
144 The tone was set in the 1870s: Kenyon, *Dictatorland,* xiii.
144 Their diamond and gold discoveries: Mavhunga, "Energy, Industry, and Transport in South-Central Africa's History."
144 For the same reason, Henry Stanley: Jeal, *Explorers of the Nile,* 1.
144 Often couched as scientific activities: Westermann, "Geology and World Politics: Mineral Resource Appraisals as Tools of Geopolitical Calculation, 1919–1939."
144 Suddenly, Africa was well beyond appealing: Mavhunga, "Energy, Industry, and Transport in South-Central Africa's History."
144 Seeing the trading opportunity, King Leopold II of Belgium: Reader, *Africa,* 528–32.
144 In *Heart of Darkness*, with some irony: Conrad, *Heart of Darkness,* 124.
145 Even though the conference was focused: Craven, "Between Law and History: The Berlin Conference of 1884–1885 and the Logic of Free Trade."
145 The king of Belgium had secured for himself: Reader, *Africa,* 534–42.
145 The Berlin Conference simply adopted: Salman, "The Helsinki Rules, the UN Watercourses Convention and the Berlin Rules: Perspectives on International Water Law."
146 It was also to plan the works: "General Act of the Conference of Berlin Concerning the Congo."
146 Science was not on their side: Schwartz, "Illuminating Charles Darwin's Morality."
146 In particular, natural selection appeared to provide: Haller, "The Species Problem: Nineteenth-Century Concepts of Racial Inferiority in the Origin of Man Controversy."
146 Near East exploration: Thornton, "British Policy in Persia, 1858–1890, I."
147 This eventually led him to be stationed in Persia: G. Rawlinson, *A Memoir of Major-General Sir Henry Creswicke Rawlinson,* 21.
147 He visited the rock of Behistun: H. C. Rawlinson, "Memoir on the Babylonian and Assyrian Inscriptions."
147 Even the Sydenham Crystal Palace: Holloway, "Biblical Assyria and Other Anxieties in the British Empire."
147 The British had competed with the French: Malley, "Layard Enterprise: Victorian Archaeology and Informal Imperialism in Mesopotamia."
147 From it, Jean-François Champollion deciphered: Champollion, *Lettre à M. Dacier,* 52.
147 In it were fifty preserved mummies: Edwards, "Was Ramases II the Pharaoh of the Exodus?"
148 The outbreak followed: Winterton, "The Soho Cholera Epidemic 1854."
148 John Snow, a local physician: Snow, *On the Mode of Communication of Cholera,* 31.
149 The unsanitary conditions highlighted by Snow: Broich, "Engineering the Empire: British Water Supply Systems and Colonial Societies."
149 In 1866, the Royal Commission on Water Supply: Ritvo, *The Dawn of Green,* 65.

149 "perpetually striving to amend, the course of nature": Mill, *Nature, the Utility of Religion, and Theism,* 32.

149 In fact, the landscape had been heavily modified by man: Ritvo, *The Dawn of Green,* 26.

12 The Great Utopian Synthesis

151 In those sixteen years, Dr. Sun: Bergère, *Sun Yat-sen,* 59.

151 He spent his days: Schiffrin, *Sun Yat-sen and the Origins of the Chinese Revolution,* 135.

152 The last subsistence crises of the Western world: Berger and Spoerer, "Economic Crises and the European Revolutions of 1848."

152 In fact, Marx went as far as to say: Marx, *Capital,* 284.

153 Montesquieu had used China: Montesquieu, *The Spirit of the Laws,* 126.

153 Marco Polo, who visited: Smith, *The Wealth of Nations,* 174.

153 For the same reason, Marx largely ignored Russia: Karl Marx, "The British and Chinese Treaty." *New York Daily Tribune,* October 15, 1858.

154 Volkhovsky had been persecuted: Bergère, *Sun Yat-sen,* 65.

154 Through her, Chernyshevsky revealed: Fokkema, *Perfect Worlds,* 211.

154 And people transformed along with the landscape: Chernyshevsky, *A Vital Question; or, What Is to Be Done,* 384.

155 Seventy percent of people and economic activity: Kelly et al., "Large-Scale Water Transfers in the USSR."

155 Russia was one of the first: Popescu, "Casting Bread upon the Waters," 17.

155 Electrification, which had started to spread: Coopersmith, *The Electrification of Russia,* 78.

156 When the First World War broke out: Josling et al., "Understanding International Trade in Agricultural Products."

156 They judged the capitalist transformation: Marx and Engels, "Manifesto of the Communist Party."

156 In his world, nature had been tamed: Bellamy, *Looking Backwards,* 79.

157 It was a vision for a new Jewish state: Penslar, "Between Honor and Authenticity: Zionism as Theodor's Life Project."

157 Herzl imagined Altneuland independent of coal: Herzl, *Altneuland.*

157 Their leader, Prosper Enfantin: Taboulet, "Aux origines du canal de Suez."

158 The French company entered into a concession agreement: McCullough, *The Path Between the Seas,* 61.

158 Nationalistic propaganda and a fair amount of corruption: Arendt, *The Origins of Totalitarianism,* 123.

159 Who exactly received those funds: Harding, *The Untold Story of Panama,* 47.

159 The treaty that emerged: Major, "Who Wrote the Hay–Bunau-Varilla Convention?"

160 The design had been used by Thomas Telford: McCullough, *The Path Between the Seas,* 28.

161 The deflagration blew up Gamboa dike: "Wilson Opens Panama Canal," *Aurora Democrat,* October 17, 1913.

161 The Panama Canal had connected the oceans: Palka, "A Geographic Overview of Panama."

161 In visiting the Gatun lock, Van Ingen saw: Van Ingen, "The Making of a Series of Murals at Panama."
162 And now it was Panama: Belhoste, "Les Origines de L'École Polytechnique."
162 The forests surrounding the lake: Carse, "Nature as Infrastructure: Making and Managing the Panama Canal Watershed."
163 "this project will be more profitable": Sun, *The International Development of China,* 33.

13 Setting the Stage for Revolution

167 By 1900, it was a free-trade zone: "General Act of the Conference of Berlin Concerning the Congo."
168 human population had grown: Kremer, "Population Growth and Technological Change."
168 Indeed, by 2000, over six billion people: Bongaarts, "Human Population Growth and the Demographic Transition."
168 But to satisfy the demands of mass consumption: Smil, *Energy and Civilization.*
169 The two decades leading up to the First World War: Wimmer and Min, "From Empire to Nation-State."
170 So, in 1902, British engineers: Robinson et al., "The High Dam at Aswan," 237.
170 Then, in 1910, Murdoch MacDonald: Hurst, "Progress in the Study of the Hydrology of the Nile in the Last Twenty Years."
170 Dams would store water in Lakes Victoria and Albert: Glennie, "The Equatorial Nile Project."
170 The water would then bypass the Sudd: Jonglei Investigation Team, "The Equatorial Nile Project and Its Effects in the Sudan."
170 Despite the success of the nineteenth century's Homestead Act: Pisani, *Water and American Government,* xi.
171 President Theodore Roosevelt signed: McGerr, *A Fierce Discontent,* xiii.
171 The expectation was that irrigation projects: Vincent et al., *Federal Land Ownership.*
171 Over time, reclamation became: Pisani, "State vs. Nation: Federal Reclamation and Water Rights in the Progressive Era."
171 The act was accompanied: Gates, "Homesteading in the High Plains."
171 In the first two decades of the twentieth century: Samson et al., "Great Plains Ecosystems: Past, Present, and Future."
172 In 1878 three-quarters of the railway workers: Snowden, *The Conquest of Malaria,* 8.
172 By 1915, 770,000 hectares: Porisini, "Le bonifiche nella politica economica dei governi Cairoli e Depretis."
172 A century later, human work accounted: Smil, "Energy in the Twentieth Century."
173 between the end of the nineteenth century: Cintrón, *Historical Statistics of the Electric Utility Industry.*
173 In the seventeenth century, the Tokugawa shogunate: Totman, *Early Modern Japan,* 11.
174 "our river systems are better adapted": Roosevelt, "Message from the President of the United States."
175 In the next decade hydropower grew: Severnini, *The Power of Hydroelectric Dams,* 65.
175 By the time Russia entered the First World War: Coopersmith, *The Electrification of Russia,* 78.

175 But to fully industrialize: Mori, "Le guerre parallele. L'industria elettrica in Italia nel periodo della grande guerra."
175 Italy's northern regions had steep rivers: Nitti, *L'Italia all'alba del secolo XX*, 169.
175 Reservoirs, mostly for water supply: Temporelli and Cassinelli, *Gli acquedotti genovesi*, 82.
176 This, as it turned out, proved: Pavese, *Cento Anni di Energia: Centrale Bertini*, 18.
176 The company proposed to electrify it: Conti, "Alle origini del sistema elettrico toscano."
176 investment capital came from Germany: Bruno, "Capitale straniero e industria elettrica nell'Italia meridionale."
176 By 1900, demand had already caught up: Pavese, *Cento Anni di Energia: Centrale Bertini*, 41.
176 Two decades later that number had grown: Ministero per la Costituente, *Rapporto della Commissione Tecnica, Industria*, 86.
178 In 1844 he had tried to answer: Dupuit, "On the Measurement of the Utility of Public Works."
179 The bridge infringed their access to the sea: U.S. Supreme Court, *Willamette Iron Bridge Co. v. Hatch*, 125 U.S. 1 (1888).
179 Section 9 of the 1899 Rivers and Harbors Act: Barker, "Sections 9 and 10 of the Rivers and Harbors Act of 1899."
179 It would make recommendations: Rivers and Harbor Act of 1902, 32 Stat. 372-73 (1903).
179 Over the subsequent two decades: River and Harbor Act, 41 Stat. 1009-10 (1920).

14 Crisis and Its Discontent

181 In an infamous 1914 editorial, H. G. Wells: Wells, *The War That Will End War*.
181 The destruction of Europe traveled along: Johnson, *Topography and Strategy in the War*, 1.
182 Similarly, Britain's dependence on commodities: Offer, *The First World War: An Agrarian Interpretation*, 468.
183 The recession of 1920: Denman and McDonald, *Unemployment Statistics from 1881 to the Present Day*, 5.
183 After the war and through most of the twenties: Lebergott, "Annual Estimates of Unemployment in the United States."
183 Most new settlers came: Libecap and Hansen, "'Rain Follows the Plow' and Dry-farming Doctrine."
183 They spread like wildfire: Baker, "Turbine-Type Windmills of the Great Plains and Midwest."
184 In hindsight he was right: Keynes, *The Economic Consequences of the Peace*, 52.
184 Rather, they drew inspiration: Lenin, "The State and Revolution," 381.
184 Wartime Germany was the archetype: Scott, *Seeing like a State*, 100.
185 Those who did not, or could not, understand: Lenin, "What Is to Be Done?," 347.
185 If the revolution was to succeed: Lenin, "Our Foreign and Domestic Position," 408.
185 The scale of what the Soviet Union: Richter, "Nature Mastered by Man: Ideology and Water in the Soviet Union."

185 Hydropower was at its heart: Coopersmith, *The Electrification of Russia,* 148.
186 American firms supplied technology and skills: Melnikova-Raich, "The Soviet Problem with Two 'Unknowns.'"
186 The disruption of international trade: Preti, "La politica agraria del fascismo: Note introduttive."
187 The Italian population had turned: Einaudi, "Il dogma della sovranità," 2.
187 Liberalization had struggled: D'Antone, "Politica e cultura agraria: Arrigo Serpieri."
187 The king, unwilling to test: Guerriero, "La Generazione di Mussolini."
187 His political education had been shaped by unionism: Delzell, "Remembering Mussolini."
188 The waters of the country became: Salvemini, *Le origini del fascismo in Italia: Lezioni di Harvard,* 452.
188 His capable propaganda machine: Diggins, *Mussolini and Fascism: The View from America,* 37.
188 Between 1925 and 1928: Migone, *The United States and Fascist Italy.*
189 No matter how intellectually inconsistent: De Felice, *Mussolini e Hitler: I rapport segreti, 1922–1933,* 17.
190 This is when Dr. Sun retreated: Sun, *The International Development of China,* v.
191 The American minister to Beijing: Bergère, *Sun Yat-sen,* 280.
191 "The disposition towards public affairs": Keynes, *The End of Laissez-Faire.*
192 Where the state could act: Keynes, *The General Theory of Employment, Interest and Money,* 383.
193 Having experienced the gains: Keynes, *The Economic Consequences of the Peace,* 3.
193 That opened the door: Berend, *An Economic History of Twentieth-Century Europe,* 1.
193 In the United Kingdom: Tanzi and Schuknecht, *Public Spending in the 20th Century,* 3.
193 The arrangement had seemed to make sense: Johnson, "Freedom of Navigation for International Rivers: What Does It Mean?"
194 They believed the Belgian government: Permanent Court of International Justice, December 12th, 1934. The Oscar Chinn Case. Series A./B., Fascicule 63: 65–90.

15 Industrializing Modernity

195 Manufacturing was sensitive to the cost of electricity: Schramm, "The Effects of Low-Cost Hydro Power on Industrial Location."
195 With advances in turbine technology: Billington et al., *The History of Large Federal Dams,* 62.
195 The Columbia, for example: Severnini, *The Power of Hydroelectric Dams,* 77.
195 Because technology to transmit: Severnini, *The Power of Hydroelectric Dams,* 77.
196 Water and its geography provided the master plan: Billington and Jackson, *Big Dams of the New Deal Era,* 3.
196 The Congo River alone was one-quarter: G. O. Smith, *World Atlas of Commercial Geology,* 3.
196 For all its fame, the Colorado: Woodhouse et al., "Updated Streamflow Reconstructions for the Upper Colorado River Basin."
196 Without expensive and continuous maintenance: Hiltzik, *Colossus,* 19.
197 The issue of who had authority over Sudan: Crabitès, "The Nile Waters Agreement."

197 The international commission of three experts: Gebbie et al., *Report of the Nile Projects Commission*, 53–58.
197 The negotiation was about supporting: Woodhouse et al., "The Twentieth-Century Pluvial in the Western United States."
198 Cory knew that the choices: Gebbie et al., *Report of the Nile Projects Commission*, 59–77.
198 In the end Anglo-Egyptian Sudan and Egypt: Crabitès, "The Nile Waters Agreement."
198 This meant that federal coordination was required: Hiltzik, *Colossus*, 61.
198 The Missouri, the upper Mississippi: Henry, "Frankenfield on the 1927 Floods in the Mississippi Valley."
198 It covered over a million hectares: Barry, *Rising Tide*, 528.
198 Its damages were equivalent to a third: Lohof, "Herbert Hoover, Spokesman of Humane Efficiency: The Mississippi Flood of 1927."
199 He was always in search of water: Hiltzik, *Colossus*, 99.
200 His instruments of destruction: Joseph Stalin, Speech Delivered at a Meeting of Voters of the Stalin Electoral District, Moscow. February 9, 1946.
200 the Volga, the longest river in Europe: Hooson, "The Middle Volga: An Emerging Focal Region in the Soviet Union."
200 Kuybyshev Dam created one of the largest reservoirs: Richter, "Nature Mastered by Man: Ideology and Water in the Soviet Union."
200 Sixty-five percent of the agricultural land: Tolmazin, "Recent Changes in Soviet Water Management."
200 Lenin had ordered three hundred thousand hectares: Teichmann, "Canals, Cotton, and the Limits of De-colonization in Soviet Uzbekistan."
200 The political objective of controlling rural peasants: Merl and Templer, "Why Did the Attempt Under Stalin to Increase Agricultural Productivity Prove to Be Such a Fundamental Failure?"
201 As state pressure and violence mounted: Davies et al., "Stalin, Grain Stocks and the Famine of 1932–33."
201 California was deemed to be the one place: Peterson, "US to USSR: American Experts, Irrigation, and Cotton in Soviet Central Asia, 1929–32."
201 Famously, in 1934, at the Seventeenth Party Congress: Smolinski, "The Scale of Soviet Industrial Establishments."
201 Just a month before the Soviet Union: Bourke-White, *Portrait of Myself*, 141.
202 When the global financial world collapsed: Giordano et al., "Italy's Industrial Great Depression: Fascist Price and Wage Policies."
202 The credit system collapsed: Bel, "The First Privatization: Selling SOEs and Privatizing Public Monopolies in Fascist Italy."
202 The banking and industrial model: Preti, "La politica agraria del fascismo: Note introduttive."
203 The regime had set high prices for grains: Corner, "Considerazioni sull'agricoltura capitalistica durante il fascismo."
204 It included all water development: Isenburg, *Acque e Stato: Energia, Bonigiche, Irrigazione in Italia fra 1930 e 1950*.
204 The grand scheme of the Fascist regime: Salvemini, "Can Italy Live at Home?"
204 Sixteen thousand kilometers of canals: Caprotti, "Malaria and Technological Networks."
204 Towards the end of the conflict: Snowden, *The Conquest of Malaria*, 6.

205 He eventually consolidated his hold: Dawn, "The Origins of Arab Nationalism."
205 The only other viable, long-term option: T. C. Jones, "State of Nature: The Politics of Water in the Making of Saudi Arabia."
206 The commission went nowhere: Reimer, "The King-Crane Commission at the Juncture of Politics and Historiography."
206 While regional production had improved: Woertz, *Oil for Food,* 35.
206 It was an economic development mission: Twitchell, *Saudi Arabia,* 139.
206 The Anglo-Iranian Oil Company: Luciani, "Oil and Political Economy in the International Relations of the Middle East," 111.
206 In 1933, the Standard Oil Company of California: Fitzgerald, "The Iraq Petroleum Company, Standard Oil of California, and the Contest for Eastern Arabia."
206 The company extended a loan to Saudi Arabia: Twitchell, *Saudi Arabia,* 148.
207 By July, seven sequential storms: Courtney, *The Nature of Disaster in China,* 5.

16 FDR's Modernization Project

210 As other countries retaliated: Josling et al., "Understanding International Trade in Agricultural Products."
210 This was not deficit spending in a Keynesian sense: Fishback and Kachanovskaya, "The Multiplier for Federal Spending in the States During the Great Depression."
211 The Depression deepened: Fishback, "US Monetary and Fiscal Policy in the 1930s."
211 In the early thirties: Woodhouse and Overpeck, "2000 Years of Drought Variability in the Central United States."
211 When cold air: B. I. Cook et al., "Dust and Sea Surface Temperature Forcing of the 1930s 'Dust Bowl' Drought."
211 Rivers were the answer: Kitchens and Fishback, "Flip the Switch: The Impact of the Rural Electrification Administration, 1935–1940."
211 And they would be put to work: Gordon, *The Rise and Fall of American Growth,* 52.
212 But her beginnings were in industrial photography: Vials, "The Popular Front in the American Century."
212 It looked like a massive defensive wall: Billington et al., *The History of Large Federal Dams,* 235.
212 "nation, conceived in adventure": Luce, "The American Century."
212 Roosevelt hoped that having government-developed projects: Franklin D. Roosevelt, October 22, 1932. Extemporaneous Remarks, Knoxville, Tennessee.
212 They were economic development investments: Franklin D. Roosevelt, January 15, 1940. Tennessee Valley Authority Message to Congress.
213 Because of its high variability: Miller and Reidinger, *Comprehensive River Development,* 9.
213 It was enormous: Lilienthal, *TVA: Democracy on the March,* 11.
213 It was an ideology: Neuse, "TVA at Age Fifty—Reflections and Retrospect."
213 "transcends mere power development": Franklin D. Roosevelt, April 10, 1933. Tennessee Valley Authority Message to Congress.
213 "a corporation clothed with the power": Franklin D. Roosevelt, April 10, 1933. Tennessee Valley Authority Message to Congress.
214 Without appropriate political safeguards: Scott, *Seeing like a State,* 93.

214 As a result, its board: Lilienthal, *TVA: Democracy on the March*, 67.
214 He became its chairman in 1941: Neuse, *David E. Lilienthal*, xv.
214 His book *TVA: Democracy on the March*: Lilienthal, *TVA: Democracy on the March*, xi.
214 The empowerment of the beneficiaries: Lilienthal, *TVA: Democracy on the March*, 80.
215 It was not just a regional authority: Lilienthal, *TVA: Democracy on the March*, 149.
215 TVA also managed fourteen more dams: Kitchens, "The Role of Publicly Provided Electricity in Economic Development."
215 During flood season the system: Miller and Reidinger, *Comprehensive River Development*, 35.
215 Modern water management had fueled: Kline and Moretti, "Local Economic Development, Agglomeration Economies, and the Big Push."
216 Then, two economists: Domar, "Capital Expansion, Rate of Growth, and Employment."
216 The Harrod-Domar model: Alacevich, *The Political Economy of the World Bank*, 26.
216 In truth, and seen with the benefit of hindsight: Bhatia and Scatasta, "Methodological Issues."
217 Over time, the distance: Molle, "River-Basin Planning and Management."
218 Indeed, China was one of the countries: Lilienthal, *TVA: Democracy on the March*, 203.
218 While China was still reeling: Courtney, *The Nature of Disaster in China*, 181.
218 The Chinese Nationalist government: Sneddon, *Concrete Revolution*, 36.
218 He was known as the first: Wolman and Lyles, "John Lucian Savage, 1879–1967."
219 There was even talk: Ekbladh, "'Mr. TVA': Grass-Roots Development, David Lilienthal, and the Rise and Fall of the Tennessee Valley Authority."
219 The idea of a Damodar Valley Corporation: D'Souza, *Drowned and Dammed*, 194.
220 It was part of a remarkable cascade of dams: Sneddon, *Concrete Revolution*, 38.
220 And he believed it would bring: Perkins, *The Roosevelt I Knew*, 88.
220 Aramco had struck commercially viable oil in 1938: T. C. Jones, "State of Nature: The Politics of Water in the Making of Saudi Arabia," 235.
220 Oil was crucial to build wealth: Luciani, "Oil and Political Economy in the International Relations of the Middle East."
221 It was the first attempt at farm mechanization: Woertz, *Oil for Food*, 64.
221 The principal theme of those developments: Hart, *Saudi Arabia and the United States*, 29.
221 The recommendations of the mission: T. C. Jones, "State of Nature: The Politics of Water in the Making of Saudi Arabia," 242.
221 The technical expertise of Aramco's staff: Crary, "Recent Agricultural Developments in Saudi Arabia."
221 But its politics were going to be another matter: Woertz, *Oil for Food*, 64.
221 Then, the same day, it was Haile Selassie: Gardner, *Three Kings*, 27.
221 After him, the president hosted: T. C. Jones, "America, Oil, and War in the Middle East," 208.
221 Roosevelt, on the other hand: Gardner, *Three Kings*, 33.
222 And it was clear that Saudi Arabia's: Yergin, *The Prize*, 385.
222 By this point, Roosevelt believed: Lippman, "The Day FDR Met Saudi Arabia's Ibn Saud."
222 In a prophetic moment, Roosevelt told: Perkins, *The Roosevelt I Knew*, 88.
222 Even the Marshall Plan was described: Macekura, "The Point Four Program and U.S. International Development Policy."

222 The TVA was so popular: Ekbladh, "'Mr. TVA': Grass-Roots Development, David Lilienthal, and the Rise and Fall of the Tennessee Valley Authority."

222 At his second inaugural address: Harry S. Truman, January 20, 1949. Inaugural Address.

223 It would dominate much of the second half: Gilman, *Mandarins of the Future*, 1.

17 Cold War

224 In his election speech: Joseph Stalin, Speech Delivered at a Meeting of Voters of the Stalin Electoral District, Moscow. February 9, 1946.

225 Italy took American support: Sapelli, *Storia Economica Contemporanea*, 1.

225 Most of Italy's proximate markets: Isenburg, *Acque e Stato: Energia, Bonigiche, Irrigazione in Italia fra 1930 e 1950*, 26.

225 Economist Arthur Lewis proposed: Gollin, "The Lewis Model: A 60-Year Retrospective."

226 "no country can fail to develop": Lewis, "Economic Development with Unlimited Supply of Labour."

226 In 1957 Karl Wittfogel: Wittfogel, *Oriental Despotism.*

226 Wittfogel adopted Weber's analysis: Wallimann et al., "Misreading Weber: The Concept of 'Macht.'"

226 He believed that social development: Walder, review of *The Science of Society: Toward an Understanding of the Life and Work of Karl August Wittfogel*, by G. L. Ulmen.

226 As a work of sociology: Alfonso A. Narvaez, "Karl A. Wittfogel, Social Scientist Who Turned On Communists, 91." *New York Times*, May 26, 1988.

228 That was not to be: Gilmartin, *Blood and Water*, 204.

228 The Indus is different from other rivers: Bookhagen and Burbank, "Toward a Complete Himalayan Hydrological Budget."

228 Glacial and snowmelt contribute: Immerzeel et al., "Climate Change Will Affect the Asian Water Towers."

228 Any extensive agriculture in the Indus: Lambrick, "The Indus Flood-Plain and the 'Indus' Civilization."

228 Flood irrigation was the most common form: Giosan et al., "Fluvial Landscapes of the Harappan Civilization."

228 But the few canals: Gulhati, *Indus Waters Treaty*, 55.

229 In retrospect, the decision appears: Gulhati, *Indus Waters Treaty*, 64f.

229 Be that as it may, the cutoff: Gilmartin, *Blood and Water*, 207.

229 Pakistan tried to argue: Gilmartin, *Blood and Water*, 211.

229 It could only be solved: Gulhati, *Indus Waters Treaty*, 60.

231 One above all: Draskoczy, "The *Put'* of *Perekovka*: Transforming Lives at Stalin's White Sea–Baltic Canal."

231 Cotton had been produced: Kandiyoti, *The Cotton Sector in Central Asia Economic Policy and Development Challenges*, 1.

231 During the Second World War production: Jasny, "Soviet Agriculture and the Fourth Five-Year Plan."

231 In fact, he was singularly obsessed: Teichmann, "Canals, Cotton, and the Limits of De-colonization in Soviet Uzbekistan."

231 It was a vast exercise in resettling forced labor: Pohl, "A Caste of Helot Labourers: Special Settlers and the Cultivation of Cotton in Soviet Central Asia."

232 Mussolini did the same: Brain, "The Great Stalin Plan for the Transformation of Nature."
232 All together they were known as: Shaw, "Mastering Nature Through Science."
232 In 1953 he noted how the results: Nikita Khrushchev, "On Measures for the Further Development of Soviet Agriculture," September 3, 1953.
232 Under Khrushchev, and Brezhnev after him: Chida, "Science, Development and Modernization in the Brezhnev Time."
232 Farms were so inefficient: Kelly et al., "Large-Scale Water Transfers in the USSR."
233 After leaving the Tennessee Valley Authority: Gulhati, *Indus Waters Treaty*, 91.
233 Eventually, a truce was called: Guha, *India after Gandhi*, 74.
233 Lilienthal described Kashmir: Lilienthal, "Another Korea in the Making?"
235 These are the "water towers" of Africa: Viviroli et al., "Assessing the Hydrological Significance of the World's Mountains."
235 The nation-states of the post-colonial era: Herbst, "The Creation and Maintenance of National Boundaries in Africa."
235 This set of borders meant: Collier, "Africa: Geography and Growth," 235.
235 The challenge became even more evident: Mavhunga, "Energy, Industry, and Transport in South-Central Africa's History."
235 The federation ostensibly stood for: Hyam, "The Geopolitical Origins of the Central African Federation."
236 It is a source of perennial water: Moore et al., "The Zambezi River."
236 Long-distance transmission: G. J. Williams, "The Changing Electrical Power Industry of the Middle Zambezi Valley."
236 That is how Kariba Dam: Burdette, "Industrial Development in Zambia, Zimbabwe and Malawi: The Primacy of Politics," 96.
236 It was a daring bet on state-led development: Tischler, *Light and Power for a Multiracial Nation*, 1.
237 It was the fastest and most terminal destruction: Hughes, "Whites and Water: How Euro-Africans Made Nature at Kariba Dam."

18 The Great Acceleration

238 If anyone ever had any doubts that ideas: Lilienthal, "Another Korea in the Making?"
238 The head of the World Bank, Eugene Black: Alacevich, *The Political Economy of the World Bank*, 2.
238 Black sent a letter to both prime ministers: Gulhati, *Indus Waters Treaty*, 95.
239 The bank's engineering advisor: The World Bank/IFC Archives, Oral History Program. Transcript of interview with General Raymond Wheeler. Oral History Research Office, Columbia University, July 14, 1961.
239 When describing the Indus: Gulhati, *Indus Waters Treaty*, 18.
239 It was a partition of rivers: Gilmartin, *Blood and Water*, 216.
240 For this reason, the agreement: Salman, "The Baglihar Difference and Its Resolution Process—a Triumph for the Indus Waters Treaty?"
240 If that failed, the countries: Briscoe, "Troubled Waters: Can a Bridge Be Built over the Indus?"
240 When the process started: Gulhati, *Indus Waters Treaty*, 39.

240 It commands an area: Badruddin, *An Overview of Irrigation in Pakistan.*
241 It starts in the high plateau: Taddese et al., *The Water of the Awash River Basin a Future Challenge to Ethiopia.*
242 The presence of a railway line to Djibouti: Wood, "Regional Development in Ethiopia."
242 Their grazing land was diminished: Harbeson, "Territorial and Development Politics in the Horn of Africa."
243 Writers like W. E. B. DuBois: Adi, *Pan-Africanism,* vii.
243 On November 11, 1965, the white nationalists: Tischler, *Light and Power for a Multiracial Nation,* 3.
243 The British had planned: Scarritt and Nkiwane, "Friends, Neighbors, and Former Enemies: The Evolution of Zambia-Zimbabwe Relations in a Changing Regional Context."
243 Its copper economy depended: Anglin, "Zambian Crisis Behavior: Rhodesia's Unilateral Declaration of Independence."
244 By then Zambia was supplying energy: Williams, "The Changing Electrical Power Industry of the Middle Zambezi Valley."
244 The merging of nationalism: Tsomondo, "From Pan-Africanism to Socialism: The Modernization of an African Liberation Ideology."
244 Kariba Dam—the technology: Mavhunga, "Energy, Industry, and Transport in South-Central Africa's History."
244 On the back of the experience: Showers, "Beyond Mega on a Mega Continent: Grand Inga on Central Africa's Congo River."
244 In his 1961 book *I Speak of Freedom*: Nkrumah, *I Speak of Freedom,* xi.
244 As for many other leaders on the continent: Miescher and Tsikata, "Hydro-power and the Promise of Modernity and Development in Ghana."
245 In the end, as often happens: Tignor, *W. Arthur Lewis and the Birth of Development Economics,* 193.
245 The uniqueness of this site: Showers, "Beyond Mega on a Mega Continent: Grand Inga on Central Africa's Congo River."
245 But Lumumba backed out after pressure: Kaplan, "The United States, Belgium, and the Congo Crisis of 1960."
246 As it happened, 1961 was the centennial: Rosenblith, *Jerry Wiesner: Scientist, Statesman, Humanist,* 286.
246 It required skill, understanding: Hassan and Lai, *Ideas and Realities: Selected Essays of Abdus Salam,* 161.
246 The next day, Revelle, Wiesner, and Salam: Rosenblith, *Jerry Wiesner: Scientist, Statesman, Humanist,* 288.
247 Thomas was the director of Harvard's water program: Revelle, "Oceanography, Population Resources and the World," 45.
247 The task force went to Pakistan several times: Interview of Roger Revelle by Earl Droessler on February 3, 1989, Niels Bohr Library & Archives, American Institute of Physics, College Park, MD USA. Date accessed: July 31, 2020. www.aip.org/history-programs/niels-bohr-library/oral-histories/5051.
247 The team proposed dividing the landscape: Revelle, "Mission to the Indus."
247 The program expanded in the sixties: Gilmartin, *Blood and Water,* 235.
248 Ultimately there were about fifteen thousand tube wells: van Steenbergen and Oliemans, "A Review of Policies in Groundwater Management in Pakistan."

248 By 2006 that number had grown to 60 percent: Gilmartin, *Blood and Water*, 239.
248 The problem was not just salt: Revelle, "Oceanography, Population Resources and the World," 53.
248 The problem was not technical but socioeconomic: Revelle, "Mission to the Indus."
248 Saving the Indus in Pakistan: H. A. Thomas, "Roger Revelle: President-Elect, 1973."
249 Within the U.S. the mainstream had turned: Sneddon, *Concrete Revolution*, 79.
250 The committee had been the result: The World Bank/IFC Archives, Oral History Program, July 14, 1961. Transcript of interview with General Raymond Wheeler. Oral History Research Office, Columbia University, 27.
250 American experts, led by the geographer Gilbert F. White: White, "The Mekong River Plan."
250 Lyndon Johnson hoped the plan: Lilienthal, *The Journals of David E. Lilienthal*, 508.

19 The End of an Era

252 "A rocky dam shall stand": Mao, *The Writings of Mao Zedong*, 82.
253 He announced that China would surpass Britain: Chen, "Cold War Competition and Food Production in China, 1957–1962."
253 It was hardly a sector: Meng et al., "The Institutional Causes of China's Great Famine, 1959–61."
253 The tragic failures of the Great Leap Forward: Chen, "Cold War Competition and Food Production in China, 1957–1962."
254 One estimate suggested: Dikötter, *Mao's Great Famine*, 25.
254 The implementation through the provincial leadership: Chen, "Cold War Competition and Food Production in China, 1957–1962."
254 For a system that was already so vulnerable: Li and Yang, "The Great Leap Forward: Anatomy of a Central Planning Disaster."
254 Famine and starvation followed: Ó Gráda, "Great Leap into Famine: A Review Essay."
254 Because of the poor state of agricultural infrastructure: Kueh, *Agricultural Instability in China, 1931–1990*, 1.
254 In 1960 and 1961 at least some of the drop in harvest: Ó Gráda, "Great Leap into Famine: A Review Essay."
256 Besides, its interests were rapidly shifting: Sneddon, *Concrete Revolution*, 87.
256 According to Ryszard Kapuściński: Kapuscinski, *The Emperor*, 130.
256 Scenes of starvation of this kind: Bewket and Conway, "A Note on the Temporal and Spatial Variability of Rainfall in the Drought-Prone Amhara Region of Ethiopia."
256 Most independence-minded, anti-imperialist movements: Tsomondo, "From Pan-Africanism to Socialism: The Modernization of an African Liberation Ideology."
258 It was to be the tallest dam in the world: Reberschak, "Una storia del 'genio italiano': il Grande Vajont," 23.
259 However, the assumption was: Reberschak, "Una storia del 'genio italiano': il Grande Vajont," 39.

259 But the slide had such momentum: Hendron and Patton, *The Vajont Slide, A Geotechnical Analysis Based on New Geologic Observations of the Failure Surface,* 8.
260 A full third of the town: Reberschak, "Una storia del 'genio italiano': il Grande Vajont," 316.
260 But the consequences of this fateful event: Sapelli, *Storia Economica Contemporanea,* 1.
261 At those prices, commercial trades overwhelmed: Sekhar, "Surge in World Wheat Prices: Learning from the Past."
261 The crisis, coupled with the oil spike: Kelly et al., "Large-Scale Water Transfers in the USSR."
261 The windswept, exposed, salt-covered bottom: Micklin, "Desiccation of the Aral Sea."
262 The company Gidroproekt had proposed: Tolmazin, "Recent Changes in Soviet Water Management."
262 The matter was ultimately resolved: Prishchepov et al., "Effects of Institutional Changes on Land Use."

20 A World of Scarcity

268 He chose to play the U.S.A. and the USSR: Gerges, *Making of the Arab World,* 18.
268 That, of course, meant electrification: Waterbury, *The Egypt of Nasser and Sadat,* 57.
268 This wasn't exactly a new problem: Vitalis, "The 'New Deal' in Egypt."
268 The idea had even entered: Feiner, "The Aswan Dam Development Project."
268 But the scale of Nasser's ambition: Vitalis, "The 'New Deal' in Egypt."
268 In 1947, Adrien Daninos, a Greek-Egyptian engineer: El Mallakh, "Some Economic Aspects of the Aswan High Dam Project in Egypt."
268 Daninos had been one of those many foreigners: Shokr, "Hydropolitics, Economy, and the Aswan High Dam in Mid-Century Egypt."
269 To secure peace in the region: Gardner, *Three Kings,* 227.
269 They called it Project Alpha: Alterman, "American Aid to Egypt in the 1950s."
269 It should have been easy for the United States: Borzutzky and Berger, "Dammed If You Do, Dammed If You Don't: The Eisenhower Administration and the Aswan Dam."
269 The perception that the Americans: Gardner, *Three Kings,* 229.
269 The mistrust came to a head when, in 1955: Gardner, *Three Kings,* 233.
270 As the U.S. withdrew its support: Borzutzky and Berger, "Dammed If You Do, Dammed If You Don't: The Eisenhower Administration and the Aswan Dam."
270 Nasser instantly sealed the deal: Goldman, "A Balance Sheet of Soviet Foreign Aid."
270 Surprising the world, at a speech: "Discours de Gamal Abdel Nasser sur la nationalisation de la Compagnie du canal de Suez (Alexandrie, 26 juillet 1956)," in *Notes et études documentaires: Écrits et Discours du colonel Nasser* (Paris: La Documentation française, 1956), 16–21.
270 Naturally, the decision to build the dam: Mansfield, *A History of the Middle East,* 277.
271 A few details aside, the agreement: Abdalla, "The 1959 Nile Waters Agreement in Sudanese-Egyptian Relations."
271 Payments had to be in dollars: Woertz, *Oil for Food,* 19.
271 The most important international source: Woertz, *Oil for Food,* 9.

271 The age of oil had arrived: Sapelli, *Storia Economica Contemporanea*, 1.

272 Israel declared this an infringement: Wolf and Ross, "The Impact of Scarce Water Resources on the Arab-Israeli Conflict."

273 The country's agriculture became addicted to them: Beaumont, "Water and Development in Saudi Arabia."

274 Farmers produced so much wheat: Woertz, *Oil for Food*, 81.

274 When, at the end of 1979: T. C. Jones, "Rebellion on the Saudi Periphery: Modernity, Marginalization, and the Shi'a Uprising of 1979."

275 In 1979, the Chinese government: Zhang and Carter, "Reforms, the Weather, and Productivity Growth in China's Grain Sector."

275 A growth led by exports: Straub et al., *Infrastructure and Economic Growth in East Asia.*

276 China grew faster than any other country: Lin, *New Structural Economics*, 13.

276 China today accounts for more than a quarter: International Hydropower Association (IHA), *2019 Hydropower Status Report*, 98.

276 Annual investment in water security: Liu and Yang, "Water Sustainability for China and Beyond."

276 The one awe-inspiring symbol: Edmonds, "The Sanxia (Three Gorges) Project: The Environmental Argument Surrounding China's Super Dam."

277 The alternative is to transport supplies: Beckley, "China and Pakistan: Fair-Weather Friends."

278 In Pakistan this was seen as a development program: Sial, "The China-Pakistan Economic Corridor."

279 Meanwhile, India was looking to build: Briscoe, "Troubled Waters: Can a Bridge Be Built over the Indus?"

279 When Kishanganga was inaugurated: Naveed Siddiqui, "Pakistan Expresses Concerns over Inauguration of Kishanganga Dam Project by India." *Dawn*, May 18, 2018.

279 On the basis of a provision of the treaty: "Kishanganga Dam Issue: World Bank Asks Pakistan to Accept India's Demand of 'Neutral Expert,'" *The Times of India*, June 5, 2018.

279 The cascade of infrastructure: Magee, "The Dragon Upstream: China's Role in Lancang-Mekong Development."

280 They point to the renewable energy: Liebman, "Trickle-down Hegemony? China's 'Peaceful Rise' and Dam Building on the Mekong."

280 From the Chinese perspective: Fan et al., "Environmental Consequences of Damming the Mainstream Lancang-Mekong River: A Review."

21 A Planetary Experiment

283 In the first months of 2010, La Niña: Trenberth and Fasullo, "Climate Extremes and Climate Change: The Russian Heat Wave and Other Climate Extremes of 2010."

283 Microwave satellite images show: Ralph and Dettinger, "Storms, Floods, and the Science of Atmospheric Rivers."

283 When added up from top to bottom: Gimeno et al., "Atmospheric Rivers: A Mini-Review."

283 With instability came the storms: B. Moore et al., "Physical Processes Associated

with Heavy Flooding Rainfall in Nashville, Tennessee, and Vicinity During 1–2 May 2010."

284 The dam overtopped and water had to be released: Subcommittee of the Committee on Appropriations, *Special Hearing*, July 22, 2010, 14–17.

284 Little could be done at that point: U.S. Army Corps of Engineers, *Cumberland and Duck River Basins: May 2010 Post Flood Technical Report*, 36.

284 So was the amount of water: Martius et al., "The Role of Upper-Level Dynamics and Surface Processes for the Pakistan Flood of July 2010."

284 Those cyclones acted as obstacles: Altenhoff et al., "Linkage of Atmospheric Blocks and Synoptic-Scale Rossby Waves."

285 It was catastrophic: Martius et al., "The Role of Upper-Level Dynamics and Surface Processes for the Pakistan Flood of July 2010."

285 Floods washed through the Indus basin: Webster et al., "Were the 2010 Pakistan Floods Predictable?"

285 Through the first part of August: Polastro et al., *Inter-Agency Real Time Evaluation of the Humanitarian Response to Pakistan's 2010 Flood Crisis*, 89.

285 Almost two thousand people had died: United Nations Secretary-General's Remarks to General Assembly meeting on "Strengthening of the Coordination of Humanitarian and Disaster Relief Assistance of the United Nations, including Special Economic Assistance," August 19, 2010.

286 In the summer of 2010, the unusually low pressure: Lau and Kim, "The 2010 Pakistan Flood and Russian Heat Wave."

286 Between July 25 and August 8: Dole et al., "Was There a Basis for Anticipating the 2010 Russian Heat Wave?"

286 The extraordinary heat and air pollution: Barriopedro et al., "The 2009/10 Drought in China: Possible Causes and Impacts on Vegetation."

286 In 2010, Russia had completed: Josling et al., "Understanding International Trade in Agricultural Products."

287 Ultimately 17 percent of the crop area: Welton, *The Impact of Russia's 2010 Grain Export Ban*.

287 The expectations of buyers and sellers: Robert Paalberg, "How Grain Markets Sow the Spikes of Fear," *The Financial Times*, August 19, 2010.

287 Despite some localized floods: Barriopedro et al., "The 2009/10 Drought in China: Possible Causes and Impacts on Vegetation."

287 The shopping spree had consequences: Sternberg, "Chinese Drought, Wheat, and the Egyptian Uprising: How a Localized Hazard Became Globalized."

288 Those countries are highly vulnerable: Enghiad et al., "An Overview of Global Wheat Market Fundamentals in an Era of Climate Concerns."

288 Pressure grew on public finances: Abderrahim and Castel, *Inflation in Tunisia*, 11.

288 Egypt was the largest wheat importer: Breisinger et al., *Beyond the Arab Awakening*, 12.

289 During the eighties and nineties, Gaddafi: Adi, *Pan-Africanism*, 213.

290 Those events shared basic traits: Matthiesen, *Sectarian Gulf*.

290 Between 2000 and 2017, China's loans: Atkins et al., *Challenges of and Opportunities from the Commodity Price Slump*.

291 Indeed, Deng Xiaoping's China: Brautigam, *The Dragon's Gift*, 1.

Bibliography

Abdalla, Ismail H. "The 1959 Nile Waters Agreement in Sudanese-Egyptian Relations." *Middle Eastern Studies* 7 (October 1971): 329–41.

Abderrahim, Kaouther, and V. Castel. *Inflation in Tunisia: Perception and Reality in a Context of Transition.* African Development Bank Report, 2012.

Adams, Robert McC. *Heartland of Cities.* Chicago: University of Chicago Press, 1981.

Adi, Hakim. *Pan-Africanism: A History.* London: Bloomsbury Academic, 2018.

Alacevich, Michele. *The Political Economy of the World Bank.* Stanford Economics and Finance/World Bank, 2009.

Alexander, Demandt. *Der Fall Roms: Die Auflösung des römischen Reiches im Urteil der Nachwelt.* Munich: C. H. Beck, 1984.

Altaweel, Mark. "Simulating the Effects of Salinization on Irrigation Agriculture in Southern Mesopotamia." In *Models of Mesopotamian Landscapes: How Small-Scale Processes Contributed to the Growth of Early Civilizations,* edited by T. K. Wilkinson, M. Gibson, and M. Widell. BAR International Series 2552, 2013: 219–38.

Altenhoff, Adrian M., O. Martius, M. Croci-Maspoli, C. Schwierz, and H. C. Davies. "Linkage of Atmospheric Blocks and Synoptic-Scale Rossby Waves: A Climatological Analysis." *Tellus A: Dynamic Meteorology and Oceanography* 60 (2008): 1053–63.

Alterman, Jon B. "American Aid to Egypt in the 1950s: From Hope to Hostility." *Middle East Journal* 52 (Winter 1998): 51–69.

An, Cheng-Bang, L. Tang, L. Barton, and F.-H. Chen. "Climate Change and Cultural Response Around 4000 Cal Yr B.P. in the Western Part of Chinese Loess Plateau." *Quaternary Research* 63 (May 2005): 347–52.

Anglin, Douglas G. "Zambian Crisis Behavior: Rhodesia's Unilateral Declaration of Independence." *International Studies Quarterly* 24 (December 1980): 581–616.

Anzelark, Daniel. *Water and Fire: The Myth of the Flood in Anglo-Saxon England.* Manchester, UK: Manchester University Press, 2006.

Arendt, Hannah. *The Origins of Totalitarianism.* London: Penguin Classics, 2017.

Aristotle. *The Politics.* Trans. T. A. Sinclair. London: Penguin Classics, 1981.

Arnold, Ellen F. "Engineering Miracles: Water Control, Conversion and the Creation of a Religious Landscape in the Medieval Ardennes." *Environment and History* 13 (November 2007): 477–502.

Ashby, Thomas. *The Aqueducts of Ancient Rome.* Oxford: Oxford University Press, 1935.

Asthana, S. N. *The Cultivation of the Opium Poppy in India.* United Nations Office on Drugs and Crime, 1954.

Astour, Michael C. "New Evidence on the Last Days of Ugarit." *American Journal of Archaeology* 69 (July 1965): 253–58.

Atchi Reddy, M. "Travails of an Irrigation Canal Company in South India, 1857–1882." *Economic and Political Weekly* 25 (March 24, 1990): 619–21, 623–28.

Atkins, Lucas, D. Brautigam, Y. Chen, and J. Hwang. *Challenges of and Opportunities from the Commodity Price Slump.* Washington, DC: Johns Hopkins University SAIS–China Africa Research Initiative, 2017.

Bacon, Francis. *The Essayes or Counsels Civill and Morall.* Edited by M. Kiernan. Cambridge, MA: Harvard University Press, 1985.

Badruddin, M. *An Overview of Irrigation in Pakistan.* International Water Management Institute, 1993.

Baena, Laura M. "Negotiating Sovereignty: The Peace Treaty of Münster, 1648." *History of Political Thought* 28 (Winter 2007): 617–41.

Bailey, Robert G. *Description of the Ecoregions of the United States.* Washington, DC: U.S. Department of Agriculture, Forest Service, 1996.

Baker, T. Lindsay. "Turbine-Type Windmills of the Great Plains and Midwest." *Agricultural History* 54 (January 1980): 38–51.

Barker, Neil J. "Sections 9 and 10 of the Rivers and Harbors Act of 1899: Potent Tools for Environmental Protection." *Ecology Law Quarterly* 6 (1976): 109–59.

Barriopedro, David, E. M. Fischer, J. Luterbacher, R. M. Trigo, and R. García-Herrera. "The Hot Summer of 2010: Redrawing the Temperature Record Map of Europe." *Science* 332 (April 2011): 220–24.

Barriopedro, David, C. M. Gouveia, R. M. Trigo, and L. Wang. "The 2009/10 Drought in China: Possible Causes and Impacts on Vegetation." *Journal of Hydrometeorology* 13 (August 2012): 1251–67.

Barry, John M. *Rising Tide: The Great Mississippi Flood of 1927 and How It Changed America.* New York: Simon & Schuster, 1997.

Barthas, Jérémie. "Machiavelli, Public Debt, and the Origins of Political Economy: An Introduction." In *The Radical Machiavelli: Politics, Philosophy and Language,* edited by F. Del Lucchese, F. Frosini, and V. Morfino, 273–305. Leiden: Brill, 2015.

Bartolomei, Cristiana, and A. Ippolito. "The Silk Mill 'alla Bolognese.'" In *Explorations in the History of Machines and Mechanisms,* edited by C. Lopez-Cajun and M. Ceccarelli, 31–38. Switzerland: Springer, 2016.

Bauer, Brian S. "Introduction." In *Accounts of the Fables and Rites of the Incas by Cristóbal de Molina,* translated and edited by Brian S. Bauer, Vania Smith-Oka, and Gabriel E. Cantarutti, xiv–xxxv. Austin: University of Texas Press, 2011.

Beaumont, Peter. "Water and Development in Saudi Arabia." *Geographical Journal* 143 (March 1977): 42–60.

Beckley, Michael. "China and Pakistan: Fair-Weather Friends." *Yale Journal of International Affairs* 7 (March 2012): 9–22.

Bel, Germà. "The First Privatization: Selling SOEs and Privatizing Public Monopolies in Fascist Italy (1922–1925)." *Cambridge Economic Journal* 35 (February 2011): 937–56.

Belhoste, Bruno. "Les Origines de L'École Polytechnique: Des anciennes écoles d'ingénieurs à l'École centrale des Travaux publics." *Histoire de l'éducation* 42 (1989): 13–53.

Bell, Duncan. "John Stuart Mill on Colonies." *Political Theory* 38 (February 2010): 34–64.

Bellamy, Edward. *Looking Backward, 2000–1887.* Toronto: William Bryce Publisher, 1888.

Berend, Ivan T. *An Economic History of Twentieth-Century Europe: Economic Regimes from Laissez-Faire to Globalization.* Cambridge: Cambridge University Press, 2006.

Berger, Helge, and M. Spoerer. "Economic Crises and the European Revolutions of 1848." *Journal of Economic History* 61 (June 2001): 293–326.

Bergère, Marie-Claire. *Sun Yat-sen.* Translated by J. Lloyd. Stanford, CA: Stanford University Press, 1998.

Bewket, Woldeamlak, and D. Conway. "A Note on the Temporal and Spatial Variability of Rainfall in the Drought-Prone Amhara Region of Ethiopia." *International Journal of Climatology* 27 (September 2007): 1467–77.

Bhatia, Ramesh, and M. Scatasta. "Methodological Issues." In *Indirect Economic Impacts of Dams: Case Studies from India, Egypt and Brazil,* edited by R. Bhatia, R. Cestti, M. Scatasta, and R. P. S. Malik, 35–86. New Delhi: Academic Foundation, 2008.

Bierhorst, John. *Mythology of the Lenape: Guide and Text.* Tucson: University of Arizona Press, 1995.

Billington, David P., and D. C. Jackson. *Big Dams of the New Deal Era: A Confluence of Engineering and Politics.* Norman: University of Oklahoma Press, 2006.

Billington, David P., D. C. Jackson, and M. V. Melosi. *The History of Large Federal Dams: Planning, Design, and Construction in the Era of Big Dams.* Denver: U.S. Department of the Interior, Bureau of Reclamation, 2005.

Bindoff, Stanley T. *The Scheldt Question to 1839.* London: Routledge, 1945.

Birrell, Anne. "The Four Flood Myth Traditions of Classical China." *T'oung Pao* 83 (October 1997): 213–59.

Black, Jeremy A., G. Cunningham, E. Fluckiger-Hawker, E. Robson, and G. Zólyomi. "The Electronic Text Corpus of Sumerian Literature." http://etcsl.orinst.ox.ac.uk/, accessed August 1, 2020.

Blanshai, Sarah R. *Politics and Justice in Late Medieval Bologna.* Leiden: Brill, 2010.

Bocquet-Appel, Jean-Pierre. "The Agricultural Demographic Transition During and After the Agriculture Inventions." *Current Anthropology* 52 (October 2011): S497—S510.

Bogart, Dan. "Did the Glorious Revolution Contribute to the Transport Revolution? Evidence from Investment in Roads and Rivers." *Economic History Review* 64 (July 2011): 1073–1112.

Bongaarts, John. "Human Population Growth and the Demographic Transition." *Philosophical Transactions of the Royal Society B Biological Sciences* 364 (October 2009): 2985–90.

Bookhagen, Bodo, and D. W. Burbank. "Toward a Complete Himalayan Hydrological Budget: Spatiotemporal Distribution of Snowmelt and Rainfall and Their Impact on River Discharge." *Journal of Geophysical Research* 115 (August 2010): F03019.

Booth, Charlotte. *The Hyksos Period in Egypt.* London: Shire Egyptology, 2005.

Borzutzky, Silvia, and D. Berger. "Dammed If You Do, Dammed If You Don't: The Eisenhower Administration and the Aswan Dam." *Middle East Journal* 64 (2010): 84–102.

Botero, Carlos A., B. Gardner, K. R. Kirby, J. Bulbulia, M. C. Gavin, and R. D. Gray. "The Ecology of Religious Beliefs." *Proceedings of the National Academy of Sciences* 111 (November 2014): 16784–89.

Boucheron, Patrick. "Water and Power in Milan, c. 1200–1500." *Urban History* 28 (August 2001): 180–93.

Bourke-White, Margaret. *Portrait of Myself.* New York: Simon & Schuster, 1963.

Bowring, Julie. "Between the Corporation and Captain Flood: The Fens and Drainage After 1663." In *Custom, Improvement and the Landscape in Early Modern Britain,* edited by R. W. Hoyle, 235–61. Farnham, UK: Ashgate, 2011.

Brain, Stephen. "The Great Stalin Plan for the Transformation of Nature." *Environmental History* 15 (October 2010): 670–700.

Brautigam, Deborah. *The Dragon's Gift: The Real Story of China in Africa.* Oxford: Oxford University Press, 2009.

Breisinger, Clemens, O. Ecker, A. Perrihan, and B. Yu. *Beyond the Arab Awakening: Policies and Investments for Poverty Reduction and Food Security.* International Food Policy Research Institute, 2012.

Bresson, Alain. *The Making of the Ancient Greek Economy: Institutions, Markets, and Growth in the City-States.* Princeton, NJ: Princeton University Press, 2015.

Brezis, Elise S. "Foreign Capital Flows in the Century of Britain's Industrial Revolution: New Estimates, Controlled Conjectures." *Economic History Review* 48 (February 1995): 46–67.

Briscoe, John. "Troubled Waters: Can a Bridge Be Built over the Indus?" *Economic and Political Weekly* 45 (December 11–17, 2010): 28–32, 5.

Brody, Aaron. "From the Hills of Adonis Through the Pillars of Hercules: Recent Advances in the Archaeology of Canaan and Phoenicia." *Near Eastern Archaeology* 65 (March 2002): 69–80.

Broich, John. "Engineering the Empire: British Water Supply Systems and Colonial Societies, 1850–1900." *Journal of British Studies* 46 (April 2007): 346–65.

Brooks Hedstrom, Darlene L. *The Monastic Landscape of Late Antique Egypt: An Archaeological Reconstruction.* Cambridge: Cambridge University Press, 2017.

Broshi, Magen, and I. Finkelstein. "The Population of Palestine in Iron Age II." *Bulletin of the American Schools of Oriental Research* 287 (August 1992): 47–60.

Brown, J. B. "Politics of the Poppy: The Society for the Suppression of the Opium Trade, 1874–1916." *Journal of Contemporary History* 8 (July 1973): 97–111.

Brown, Jonathan. *Agriculture in England: A Survey of Farming, 1870–1947.* Manchester, UK: Manchester University Press, 1987.

Brown, Peter. *Augustine of Hippo: A Biography.* Berkeley: University of California Press, 2000.

Brumfield, Sara. "Imperial Methods: Using Text Mining and Social Network Analysis to Detect Regional Strategies in the Akkadian Empire." PhD diss., UCLA, 2013.

Bruno, Giovanni. "Capitale straniero e industria elettrica nell'Italia meridionale (1895–1935)." *Studi Storici* 28 (October–December 1987): 943–84.

Büntgen, Ulf, and N. Di Cosmo. "Climatic and Environmental Aspects of the Mongol Withdrawal from Hungary in 1242 CE." *Nature Scientific Reports* 6 (May 2016): 25606.

Büntgen, Ulf, W. Tegel, K. Nicolussi, M. McCormick, D. Frank, V. Trouet, J. O. Kaplan, F. Herzig, K.-U. Heussner, H. Wanner, J. Luterbacher, and J. Esper. "2500 Years of European Climate Variability and Human Susceptibility." *Science* 331 (February 2011): 578–82.

Burdette, Marcia M. "Industrial Development in Zambia, Zimbabwe and Malawi: The Primacy of Politics." In *Studies in the Economic History of Southern Africa,* vol. 1, *The Front-Line States,* edited by Z. A. Konczacki, J. L. Parpart, and T. M. Shaw, 75–126. London: Frank Cass & Co, 1990.

Butzer, Karl W. *Early Hydraulic Civilization in Egypt. A Study in Cultural Ecology.* Chicago: University of Chicago Press, 1976.

Cain, Peter. "British Free Trade, 1850–1914: Economics and Policy." *Recent Findings of Research in Economic & Social History* 29 (Autumn 1999): 1–4.

Capinera, John L. "Insects in Art and Religion: The American South West." *American Entomologist* 39 (Winter 1993): 221–30.

Caprotti, Federico. "Malaria and Technological Networks: Medical Geography in the Pontine Marshes, Italy, in the 1930s." *Geographical Journal* 172 (June 2006): 145–55.

Carneiro, Robert L. "A Theory of the Origin of the State." *Science* 169 (August 1970): 733–38.

Carse, Ashley. "Nature as Infrastructure: Making and Managing the Panama Canal Watershed." *Social Studies of Science* 42 (April 2012): 539–63.

Cavallar, Osvaldo. "River of Law: Bartolus's Tiberiadis (De Alluvione)." In *A Renaissance of Conflicts: Visions and Revisions of Law and Society in Italy and Spain,* edited by J. A. Marino and T. Kuehn, 31–116. Toronto: Center for Reformation and Renaissance Studies, 2004.

Chadwick, John. *The Mycenaean World.* Cambridge: Cambridge University Press, 1976.

Champollion, Jean-François. *Lettre à M. Dacier.* Paris: Firmin Didot, 1822.

Chen, Yixin. "Cold War Competition and Food Production in China, 1957–1962." *Agricultural History* 83 (Winter 2009): 51–78.

Chen, Yunzhen, J. P. M. Syvitski, S. Gao, I. Overeem, and A. J. Kettner. "Socio-economic Impacts on Flooding: A 4000-Year History of the Yellow River, China." *AMBIO* 41 (November 2012): 682–98.

Chernoff, Miriam C., and S. M. Paley. "Dynamics of Cereal Production at Tell el Ifshar, Israel During the Middle Bronze Age." *Journal of Field Archaeology* 25 (July 1998) 397–416.

Chernyshevsky, Nikolay G. *A Vital Question; or, What Is to Be Done.* Translated by N. H. Dole and S. S. Skydelsky. New York: T. Y. Crowell, 1886.

Cheyette, Fredric L. "The Disappearance of the Ancient Landscape and the Climatic Anomaly of the Early Middle Ages: A Question to Be Pursued." *Early Medieval Europe* 16 (March 2008): 127–65.

Chida, Tetsuro. "Science, Development and Modernization in the Brezhnev Time: The Water Development in the Lake Balkhash Basin." *Cahiers du Monde Russe* 54 (2013): 239–64.

Christensen, Katherine. "Introduction." In *Gratian: The Treatise on Laws,* translated by A. Thompson. Washington, DC: Catholic University of America Press, 1993.

Cicero. *On Obligations.* Translated by P. G. Walsh. Oxford: Oxford World's Classics, 2000.

Cicero. *The Republic and the Laws.* Translated by N. Rudd. Oxford: Oxford World's Classics, 2008.

Cintrón, Lizette. *Historical Statistics of the Electric Utility Industry Through 1992.* Washington, DC: Edison Electric Institute, 1995.

Cipolla, Carlo. *Storia Economica dell'Europa Pre-Industriale.* Bologna: Il Mulino, 1974.

Ciriacono, Salvatore. *Building on Water: Venice, Holland, and the Construction of the European Landscape in Early Modern Times.* Translated by J. Scott. New York: Berghahn Books, 2006.

Clark, Peter U., A. S. Dyke, J. D. Shakun, A. E. Carlson, J. Clark, B. Wohlfarth, J. X. Mitrovica, S. W. Hostetler, and A. M. McCabe. "The Last Glacial Maximum." *Science* 325 (August 2009): 710–14.

Cole, Steven W., and H. Gasche. "Second- and First-Millennium BC Rivers in Northern Babylonia." In *Changing Watercourse in Babylonia: Towards a Reconstruction of the Ancient Environment in Lower Mesopotamia,* vol. 1, edited by H. Gasche and M. Tanret, 4–7. University of Ghent and Oriental Institute of the University of Chicago, 1998.

Colish, Marcia L. "Machiavelli's Art of War: A Reconsideration." *Renaissance Quarterly* 51 (1998): 1151–68.

Collier, Paul. "Africa: Geography and Growth." In *Proceedings—Economic Policy Symposium—Jackson Hole,* 235–52. Federal Reserve Bank of Kansas City, 2006.

Columella. *De Re Rustica,* vol. 1. Translated by H. B. Ash. Loeb Classical Library edition, 1941.

Conrad, Joseph. *Heart of Darkness*. Peterborough, ON: Broadview Press, 1999.

Conti, Fulvio. "Alle origini del sistema elettrico toscano: Strategie d'impresa e concentrazioni industriali (1890–1920)." *Studi Storici* 32 (January–March 1991): 137–60.

Conway, Declan. "The Climate and Hydrology of the Upper Blue Nile River." *Geographical Journal* 166 (March 2000): 49–62.

Cook, Benjamin I., R. L. Miller, and R. Seager. "Dust and Sea Surface Temperature Forcing of the 1930s 'Dust Bowl' Drought." *Geophysical Research Letters* 35 (April 2008): L08710.

Cook, Edward R., K. J. Anchukaitis, B. M. Buckley, R. D. D'Arrigo, G. C. Jacoby, and W. E. Wright. "Asian Monsoon Failure and Megadrought During the Last Millennium." *Science* 328 (April 2010); 486–89.

Cooper, Jerrold S. *Reconstructing History from Ancient Inscriptions: The Lagash-Umma Border Conflict.* Malibu, CA: Undena Publications, 1983.

Coopersmith, Jonathan. *The Electrification of Russia, 1880–1926.* Ithaca, NY: Cornell University Press, 1992.

Cornelisse, Charles L. E. "The Economy of Peat and Its Environmental Consequences in Holland During the Late Middle Ages." In *Water Management, Communities, and Environment: The Low Countries in Comparative Perspective, c.1000—c.1800,* edited by H. Greefs and M. 't Hart, 95–122. Ghent: Academia Press, 2006.

Corner, Paul. "Considerazioni sull'agricoltura capitalistica durante il fascismo." *Quaderni storici* 10 (May–December 1975): 519–29.

Courtney, Chris. *The Nature of Disaster in China: The 1931 Yangzi River Flood.* Cambridge: Cambridge University Press, 2018.

Crabitès, Pierre. "The Nile Waters Agreement." *Foreign Affairs* 8 (October 1929): 145–49.

Crary, Douglas D. "Recent Agricultural Developments in Saudi Arabia." *Geographical Review* 41 (July 1951): 366–83.

Craven, Matthew. "Between Law and History: The Berlin Conference of 1884–1885 and the Logic of Free Trade." *London Review of International Law* 3 (March 2015): 31–59.

Cregan-Reid, Vybarr. *Discovering Gilgamesh: Geology, Narrative and the Historical Sublime in Victorian Culture.* Manchester: Manchester University Press, 2013.

Crouch, Dora P. *Geology and Settlement: Greco-Roman Patterns.* Oxford: Oxford University Press, 2003.

Crouch, Dora P. *Water Management in Ancient Greek Cities.* Oxford: Oxford University Press, 1993.

Cullen, Heidi M., P. B. DeMenocal, S. Hemming, G. Hemming, F. H. Brown, T. Guilderson, F. Sirocko. "Climate Change and the Collapse of the Akkadian Empire: Evidence from the Deep Sea." *Geology* 28 (2000): 379–82.

Cutler, David, and G. Miller. "Water, Water Everywhere: Municipal Finance and Water Supply in American Cities." NBER Working Paper 11096, 2005.

Cyberski, Jerzy, M. Grześ, M. Gutry-Korycka, E. Nachlik, and Z. W. Kundzewicz. "History of Floods on the River Vistula." *Hydrological Sciences Journal* 51 (June 2006) 799–817.

Dalley, Stephanie, trans. *Myths from Mesopotamia: Creation, the Flood, Gilgamesh, and Others.* Oxford: Oxford University Press, 1989.

Dameron, George. "The Church as Lord." In *The Oxford Handbook of Medieval Christianity,* edited by J. H. Arnold, 457–67. Oxford: Oxford Univeristy Press, 2014.

D'Antone, Lea. "Politica e cultura agraria: Arrigo Serpieri." *Studi Storici* 20 (July–September 1979): 609–42.

David, Nicholas. "The Scheldt Trade and the 'Ghent War' of 1379–1385." *Bulletin de la Commission royale d'Histoire* 144 (1978): 189–359.

Davies, Arthur. "The First Discovery and Exploration of the Amazon in 1498–99." *Transactions and Papers* (Institute of British Geographers) 22 (1956): 87–96.

Davies, Robert W., M. B. Tauger, and S. G. Wheatcroft. "Stalin, Grain Stocks and the Famine of 1932–33." *Slavic Review* 54 (1995): 642–57.

Dawn, C. Ernest. "The Origins of Arab Nationalism." In *The Origins of Arab Nationalism*, edited by R. Khalidi, L. Anderson, M. Muslih, and R. S. Simon, 3–30. New York: Columbia University Press, 1991.

Day, John W., Jr., J. D. Gunn, W. J. Folan, A. Yáñez-Arancibia, and B. P. Horton. "Emergence of Complex Societies After Sea Level Stabilized." *EOS* 88 (June 2007): 169–76.

De Felice, Renzo. *Mussolini e Hitler: I rapport segreti, 1922–1933*. Firenze: Le Monnier, 1975.

De Tocqueville, Alexis. *Democracy in America*, vol. 1. Edited by E. Nolla. Translated by J. T. Schleifer. Carmel, IN: Liberty Fund, 2010.

De Vries, Jan. *Barges and Capitalism, Passenger Transportation in the Dutch Economy, 1632–1839*. Utrecht: HES Publishers, 1981.

De Vries, Jan. *The Dutch Rural Economy in the Golden Age, 1500–1700*. New Haven, CT: Yale University Press, 1974.

De Vries, Jan. "The Economic Crisis of the Seventeenth Century After Fifty Years." *Journal of Interdisciplinary History* 40 (Autumn 2009): 151–94.

De Vries, Jan. *The Economy of Europe in an Age of Crisis, 1600–1750*. Cambridge: Cambridge University Press, 1976.

Degroot, Dagomar. *The Frigid Golden Age. Climate Change, the Little Ice Age, and the Dutch Republic, 1560–1720*. Cambridge: Cambridge University Press, 2018.

Delzell, Charles F. "Remembering Mussolini." *Wilson Quarterly* 12 (Spring 1988): 118–35.

Demosthenes. *Orations*. Translated by N. W. DeWitt and N. J. DeWitt. Cambridge, MA: Harvard University Press, 1949.

Denman, James, and P. McDonald. *Unemployment Statistics from 1881 to the Present Day*. Labour Market Trends, 1996.

Dever, William G. "Histories and Non-Histories of Ancient Israel: The Question of the United Monarchy." In *In Search of Pre-Exilic Israel*, edited by J. Day, 65–94. London: Bloomsbury, 2005.

Di Cosimo, Nicola. "China-Steppe Relations in Historical Perspective." In *Complexity and Interaction Along the Eurasian Steppe Zone in the First Millennium CE*, edited by J. Bemmann and M. Schmauder, 49–72. Bonn: Universität Bonn, 2015.

Di Matteo, Federico. *Villa di Nerone a Subiaco: il complesso dei Simbruina stagna*. Rome: L'Erma di Bretschneider, 2005.

Diggins, John P. *Mussolini and Fascism: The View from America*. Princeton, NJ: Princeton University Press, 2015.

Dikötter, Frank. *Mao's Great Famine: The History of China's Most Devastating Catastrophe, 1958–1962*. London: Bloomsbury, 2010.

Dodgen, Randall A. *Controlling the Dragon: Confucian Engineers and the Yellow River in Late Imperial China*. Honolulu: University of Hawaii Press, 2001.

Dole, Randall, M. Hoerling, J. Perlwitz, J. Eischeid, P. Pegion, T. Zhang, X-W Quan, T. Xu, and D. Murray. "Was There a Basis for Anticipating the 2010 Russian Heat Wave?" *Geophysical Research Letters* 38 (March 2011): L06702.

Domar, Evsey D. "Capital Expansion, Rate of Growth, and Employment." *Econometrica* 14 (April 1946): 137–47.

Dong, Guanghui, X. Jia, C. An, F. Chen, Y. Zhao, S. Tao, and M. Ma. "Mid-Holocene Climate Change and Its Effect on Prehistoric Cultural Evolution in Eastern Qinghai Province, China." *Quaternary Research* 77 (2012): 23–30.

Doyle, Martin. *The Source: How Rivers Made America and America Remade Its Rivers.* New York: W. W. Norton, 2018.

Drake, Brandon L. "The Influence of Climatic Change on the Late Bronze Age Collapse and the Greek Dark Ages." *Journal of Archaeological Science* 39 (June 2012): 1862–70.

Draskoczy, Julie. "The *Put'* of *Perekovka*: Transforming Lives at Stalin's White Sea–Baltic Canal." *Russian Review* 71 (January 2012): 30–48.

D'Souza, Rohan. *Drowned and Dammed: Colonial Capitalism and Flood Control in Eastern India.* Oxford: Oxford University Press, 2006.

Duncan-Jones, Richard P. "Giant Cargo-Ships in Antiquity." *Classical Quarterly* 27 (December 1977): 331–32.

Dünkeloh, Armin, and J. Jacobeit. "Circulation Dynamics of Mediterranean Precipitation Variability." *International Journal of Climatology* 23 (December 2003): 1843–66.

Dupuit, Jules. "On the Measurement of the Utility of Public Works." *International Economic Papers* 2 (1952): 83–110.

Edgerton-Tarpley, Kathryn. *Tears from Iron: Cultural Responses to Famine in Nineteenth-Century China.* Berkeley: University of California Press, 2008.

Edmonds, Richard L. "The Sanxia (Three Gorges) Project: The Environmental Argument Surrounding China's Super Dam." *Global Ecology and Biogeography Letters* 2 (July 1992) 105–25.

Edwards, Amelia. "Was Rameses II the Pharaoh of the Exodus?" *Knowledge* 2 (1882): 1.

Einaudi, Luigi. "Il dogma della sovranità e l'idea della società delle nazioni." *Corriere della Sera,* 28 December 1918.

Eisenstadt, Shmuel N. *The Political Systems of Empires.* Piscataway, NJ: Transaction Publishers, 1963.

Ekbladh, David. "'Mr. TVA': Grass-Roots Development, David Lilienthal, and the Rise and Fall of the Tennessee Valley Authority as a Symbol for U.S. Overseas Development, 1933–1973." *Diplomatic History* 26 (July 2002): 335–74.

El Mallakh, Ragaei. "Some Economic Aspects of the Aswan High Dam Project in Egypt." *Land Economics* 35 (February 1959): 15–23.

Elden, Stuart. "Bartolus of Sassoferrato and the Emergence of Territorial Sovereignty." In *Proceedings of 14th International Congress of Historical Geographers,* edited by Akihiro Kinda, T. Komeie, S. Minamide, T. Mizoguchi, and K. Uesugi, 48–49. Kyoto: Kyoto University Press, 2010.

Emanuel, Kerry A. "The Power of a Hurricane: An Example of Reckless Driving on the Information Superhighway" *Weather* 54 (April 1999): 107–108.

Enghiad, Aliakbar, D. Ufer, A. M. Countryman, and D. D. Thilmany. "An Overview of Global Wheat Market Fundamentals in an Era of Climate Concerns." *International Journal of Agronomy* 2017 (July 2017): 1–15.

Epstein, Steven A. *An Economic and Social History of Later Medieval Europe, 1000–1500.* Cambridge: Cambridge University Press, 2009.

Erdkamp, Paul. *The Grain Market in the Roman Empire: A Social, Political and Economic Study.* Cambridge: Cambridge University Press, 2009.

Erickson, Clark L. "An Artificial Landscape-Scale Fishery in the Bolivian Amazon." *Nature* 408 (December 2000): 190–93.

Erickson, Clark L. "The Domesticated Landscapes of the Bolivian Amazon." In *Time and Complexity in Historical Ecology,* edited by W. Balée and C. L. Erickson, 235–78. New York: Columbia University Press, 2006.

Evans, Harry B. "Agrippa's Water Plan." *American Journal of Archaeology* 86 (July 1982): 401–11.

Everett, Edward. *Address of Hon. Edward Everett, at the Consecration of the National Cemetery at Gettysburg, 19th November, 1863, with the dedicatory speech of President Lincoln, and the other exercises of the occasion.* Boston: Little Brown and Company, 1864.

Fan, Hui, D. He, and H.Wang. "Environmental Consequences of Damming the Mainstream Lancang-Mekong River: A Review." *Earth-Science Reviews* 146 (July 2015): 77–91.

Farrand, Max, ed. *Records of the Federal Convention of 1787,* vol. 1. New Haven, CT: Yale University Press, 1911.

Farrand, Max, ed. *Records of the Federal Convention of 1787,* vol. 2. New Haven, CT: Yale University Press, 1911.

Faust, Avraham, and E. Weiss. "Judah, Philistia, and the Mediterranean World: Reconstructing the Economic System of the Seventh Century B.C.E." *Bulletin of the American Schools of Oriental Research* 338 (May 2005): 71–92.

Feiner, Leon. "The Aswan Dam Development Project." *Middle East Journal* 6 (Autumn 1952): 464–67.

Finkelstein, Israel. *The Forgotten Kingdom: The Archaeology and History of Northern Israel.* Atlanta, GA: Society of Biblical Literature, 2013.

Finkelstein, Israel, and N. A. Silberman. *The Bible Unearthed: Archaeology's New Vision of Ancient Israel and the Origin of Its Sacred Texts.* New York: The Free Press, 2001.

Finkelstein, Jacob J. "Mesopotamia." *Journal of Near Eastern Studies* 21 (April 1962): 73–92.

Finné, Martin, K. Holmgren, C.-C. Shen, H.-M. Hu, M. Boyd, and S. Stocker. "Late Bronze Age Climate Change and the Destruction of the Mycenaean Palace of Nestor at Pylos." *PLOS One* 12 (2017): e0189447.

Fishback, P. V. "US Monetary and Fiscal Policy in the 1930s." *Oxford Review of Economic Policy* 26 (Autumn 2010): 385–413.

Fishback, Price V., and V. Kachanovskaya. "The Multiplier for Federal Spending in the States During the Great Depression." *Journal of Economic History* 75 (November 2015): 125–62.

Fiske, John. *The Critical Period of American History: 1783–1789.* 3rd ed. Boston: Houghton Mifflin, 1899.

Fitzgerald, Edward P. "The Iraq Petroleum Company, Standard Oil of California, and the Contest for Eastern Arabia, 1930–1933." *International History Review* 13 (August 1991): 441–65.

Fokkema, Douwe. *Perfect Worlds: Utopian Fiction in China and the West.* Amsterdam: Amsterdam University Press, 2011.

Forrest, William G. *The Emergence of the Greek Democracy, 800–400 BC.* London: World University Library, 1976.

Foster, Benjamin R. *Before the Muses: An Anthology of Akkadian Literature.* Vol. 1, *Archaic, Classical, Mature.* Bethesda, MD: CDL Press, 1996.

Foxhall, Lin. *Olive Cultivation in Ancient Greece: Seeking the Ancient Economy.* Oxford: Oxford University Press, 2007.

French, Alf. "The Economic Background to Solon's Reforms." *Classical Quarterly* 1956 (May 1956): 11–25.

Gallant, Thomas W. "Agricultural Systems, Land Tenure, and the Reforms of Solon." *Annual of the British School at Athens* 77 (November 1982): 111–24.

García, Erik V. "The Maya Flood Myth and the Decapitation of the Cosmic Caiman." *PARI Journal* 7 (Summer 2006): 1–10.

Gardner, Lloyd C. *Three Kings: The Rise of an American Empire in the Middle East After World War II.* New York: New Press, 2009.

Gardoni, Giuseppe. "Élites cittadine fra XI e XII secolo: il caso mantovano." In *Medioevo:*

Studi e documenti, II, edited by A. Castagnetti, A. Ciaralli, and G. M. Varanini, 281–350. Libreria universitaria editrice, 2007.

Gardoni, Giuseppe. "Uomini e acque nel territorio Mantovano (secoli X–XIII)." In *La Civiltà delle Acque: Tra Medioevo e Rinascimento,* edited by A. Calzona and D. Lamberini, 143–76. Firenze: Leo S. Olschki, 2010.

Gates, Paul W. "Homesteading in the High Plains." *Agricultural History* 5 (January 1977): 109–33.

Gebbie, F. S. J., H. F. Cory, and G. C. Simpson. *Report of the Nile Projects Commission.* Egyptian Government, 1920.

"General Act of the Conference of Berlin Concerning the Congo." *American Journal of International Law* 3 (1909): 7–25.

George, Andrew, trans. *The Epic of Gilgamesh: A New Translation.* London: Penguin, 1999.

Gerges, Fawaz A. *Making of the Arab World: Nasser, Qutb, and the Clash That Shaped the Middle East.* Princeton, NJ: Princeton University Press, 2018.

Gilman, Nils. *Mandarins of the Future: Modernization Theory in Cold War America.* Baltimore: Johns Hopkins University Press, 2003.

Gilmartin, David. *Blood and Water: The Indus River Basin in Modern History.* Berkeley: University of California Press, 2015.

Gilmour, David. "The Ends of Empire." *Wilson Quarterly* 21 (Spring 1997): 32–39.

Gimeno, Luis, L. R. Nieto, M. Vázquez, and D. A. Lavers. "Atmospheric Rivers: A Mini-Review." *Frontiers in Earth Science* 2 (March 2014): 1–6.

Giordano, Claire, G. Piga, and G. Trovato. "Italy's Industrial Great Depression: Fascist Price and Wage Policies." *Macroeconomic Dynamics* 18 (May 2013): 689–720.

Giosan, Liviu, P. D. Clift, M. G. Macklin, D. Q. Fuller, S. Constantinescu, J. A. Durcan, T. Stevens, G. A. T. Duller, A. R. Tabrez, K. Gangal, R. Adhikari, A. Alizai, F. Filip, S. VanLaningham, and J. P. M. Syvitski. "Fluvial Landscapes of the Harappan Civilization." *Proceedings of the National Academy of Sciences* 109 (June 2012): E1688—E1694.

Glennie, J. F. "The Equatorial Nile Project." *Sudan Notes and Records* 38 (1957): 67–73.

Goldman, Marshall I. "A Balance Sheet of Soviet Foreign Aid." *Foreign Affairs* 43 (January 1965): 349–60.

Goldstone, Jack A., and J. F. Haldon, 2009. "Ancient States, Empires, and Exploitation: Problems and Perspectives." In *The Dynamics of Ancient Empires, State Power from Assyria to Byzantium,* edited by I. Morris and W. Scheidel, 3–29. Oxford: Oxford University Press, 2009.

Gollin, Douglas. "The Lewis Model: A 60-Year Retrospective." *Journal of Economic Perspectives* 28 (Summer 2014): 71–88.

Goosse, Hugues, O. Arzel, J. Luterbacher, M. E. Mann, H. Renssen, N. Riedwyl, A. Timmermann, E. Xoplaki, and H. Wanner. "The Origin of the European 'Medieval Warm Period.'" *Climate of the Past* 2 (September 2006): 99–113.

Gordon, Robert J. *The Rise and Fall of American Growth.* Princeton, NJ: Princeton University Press, 2016.

Gray, Lesley J., J. Beer, M. Geller, J. D. Haigh, M. Lockwood, K. Matthes, U. Cubasch, D. Fleitmann, G. Harrison, L. Hood, J. Luterbacher, G. A. Meehl, D. Shindell, B. van Geel, and W. White. "Solar Influences on Climate." *Reviews of Geophysics* 48 (October 2010): 1–53.

Gregory of Tours. *The History of the Franks.* Translated by L. Thorpe. London: Penguin Books, 1974.

Grillo, Paolo "Cistercensi e società cittadina in età comunale: Il monastero di Chiaravalle Milanese (1180–1276)." *Studi Storici* 40 (June 1999): 357–94.

Gross, Leo. "The Peace of Westphalia, 1648–1948." *American Journal of International Law* 42 (January 1948): 20–41.

Guerriero, Stefano. "La Generazione di Mussolini" *Belfagor* 67 (May 2012): 277–87.

Guha, Ramachandra. *India after Gandhi: The History of the World's Largest Democracy.* New York: HarperPerennial, 2003.

Gulhati, Niranjan D. *Indus Waters Treaty: An Exercise in International Mediation.* Bombay: Allied Publishers, 1973.

Haldon, John. "Some Thoughts on Climate Change, Local Environment, and Grain Production in Byzantine Northern Anatolia." In *Environment and Society in the Long Late Antiquity,* edited by A. Izdebski and M. Mulryan, 200–206. Leiden: Brill, 2019.

Hallenbeck, Jan T. *Pavia and Rome: The Lombard Monarchy and the Papacy in the Eighth Century.* Transactions of the American Philosophical Society, 1982.

Haller, John S. "The Species Problem: Nineteenth-Century Concepts of Racial Inferiority in the Origin of Man Controversy." *American Anthropologist* 72 (1970): 1319–29.

Hallis, Lydia J., G. R. Huss, K. Nagashima, G. J. Taylor, S. A. Halldórsson, D. R. Hilton, M. J. Mottl, and K. J. Meech. "Evidence for Primordial Water in Earth's Deep Mantle." *Science* 350 (November 2015): 795–97.

Hamilton, Alexander. *The Papers of Alexander Hamilton.* Vol. 3, 1782–1786. Edited by H. C. Syrett. New York: Columbia University Press, 1962.

Hansen, Victor D. *The Other Greeks.* Berkeley: University of California Press, 1999.

Hanyan, Craig R. "China and the Erie Canal." *Business History Review* 35 (Winter 1961): 558–66.

Hao, Zhixin, J. Zheng, G. Wu, X. Zhang, and Q. Ge. "1876–1878 Severe Drought in North China: Facts, Impacts and Climatic Background." *Chinese Science Bulletin* 55 (September 2010): 3001–7.

Harbeson, John W. "Territorial and Development Politics in the Horn of Africa: The Afar of the Awash Valley." *African Affairs* 77 (October 1978): 479–98.

Harding, Earl. *The Untold Story of Panama.* New York: Athene Press, 1959.

Harris, Abram L. "John Stuart Mill: Servant of the East India Company." *Canadian Journal of Economics and Political Science* 30 (May 1964): 185–202.

Hart, Parker T. *Saudi Arabia and the United States: Birth of a Security Partnership.* Bloomington: Indiana University Press, 1998.

Hassan, Fekri A. 'The Dynamics of a Riverine Civilization: A Geoarchaeological Perspective on the Nile Valley, Egypt." *World Archaeology* 29 (June 1997): 51–57.

Hassan, Z., and C. H. Lai, eds. *Ideas and Realities: Selected Essays of Abdus Salam.* World Scientific Publishing, 1984.

Haw, Stephen G. *Marco Polo's China: A Venetian in the Realm of Khubilai Khan.* New York: Routledge, 2006.

Heather, Peter. *The Fall of the Roman Empire: A New History.* London: Pan Books, 2006.

Helmholz, Richard H. "Continental Law and Common Law: Historical Strangers or Companions?" *Duke Law Journal* 1990 (December 1990): 1207–28.

Helmholz, Richard H. "Magna Carta and the Ius Commune." *University of Chicago Law Review* 66 (Spring 1999): 297–371.

Helmholz, Richard H. "Magna Carta and the Law of Nature." *Loyola Law Review* 62 (Fall 2016): 869–86.

Hendron, Alfred J., Jr., and F. D. Patton. *The Vajont Slide, A Geotechnical Analysis Based on New Geologic Observations of the Failure Surface.* Vicksburg, MS: U.S. Army Corps of Engineers Waterways Experiment Station, 1985.

Henry, Alfred J. "Frankenfield on the 1927 Floods in the Mississippi Valley." *Monthly Weather Review* 55 (October 1927): 437–52.

Herbst, Jeffrey. "The Creation and Maintenance of National Boundaries in Africa." *International Organization* 43 (Autumn 1989): 673–92.

Herodotus. *The Histories.* Translated by A. de Sélincourt. London: Penguin Classics, 1972.

Herzl, Theodor. *Altneuland.* Translated by D. S. Blondheim. Federation of American Zionists, 1916.

Hiltzik, Michael. *Colossus.* New York: Simon & Schuster, 2010.

Hobsbawm, Eric. *Age of Revolution, 1789–1848.* London: Abacus, 1996.

Hodge, A. Trevor. "A Roman Factory." *Scientific American* 263 (November 1990): 106–13.

Hodgkin, Thomas. *The Letters of Cassiodorus: Book XII, "Containing twenty eight letters written by Cassiodorus in his own name as praetorian praefect."* London: Henry Frowde, 1886.

Hoffmann, Richard C. "Economic Development and Aquatic Ecosystems in Medieval Europe." *American Historical Review* 101 (June 1996): 631–69.

Holloway, Steven W. "Biblical Assyria and Other Anxieties in the British Empire." *Journal of Religion & Society* 3 (2001): 1–19.

Homer. *Odyssey.* Translated by R. Fagles. London: Penguin, 1996.

Hooson, David J. M. "The Middle Volga. An Emerging Focal Region in the Soviet Union." *Geographical Journal* 126 (December 1960): 180–90.

Horden, Peregrine, and N. Purcell. *The Corrupting Sea: A Study of Mediterranean History.* Oxford, UK: Blackwell Publishing, 2000.

Hoskins, Brian J., and K. I. Hodges. "New Perspectives on the Northern Hemisphere Winter Storm Tracks." *Journal of Atmospheric Science* 59 (March 2002): 1041–61.

Huang, Ray. *Taxation and Governmental Finance in Sixteenth-Century Ming China.* Cambridge: Cambridge University Press, 1974.

Hughes, David M. "Whites and Water: How Euro-Africans Made Nature at Kariba Dam." *Journal of Southern African Studies* 32 (March 2006): 823–38.

Hughes, J. Donald. *Environmental Problems of the Greeks and the Romans.* 2nd ed. Baltimore: Johns Hopkins University Press, 2014.

Hughes, J. Donald. "Sustainable Agriculture in Ancient Egypt." *Agricultural History 66* (Spring 1992): 12–22.

Hurst, Harold E. "Progress in the Study of the Hydrology of the Nile in the Last Twenty Years." *Geographical Journal* 70 (November 1927): 440–58.

Hyam, Ronald. "The Geopolitical Origins of the Central African Federation: Britain, Rhodesia and South Africa, 1948–1953." *Historical Journal* 30 (March 1987): 145–72.

Immerzeel, Walter W., L. P. H. van Beek, and M. F. P. Bierkens. "Climate Change Will Affect the Asian Water Towers." *Science* 328 (June 2010): 1382–85.

International Hydropower Association (IHA). *2019 Hydropower Status Report: Sector Trends and Insights.* London: International Hydropower Association, 2019.

Isenburg, Teresa. *Acque e Stato: Energia, Bonigiche, Irrigazione in Italia fra 1930 e 1950.* Milan: Franco Angeli Editore, 1981.

Israel, Jonathan L. *The Dutch Republic: Its Rise, Greatness, and Fall, 1477–1806.* Oxford, UK: Clarendon Press, 1995.

Jacobsen, Thorkild. "The Historian and the Sumerian Gods." *Journal of the American Oriental Society* 114 (April–June 1994): 145–53.

Jacobsen, Thorkild, and R. M. Adams. "Salt and Silt in Ancient Mesopotamian Agriculture." *Science* 128 (November 1958): 1251–58.

Janku, Andrea. "Drought and Famine in Northwest China: A Late Victorian Tragedy?" *Journal of Chinese History* 2 (July 2018): 373–91.

Jasny, Naum. "Soviet Agriculture and the Fourth Five-Year Plan." *Russian Review* 8 (April 1949): 135–41.

Jeal, Tim. *Explorers of the Nile: The Triumph and Tragedy of a Great Victorian Adventure.* New Haven, CT: Yale University Press, 2011.

Jerome, Saint. "Letter 127, to Principia." In *Nicene and Post-Nicene Fathers,* 2nd series, vol. 6, edited by P. Schaff and H. Wace, translated by W. H. Fremantle, 253–58. Buffalo, NY: Christian Literature Publishing Co., 1893.

Jia, Xin, G. Dong, H. Li, K. Brunson, F. Chen, M. Ma, H. Wang, C. An, and K. Zhang. "The Development of Agriculture and Its Impact on Cultural Expansion During the Late Neolithic in the Western Loess Plateau, China." *Holocene* 23 (January 2013): 85–92.

Joffe, Alexander H. "The Rise of Secondary States in the Iron Age Levant." *Journal of the Economic and Social History of the Orient* 45 (December 2002): 425–67.

Johnson, Douglas W. *Topography and Strategy in the War.* New York: Henry Holt, 1917.

Johnson, Ralph W. "Freedom of Navigation for International Rivers: What Does It Mean?" *Michigan Law Review* 62 (January 1964): 465–84.

Jones, A. H. M. "Capitatio and Iugatio." *Journal of Roman Studies* 47 (November 1957) 88–94.

Jones, A. H. M. *The Later Roman Empire, 284–602: A Social Economic and Administrative Survey.* Vol. 2. Oxford: Oxford University Press, 1964.

Jones, Toby C. "America, Oil, and War in the Middle East." *Journal of American History* 99 (June 2012): 208–18.

Jones, Toby C. "Rebellion on the Saudi Periphery: Modernity, Marginalization, and the Shiʿa Uprising of 1979." *International Journal of Middle East Studies* 38 (May 2006): 213–33.

Jones, Toby C. "State of Nature: The Politics of Water in the Making of Saudi Arabia." In *Water on Sand: Environmental Histories of the Middle East and North Africa,* edited by A. Mikhail, 231–50. Oxford: Oxford University Press, 2012.

Jonglei Investigation Team. "The Equatorial Nile Project and Its Effects in the Sudan." *Geographical Journal* 119 (March 1953) 33–48.

Jordan, William C. "The Great Famine: 1315–1322 Revisited." In *Ecologies and Economies in Medieval and Early Modern Europe,* edited by S. G. Bruce, 45–62. Leiden: Brill, 2010.

Josling, Tim, K. Anderson, A. Schmitz, and S. Tangermann. "Understanding International Trade in Agricultural Products: One Hundred Years of Contributions by Agricultural Economists." *American Journal of Agricultural Economics* 92 (April 2010): 424–46.

Kaijser, Arne. "System Building from Below: Institutional Change in Dutch Water Control Systems." *Technology and Culture* 43 (July 2002): 521–48.

Kandiyoti, Deniz, ed. *The Cotton Sector in Central Asia Economic Policy and Development Challenges.* London: School of Oriental and African Studies, 2005.

Kaniewski, David, E. Paulissen, E. Van Campo, H. Weiss, T. Otto, J. Bretschneider, and K. Van Lerberghe. "Late Second–Early First Millennium BC Abrupt Climate Changes in Coastal Syria and Their Possible Significance for the History of the Eastern Mediterranean." *Quaternary Research* 74 (September 2010): 207–215

Kaniewski, David, E. Van Campo, K. Van Lerberghe, T. Boiy, K. Vansteenhuyse, G. Jans, K. Nys, H. Weiss, C. Morhange, T. Otto, and J. Bretschneider. "The Sea Peoples, from Cuneiform Tablets to Carbon Dating." *PLOS One* 6 (June 2011): e20232.

Kaniewski, David, E. Van Campo, J. Guiot, S. Le Burel, T. Otto, and C. Baeteman. "Environmental Roots of the Late Bronze Age Crisis." *PLOS One* 8 (August 2013): e71004.

Kaniewski, David, J. Guiot, and E. Van Campo. "Drought and Societal Collapse 3200 Years Ago in the Eastern Mediterranean: A Review." *WIREs Climate Change* 6 (June/August 2015): 369–82.

Kaplan, Lawrence S. "The United States, Belgium, and the Congo Crisis of 1960." *Review of Politics* 29 (April 1967): 239–56.

Kapuscinski, Ryszard. *The Emperor.* Translated by W. R. Brand and K. Mroczkowska-Brand. London: Penguin Books, 1983.

Kaufmann, Yehezkel. "The Bible and Mythological Polytheism." Translated by Moshe Greenberg. *Journal of Biblical Literature* 70 (September 1951): 179–97.

Kay, Philip. *Rome's Economic Revolution.* Oxford: Oxford University Press, 2014.

Keay, Simon. "The Port System of Imperial Rome" In *Rome, Portus and the Mediterranean,* edited by S. Keay, 33–67. London: Archaeological Monographs of the British School at Rome, 2012.

Kelly, P. M., D. A. Campbell, P. P. Micklin, and J. R. Tarrant. "Large-Scale Water Transfers in the USSR." *GeoJournal* 7 (June 1983): 201–14.

Kemp, Barry J. *Ancient Egypt: Anatomy of a Civilization.* London: Routledge, 2006.

Kennett, Douglas J., and P. J. Kennett. "Early State Formation in Southern Mesopotamia: Sea Levels, Shorelines, and Climate Change." *Journal of Island & Coastal Archaeology* 1 (2006): 67–99.

Kenyon, Paul. *Dictatorland: The Men Who Stole Africa.* London: Head of Zeus, 2018.

Kershaw, Ian. "The Great Famine and Agrarian Crisis in England, 1315–1322." *Past & Present* 59 (May 1973): 3–50.

Kessler, David, and P. Temin. "The Organization of the Grain Trade in the Early Roman Empire." *Economic History Review* 60 (May 2007): 313–32.

Keynes, John M. *The Economic Consequences of the Peace.* London: Harcourt, Brace and Howe, 1920.

Keynes, John M. *The End of Laissez-Faire.* London: Hogarth Press, 1926.

Keynes, John M. *The General Theory of Employment, Interest and Money.* London: Macmillan, 1936.

Kilmer, Anne D. "The Mesopotamian Concept of Overpopulation and Its Solution as Reflected in the Mythology." *Orientalia* 41 (January 1972): 160–77.

Kitchens, Carl. "The Role of Publicly Provided Electricity in Economic Development: The Experience of the Tennessee Valley Authority, 1929–1955." *Journal of Economic History* 74 (June 2014): 389–419.

Kitchens, Carl, and P. Fishback. "Flip the Switch: The Impact of the Rural Electrification Administration, 1935–1940." *Journal of Economic History* 75 (December 2015): 1161–95.

Kletter, Raz. "Pots and Polities: Material Remains of Late Iron Age Judah in Relation to Its Political Borders." *Bulletin of the American Schools of Oriental Research* 314 (May 1999): 19–54.

Kline, Patrick M., and E. Moretti. "Local Economic Development, Agglomeration Economies, and the Big Push: 100 Years of Evidence from the Tennessee Valley Authority." NBER Working Paper 19293, National Bureau of Economic Research, 2013.

Knittl, Margaret A. "The Design for the Initial Drainage of the Great Level of the Fens: An Historical Whodunit in Three Parts." *Agriculture History Review* 55 (June 2007): 23–50.

Koch, Alexander, C. Brierley, M. M. Maslin, and S. L. Lewis. "Earth System Impacts of the European Arrival and Great Dying in the Americas After 1492." *Quaternary Science Reviews* 207 (March 2019): 13–36.

Koutsoyiannis, Demetris, N. Zarkadoulas, A. N. Angelakis, and G. Tchobanoglous. "Urban Water Management in Ancient Greece: Legacies and Lessons." *Journal of Water Resources Planning and Management* 134 (January/February 2008): 45–54.

Kramnick, Isaac. "Editor's Introduction." In *James Madison, Alexander Hamilton, and John Jay: The Federalist Papers,* edited by Isaac Kramnick, 11–81. London: Penguin Classics, 1987.

Krasilnikoff, Jens A. "Irrigation as Innovation in Ancient Greek Agriculture." *World Archaeology* 42 (January 2010): 108–21.

Kremer, Michael. "Population Growth and Technological Change: One Million B.C. to 1990." *Quarterly Journal of Economics* 108 (August 1993): 681–716.

Kristiansen, Kristian, and T. B. Larsson. *The Rise of Bronze Age Society: Travels, Transmissions and Transformations.* Cambridge: Cambridge University Press, 2005.

Kristiansen, Kristian, and P. Suchowska-Ducke. "Connected Histories: The Dynamics of Bronze Age Interaction and Trade, 1500–1100 BC." *Proceedings of the Prehistoric Society* 81 (December 2015): 361–92.

Kueh, Y. Y. *Agricultural Instability in China, 1931–1990: Weather, Technology, and Institutions.* Oxford: Clarendon Press, 1995.

Kummu, Matti, H. de Moel, P. J. Ward, and O. Varis. "How Close Do We Live to Water? A Global Analysis of Population Distance to Freshwater Bodies." *PLOS One* 6 (June 2011): 1–13.

Lamb, Hubert H. *Climate, History, and the Modern World.* New York: Routledge, 1995.

Lambeck, Kurt, H. Rouby, A. Purcell, Y. Sun, and M. Sambridge. "Sea Level and Global Ice Volumes from the Last Glacial Maximum to the Holocene." *Proceedings of the National Academy of Sciences* 111 (October 2014): 296–303.

Lambertenghi, Giulio P. "Codex Diplomaticus Longobardiae." In *Monumenta Historiae Patriae, Tome XIII.* Torino: Regio Typographeo, 1873: 17–18.

Lambrick, Hugh T. "The Indus Flood-Plain and the 'Indus' Civilization." *Geographical Journal* 133 (December 1967): 483–95.

Laster, Richard, D. Aronowski, and D. Livney. "Water in the Jewish Legal Tradition." In *The Evolution of the Law and Politics of Water,* edited by J. W. Dellapenna and J. Gupta, 53–66. Dordrecht, Netherlands: Springer, 2009.

Lattimore, Owen. "Origins of the Great Wall of China: A Frontier Concept in Theory and Practice." *Geographical Review* 27 (October 1937): 529–49.

Lau, K.-M., G. J. Yang, and S. H. Shen. "Seasonal and Intraseasonal Climatology of Summer Monsoon Rainfall over East Asia." *Monthly Weather Review* 116 (1988): 18–37.

Lau, William K. M. and K.-M. Kim. "The 2010 Pakistan Flood and Russian Heat Wave: Teleconnection of Hydrometeorological Extremes." *Journal of Hydrometeorology* 13 (February 2012): 392–403.

Lebergott, Stanley. "Annual Estimates of Unemployment in the United States, 1900–1954." In *The Measurement and Behavior of Unemployment,* 211–42. Universities-National Bureau, 1957.

Lenin, Vladimir I. "Our Foreign and Domestic Position and Party Tasks." In *Lenin's Collected Works,* vol. 31, edited by J. Katzer, 408–26. Moscow: Progress Publishers, 1965.

Lenin, Vladimir I. "The State and Revolution: The Marxist Theory of the State and the Tasks of the Proletariat in the Revolution." In *Lenin's Collected Works*, vol. 25, edited by S. Apresyan and J. Riordan, 381–492. Moscow: Progress Publishers, 1965.

Lenin, Vladmir I. "What Is to Be Done?" In *Lenin's Collected Works*, vol. 5, edited by V. Jerome, 347–530. Moscow: Progress Publishers, 1961.

Lev-Yadun, Simcha, A. Gopher, S. Abbo. "The Cradle of Agriculture." *Science* 288 (June 2000): 1602–3.

Lewis, W. Arthur. "Economic Development with Unlimited Supply of Labour." *Manchester School* 22 (May 1954): 139–91.

Lewit, Tamara. "'Vanishing Villas': What Happened to Élite Rural Habitation in the West in the 5th—6th C?" *Journal of Roman Archeology* 16 (2003): 260–74.

Li, Wei, and D. T. Yang. "The Great Leap Forward: Anatomy of a Central Planning Disaster." *Journal of Political Economy* 113 (August 2005): 840–77.

Libecap, Gary D., and Z. K. Hansen. "'Rain Follows the Plow' and Dryfarming Doctrine: The Climate Information Problem and Homestead Failure in the Upper Great Plains, 1890–1925." NBER Historical Working Paper No. 127, 2000.

Liebman, Alex. "Trickle-down Hegemony? China's 'Peaceful Rise' and Dam Building on the Mekong." *Contemporary Southeast Asia* 27 (August 2005): 281–304.

Lilienthal, David E. "Another Korea in the Making?" *Collier's,* August 4, 1951.

Lilienthal, David E. *The Journals of David E. Lilienthal,* vol. 6. New York: Harper & Row, 1976.

Lilienthal, David E. *TVA: Democracy on the March.* New York: Harper & Brothers, 1944.

Lin, Justin Y. *New Structural Economics: A Framework for Rethinking Development Policy.* Washington, DC: World Bank Press, 2012.

Lintott, Andrew. *The Constitution of the Roman Republic.* Oxford: Oxford University Press, 1999.

Lionello, Piero, F. Abrantes, L. Congedi, F. Dulac, M. Gacic, D. Gomis, C. Goodess, H. Hoff, H. Kutiel, J. Luterbacher, S. Planton, M. Reale, K. Schröder, M. V. Struglia, A. Toretino, M. Tsimplis, U. Ulbrich, and E. Xoplaki. "Introduction: Mediterranean Climate—Background Information." In *The Climate of the Mediterranean Region: From the Past to the Future,* edited by P. Lionello. London: Springer, 2012: xxxv–xc.

Lippman, Thomas W. "The Day FDR Met Saudi Arabia's Ibn Saud." *The Link* 38 (April–May 2005): 1–12.

Lipsey, Robert E. "U.S. Foreign Trade and the Balance of Payments, 1800–1913." NBER Working Paper No. 4710, 1994.

Littlefield, Douglas R. "The Potomac Company: A Misadventure in Financing an Early American Internal Improvement Project." *Business History Review* 58 (Winter 1984): 562–85.

Liu, Jianguo, and W. Yang. "Water Sustainability for China and Beyond." *Science* 337 (August 2012): 649–50.

Liverani, Mario. *The Ancient Near East: History, Society and Economy.* London: Routledge, 2014.

Liverani, Mario. *Israel's History and the History of Israel.* Sheffield, UK: Equinox, 2005.

Liverani, Mario. *Uruk, the First City.* Sheffield, UK: Equinox, 2006.

Lohof, Bruce A. "Herbert Hoover, Spokesman of Humane Efficiency: The Mississippi Flood of 1927." *American Quarterly* 22 (1970): 690–700.

Luce, Henry R. "The American Century." *Life,* February 17, 1941.

Luciani, Giacomo. "Oil and Political Economy in the International Relations of the Middle East." In *International Relations of the Middle East,* edited by L. Fawcett, 108–29. Oxford: Oxford University Press, 2016.

Luckenbill, Daniel D. *The Annals of Sennacherib.* Chicago: University of Chicago Press, 1924.

MacDonald, Alan R., and J. McCallum. "The Evidence for Early Seventeenth-Century Climate from Scottish Ecclesiastical Records." *Environment and History* 19 (November 2013): 487–509.

Macekura, Stephen. "The Point Four Program and U.S. International Development Policy." *Political Science Quarterly* 128 (April 2013): 127–60.

MacGrady, Glenn J. "The Navigability Concept in the Civil and Common Law: Historical Development, Current Importance, and Some Doctrines That Don't Hold Water." *Florida State University Law Review* 3 (Fall 1975): 511–615.

Machiavelli, Niccolò. *Discorsi sopra la prima Deca di Tito Livio.* Edited by M. Martelli. Letteratura Italiana Einaudi, 1971.

Madison, James. *The Papers of James Madison.* Vol. 8, *10 March 1784—28 March 1786*, edited by R. A. Rutland and W. M. E. Rachal. Chicago: University of Chicago Press, 1973.

Magee, Darrin L. "The Dragon Upstream: China's Role in Lancang-Mekong Development." In *Politics and Development in a Transboundary Watershed*, edited by J. Öjendal, S. Hansson, and S. Hellberg, 171–93. Dordrecht, Netherlands: Springer, 2011.

Major, John. "Who Wrote the Hay-Bunau-Varilla Convention?" *Diplomatic History* 8 (April 1984): 115–23.

Malara, Empio, and C. Coscarella. *Milano & navigli: un parco lineare tra il Ticino e l'Adda.* Milan: Di Baio Editore, 1990.

Malley, Shawn. "Layard Enterprise: Victorian Archaeology and Informal Imperialism in Mesopotamia." *International Journal of Middle East Studies* 40 (November 2008): 623–46.

Mann, Michael. *The Sources of Social Power.* Vol. 1, *A History of Power from the Beginning to AD 1760.* Cambridge: Cambridge University Press, 2012.

Mansfield, Peter. *A History of the Middle East.* 4th ed. New York: Penguin Books, 2013.

Manville, Philip B. *The Origins of Citizenship in Ancient Athens.* Princeton, NJ: Princeton University Press, 1990.

Mao Zedong. *The Writings of Mao Zedong, 1949–1976.* Vol. 2, *January 1956–December 1957*, edited by J. K. Leung and M. Y. M. Kau. New York: M. E. Sharpe, 1992.

Martius, Olivia, H. Sodemann, H. Joos, S. Pfahl, A. Winschall, M. Croci-Maspoli, M. Graf, E. Madonna, B. Mueller, S. Schemm, J. Sedláček, M. Sprenger, and H. Wernli. "The Role of Upper-Level Dynamics and Surface Processes for the Pakistan Flood of July 2010." *Quarterly Journal of the Royal Meteorological Society* 139 (January 2013): 1780–97.

Marx, Karl. *Capital.* Vol. 1, translated by B. Fowkes. London: Penguin Classics, 1990.

Marx, Karl, and F. Engels. "Manifesto of the Communist Party." In *Marx/Engels Selected Works*, Vol. 1, 98–137. Moscow: Progress Publishers, 1969.

Masters, Roger D. *Machiavelli, Leonardo, and the Science of Power.* Notre Dame, IN: University of Notre Dame Press, 1996.

Mathew, Sarah, and C. Perreault. "Behavioural Variation in 172 Small-Scale Societies Indicates That Social Learning Is the Main Mode of Human Adaptation." *Proceedings of the Royal Society B* 282 (July 2015): 20150061.

Matthiesen, Toby. *Sectarian Gulf: Bahrain, Saudi Arabia, and the Arab Spring That Wasn't.* Stanford, CA: Stanford Briefs, 2013.

Mavhunga, Clapperton C. "Energy, Industry, and Transport in South-Central Africa's History." *Rachel Carson Center Perspectives* 5 (2014): 9–18.

McCormick, Michael, U. Büntgen, M. A. Cane, E. R. Cook, K. Harper, P. Huybers, T. Litt, S. W. Manning, P. A. Mayewski, A. F. M. More, K. Nicolussi, and W. Tegel. "Climate Change During and After the Roman Empire: Reconstructing the Past from Scientific and Historical Evidence." *Journal of Interdisciplinary History* 43 (Autumn 2012): 169–220.

McCormick, Michael, P. E. Dutton, and P. A. Mayewski. "Volcanoes and the Climate Forcing of Carolingian Europe, A.D. 750–950." *Speculum* 82 (October 2007): 865–95.

McCullough, David. *The Path Between the Seas: The Creation of the Panama Canal, 1870–1914.* New York: Simon & Schuster, 1977.

McGerr, Michael. *A Fierce Discontent: The Rise and Fall of the Progressive Movement in America, 1870–1920.* Oxford: Oxford University Press, 2003.

McSweeney, Thomas J. "Magna Carta, Civil Law, and Canon Law." In *Magna Carta and the Rule of Law*, edited by A. Martinez, D. B. Magraw Jr., and R. E. Brownell II, 281–309. American Bar Association, 2014.

Medina, Jose T. *Relación que escribió Fr. Gaspar de Carvajal in Descubrimiento del Rio de Las Amazonas*. Seville: E. Rasco, 1894.

Meggers, Betty J. "Environmental Limitation on the Development of Culture." *American Anthropologist* 56 (October 1954): 801–24.

Melnikova-Raich, Sonia. "The Soviet Problem with Two 'Unknowns': How an American Architect and a Soviet Negotiator Jump-Started the Industrialization of Russia, Part I: Albert Kahn." *Journal of the Society for Industrial Archeology* 36 (January 2010): 57–80.

Melnikova-Raich, Sonia. "The Soviet Problem with Two 'Unknowns': How an American Architect and a Soviet Negotiator Jump-Started the Industrialization of Russia, Part II: Saul Bron." *Journal of the Society for Industrial Archeology* 37 (January 2011): 5–28.

Meng, Xin, N. Qian, and P. Yared. "The Institutional Causes of China's Great Famine, 1959–61." *Review of Economic Studies* 82 (April 2015): 1568–1611.

Merl, Stephan, and B. Templer. "Why Did the Attempt Under Stalin to Increase Agricultural Productivity Prove to Be Such a Fundamental Failure? On Blocking the Implementation of Progress in Agrarian Technology (1929–1941)." *Cahiers du Monde Russe* 57 (January 2016): 191–220.

Micklin, Philip P. "Desiccation of the Aral Sea: A Water Management Disaster in the Soviet Union." *Science* 241 (September 1988): 1170–76.

Miescher, Stephan F., and D. Tsikata. "Hydro-power and the Promise of Modernity and Development in Ghana: Comparing the Akosombo and Bui Dam Projects." *Ghana Studies* 12/13 (2009–2010): 15–53.

Migone, Gian Giacomo. *The United States and Fascist Italy: The Rise of American Finance in Europe*. Cambridge: Cambridge University Press, 2015.

Mill, John S. *Memorandum of the Improvements in the Administration of India During the Last Thirty Years, and the Petition of the East-India Company to Parliament*. London: W. H. Allen & Co., 1858.

Mill, John S. *Nature, the Utility of Religion, and Theism*. London: Watts & Co., Rationalist Press, 1904.

Miller, Barbara A., and R. B. Reidinger. *Comprehensive River Development: The Tennessee River Valley Authority*. Washington, DC: World Bank Technical Paper 416, 1998.

Ministero per la Costituente. *Rapporto della Commissione Tecnica, Industria*. Vol. 1. Rome: Istituto Poligrafico dello Stato, 1947.

Molle, Francois. "River-Basin Planning and Management: The Social Life of a Concept." *Geoforum* 40 (May 2009): 484–94.

Montesquieu, Charles-Louis de Secondat. *The Spirit of the Laws*. Edited by A. M. Cohler, B. C. Miller, and H. S. Stone. Cambridge: Cambridge University Press, 1989.

Moore, Andy E., F. P. D. Cotterill, M. P.L. Main, and H. B. Williams. "The Zambezi River." In *Large Rivers: Geomorphology and Management*, edited by A. Gupta, 311–32. Chichester, UK: John Wiley & Sons, 2007.

Moore, Benjamin J., P. J. Neiman, F. M. Ralph, and F. E. Barthold. "Physical Processes Associated with Heavy Flooding Rainfall in Nashville, Tennessee, and Vicinity During 1–2 May 2010: The Role of an Atmospheric River and Mesoscale Convective Systems." *Monthly Weather Review* 140 (February 2012): 358–78.

More, Thomas. *Utopia*. Translated by D. Baker-Smith. London: Penguin Classics, 2012.

Mori, Giorgio. "Le guerre parallele. L'industria elettrica in Italia nel periodo della grande guerra (1914–1919)." *Studi Storici* 14 (April–June 1973): 292–372.

Morison, William S. "An Honorary Deme Decree and the Administration of a Palaistra in Kephissia." *Zeitschrift für Papyrologie und Epigraphik* 131 (2000): 93–98.

Morris, Ian. "Economic Growth in Ancient Greece." *Journal of Institutional and Theoretical Economics* 160 (December 2004): 709–42.

Muller, Wolfgang P. "The Recovery of Justinian's Digest in the Middle Ages." *Bulletin of Medieval Canon Law* 20 (1990): 1–29.

Murphy, Brian P. "'A Very Convenient Instrument': The Manhattan Company, Aaron Burr, and the Election of 1800." *William and Mary Quarterly* 65 (April 2008): 233–66.

Needham, Joseph. *Science and Civilization in China.* Vol. 4, *Physics and Physical Technology. Part III: Civil Engineering and Nautics.* Cambridge: Cambridge University Press, 1971.

Nelson, Eric. "Republican Visions." In *The Oxford Handbook of Political Philosophy*, edited by J. S. Dryzek, B. Honig, and A. Phillips, 193–210. Oxford: Oxford University Press, 2008.

Nesje, Atle, and S. O. Dahl. "The 'Little Ice Age'—Only Temperature?" *The Holocene* 13 (January 2003): 139–45.

Neuse, Steven M. *David E. Lilienthal: The Journey of an American Liberal.* Knoxville: University of Tennessee Press, 1996.

Neuse, Steven M. "TVA at Age Fifty—Reflections and Retrospect." *Public Administration Review* 43 (November/December 1983): 491–99.

Newfield, Timothy P. "A Cattle Panzootic in Early Fourteenth-Century Europe." *Agricultural History Review* 57 (December 2009): 155–90.

Newfield, Timothy P. "The Climate Downturn of 536–50." In *The Palgrave Handbook of Climate History,* edited by S. White, C. Pfister, and F. Mauelshagen, 447–93. London: Palgrave Macmillan, 2018.

Newfield, Timothy P. "Mysterious and Mortiferous Clouds: The Climate Cooling and Disease Burden of Late Antiquity." In *Environment and Society in the Long Late Antiquity,* edited by A. Izdebski and M. Mulryan, 89–115. Leiden: Brill, 2019.

Niebling, William, J. Baker, L. Kasuri, S. Katz, and K. Smet. "Challenge and Response in the Mississippi River Basin." *Water Policy* 16 (March 2014): 87–116.

Nitti, Francesco S. *L'Italia all'alba del secolo XX: Discorsi ai giovani d'Italia.* Torino: Casa Editrice Nazionale Roux e Varengo, 1901.

Nkrumah, Kwame. *I Speak of Freedom: A Statement of African Ideology.* London: William Heinemann Ltd., 1961.

North, Douglass C., and B. R. Weingast. "Constitutions and Commitments: The Evolution of Institutions, Governing Public Choice in Seventeenth Century England." *Journal of Economic History* 49 (December 1989): 803–832.

Nunn, Patrick D., and N. J. Reid. "Aboriginal Memories of Inundation of the Australian Coast Dating from More Than 7,000 Years Ago." *Australian Geographer* 47 (January 2016): 11–47.

Ó Gráda, Cormac. "Great Leap into Famine: A Review Essay." *Population and Development Review* 37 (March 2011): 191–202.

Offer, Avner. *The First World War: An Agrarian Interpretation.* Oxford: Oxford University Press, 1991.

Ogilvie, Sheilagh C. "Germany and the Seventeenth-Century Crisis." *Historical Journal* 35 (June 1992): 417–41.

Paine, Thomas. *Common Sense.* London: Penguin Classics, 2012.

Palka, Eugene J. "A Geographic Overview of Panama." In *The Río Chagres, Panama: A Multidisciplinary Profile of a Tropical Watershed,* edited by R. S. Harmon, 3–18. Amsterdam: Springer, 2005.

Parker, Charles H. *Global Interactions in the Early Modern Age, 1400–1800.* New York: Cambridge University Press, 2010.

Parker, Geoffrey. "Crisis and Catastrophe: The Global Crisis of the Seventeenth Century Reconsidered." *American Historical Review* 113 (October 2008): 1053–79.

Parkinson, Richard B., trans. *The Tale of Sinuhe and Other Ancient Egyptian Poems, 1940–1640 BC.* Oxford: Oxford University Press, 1997.

Pavese, Claudio. *Cento Anni di Energia: Centrale Bertini, 1898–1998.* Milan: Edison, 1998.

Pederson, Neil, A. E. Hessl, N. Baatarbileg, K. J. Anchukaitis, and N. Di Cosmo. "Pluvials, Droughts, the Mongol Empire, and Modern Mongolia." *Proceedings of the National Academy of Sciences* 111 (March 2014): 4375–79.

Penslar, Derek J. "Between Honor and Authenticity: Zionism as Theodor's Life Project." In *On the Word of a Jew: Religion, Reliability, and the Dynamics of Trust,* edited by N. Caputo and M. B. Hart, 276–96. Bloomington: Indiana University Press, 2019.

Perkins, Frances. *The Roosevelt I Knew.* New York: Viking Press, 1946.

Peterson, Maya. "US to USSR: American Experts, Irrigation, and Cotton in Soviet Central Asia, 1929–32." *Environmental History* 21 (July 2016): 442–46.

Pini, Antonio I. "Classe Politica e Progettualità Urbana a Bologna nel XII e XIII secolo." In *Villes et sociétés urbaines au Moyen âge,* edited by M. T. Caron, 25–27. Paris: Presses de l'Université de Paris Sorbonee, 1994.

Pisani, Donald J. "State vs. Nation: Federal Reclamation and Water Rights in the Progressive Era." *Pacific Historical Review* 51 (August 1982): 265–82.

Pisani, Donald J. *Water and American Government: The Reclamation Bureau, National Water Policy and the West, 1902–1935.* Berkeley: University of California Press, 2002.

Pitts, Lynn F. "Relations Between Rome and the German 'Kings' on the Middle Danube in the First to Fourth Centuries A.D." *Journal of Roman Studies* 79 (November 1989): 45–58.

Plato. *The Laws.* Translated by T. J. Saunders. London: Penguin Classics, 1975.

Plato. *The Republic.* Translated by D. Lee. London: Penguin Classics, 1987.

Pnevmatikos, John D., and B. D. Katsoulis, 2006. "The Changing Rainfall Regime in Greece and Its Impact on Climatological Means." *Meteorological Applications* 13 (December 2006): 331–45.

Pohl, J. Otto. "A Caste of Helot Labourers: Special Settlers and the Cultivation of Cotton in Soviet Central Asia: 1944–1956." In *The Cotton Sector in Central Asia Economic Policy and Development Challenges: Proceedings of a Conference held at SOAS University of London, 3–4 November 2005,* edited by Deniz Kandiyoti, 12–28. London: School of Oriental and African Studies (SOAS), 2007.

Polastro, Riccardo, A. Nagrah, N. Steen, and F. Zafar. *Inter-Agency Real Time Evaluation of the Humanitarian Response to Pakistan's 2010 Flood Crisis.* Dara Report, 2011.

Polybius. *The Histories.* Translated by R. Waterfield. Oxford: Oxford University Press, 2010.

Popescu, Adina. "Casting Bread upon the Waters: American Farming and the International Wheat Market, 1880–1920." PhD diss., Columbia University, New York, 2014.

Porisini, Giorgio. "Le bonifiche nella politica economica dei governi Cairoli e Depretis." *Studi Storici* 15 (July–September 1974): 589–623.

Portmann, Robert W., S. Solomon, and G. C. Hegerl. "Spatial and Seasonal Patterns in Climate Change, Temperatures, and Precipitation Across the United States." *Proceedings of the National Academy of Sciences* 106 (April 2009): 7324–29.

Potter, David S. *The Roman Empire at Bay, AD 180–395.* London: Routledge, 2014.

Pournelle, Jennifer. "Marshland of Cities: Deltaic Landscapes and the Evolution of Early Mesopotamian Civilization." PhD diss., University of California, San Diego, 2003.

Powell, John W. *Report on the Lands of the Arid Region of the United States with a More Detailed Account of the Lands of Utah.* 2nd ed. Washington, DC: Government Printing Office, 1879.

Preti, Domenico. "La politica agraria del fascismo: Note introduttive." *Studi Storici* 14 (October–December 1973): 802–69.

Prishchepov, Alexander V., V. C. Radeloff, M. Baumann, T. Kuemmerle, and D. Müller. "Effects of Institutional Changes on Land Use: Agricultural Land Abandonment During the Transition from State-Command to Market-Driven Economies in Post-Soviet Eastern Europe." *Environmental Research Letters* 7 (June 2012): 024021.

Procopius. *History of the Wars, Books I and II.* Translated by H. B. Dewing. London: William Heinemann, 1914.

Purzycki, Benjamin G., C. Apicella, Q. D. Atkinson, E. Cohen, R. A. McNamara, A. Willard, D. Xygalatas, A. Norenzayan, and J. Henrich. "Moralistic Gods, Supernatural Punishment and the Expansion of Human Sociality." *Nature* 530 (February 2016): 327–30.

Putnam, Aaron E., D. E. Putnam, L. Andreu-Hayles, E. R. Cook, J. G. Palmer, E. H. Clark, C. Wang, F. Chen, G. H. Denton, D. P. Boyle, S. D. Bassett, S. D. Birkel, J. Martin-Fernandez, I. Hajdas, J. Southon, C. B. Garner, H. Cheng, and W. S. Broecker. "Little Ice Age Wetting of Interior Asian Deserts and the Rise of the Mongol Empire." *Quaternary Science Reviews* 131 (January 2016): 33–50.

Quilici Gigli, Stefania. "Su alcuni segni dell'antico paesaggio agrario presso Roma." *Quaderni del Centro di studio per l'archeologia Etrusco-Italica* 14 (1987): 152–66.

Racine, Pierre. "Poteri medievali e percorsi fluviali nell'Italia padana." *Quaderni storici* 21 (April 1986): 9–32.

Radivojević, Miljana, B. W. Roberts, E. Pernicka, Z. Stos-Gale, M. Martinón-Torres, T. Rehren, P. Bray, D. Brandherm, J. Ling, J. Mei, H. Vandkilde, K. Kristiansen, S. J. Shennan, and C. Broodbank. "The Provenance, Use, and Circulation of Metals in the European Bronze Age: The State of Debate." *Journal of Archaeological Research* 27 (June 2019): 131–85.

Ralph, F. Martin, and M. D. Dettinger. "Storms, Floods, and the Science of Atmospheric Rivers." *EOS* 92 (August 2011): 265–72.

Rawlinson, George. *A Memoir of Major-General Sir Henry Creswicke Rawlinson.* London: Longmans Green and Co., 1898.

Rawlinson, Henry C. "Memoir on the Babylonian and Assyrian Inscriptions." *Journal of the Royal Asiatic Society of Great Britain and Ireland* 14 (1851): i–civ, 1–16.

Reader, John. *Africa: A Biography of a Continent.* New York: Vintage Books, 1999.

Reberschak, Maurizio. "Una storia del 'genio italiano': il Grande Vajont." In *Il Grande Vajont,* edited by M. Reberschak. Verona: Cierre Edizioni, 2003.

Reimer, Michael. "The King-Crane Commission at the Juncture of Politics and Historiography." *Critique: Critical Middle Eastern Studies* 15 (Summer 2006): 129–50.

Reinhold, Meyer. *Marcus Agrippa: A Biography.* New York: W. F. Humphrey Press, Geneva, 1933.

Reuss, Martin. "Is It Time to Resurrect the Harvard Water Program?" *Journal of Water Resources Planning and Management* 129 (September 2003): 357–60.

Revelle, Roger. "Mission to the Indus." *New Scientist* 17 (February 1963): 340–42.

Revelle, Roger. "Oceanography, Population Resources and the World; Roger Randall

Dougan Revelle, Director of Scripps Institution of Oceanography, 1951–1964," an oral history conducted in 1984 by Sarah Sharp, University of California, Regional Oral History Office, Bancroft Library. Berkeley: University of California Press, 1988.

Rey, Sébastien. *For the Gods of Girsu: City-State Formation in Ancient Sumer.* Oxford, UK: Archaeopress, 2016.

Richard, Carl J. *The Founders and the Classics: Greece, Rome, and the American Enlightenment.* Cambridge, MA: Harvard University Press, 1994.

Richards, John F. "The Indian Empire and Peasant Production of Opium in the Nineteenth Century." *Modern Asian Studies* 15 (February 1981): 59–82.

Richerson, Peter J., and R. Boyd. "Cultural Inheritance and Evolutionary Ecology." In *Evolutionary Ecology and Human Behavior,* edited by E. A. Smith and B. Winterhalder, 61–92. New York: Aldine de Gruyter, 1992.

Richter, Bernd S. "Nature Mastered by Man: Ideology and Water in the Soviet Union." *Environment and History* 3 (February 1997): 69–96.

Rinaldi, Rossella. "Il Fiume Mobile. Il Po Mantovano tra monaci-signori, vescovi cittadini e comunità (secoli XI–XII)." In *Il Paesaggio Mantovano nelle trace materiali, nelle lettere e nelle arti,* edited by E. Camerlenghi, V. Rebonato, and S. Tammaccaro, 113–31. Florence: Leo S. Olschki Editore, 2005.

Ritvo, Harriet. *The Dawn of Green: Manchester, Thirlmere, and Modern Environmentalism.* Chicago: University of Chicago Press, 2009.

Roberts, Clayton. "The Earl of Bedford and the Coming of the English Revolution." *Journal of Modern History* 49 (1977): 600–16.

Robinson, Sherman, K. Strzepek, M. El-Said, and H. Lofgren. "The High Dam at Aswan." In *Indirect Economic Impacts of Dams: Case Studies from India, Egypt and Brazil,* edited by R. Bhatia, Rita Cestti, M. Scatasta, and R. P. S. Malik. New Delhi: Academic Foundation, 2008.

Rodes, Robert E., Jr. "The Canon Law as a Legal System—Function, Obligation, and Sanction." *Natural Law Forum,* Paper 82, 1964.

Rodwell, Mark J., and B. J. Hoskins. "Monsoons and the Dynamics of Deserts." *Journal of the Royal Meteorological Society* 122 (July 1996): 1385–1404.

Rogers, Peter. *America's Water: Federal Roles and Responsibilities.* Cambridge, MA: MIT Press, 1993.

Roosevelt, Theodore. "Message from the President of the United States, Transmitting a Preliminary Report of the Inland Waterways Commission." In *Preliminary Report of the Inland Waterways Commission,* iii–v. Washington, DC: Government Printing Office, 1908.

Rosenblith, Judy, ed. *Jerry Wiesner: Scientist, Statesman, Humanist—Memories and Memoirs.* Cambridge, MA: MIT Press, 2003.

Rothman, Mitchell S. "Studying the Development of Complex Society: Mesopotamia in the Late Fifth and Fourth Millennia BC." *Journal of Archaeological Research* 12 (January 2004): 75–119.

Rowen, Herbert H., ed. *The Low Countries in Early Modern Times.* New York: Palgrave Macmillan, 1972.

Rowland, Kate M. "The Mount Vernon Convention." *Pennsylvania Magazine of History and Biography* 11 (January 1888): 410–25.

Sallust, 2007. *Catiline's War, The Jugurthine War, Histories.* Translated by A. J. Woodman. London: Penguin Classics, 2007.

Salman, Salman M. A. "The Baglihar Difference and Its Resolution Process—a Triumph for the Indus Waters Treaty?" *Water Policy* 10 (2008): 105–17.

Salman, Salman M. A. "The Helsinki Rules, the UN Watercourses Convention and the Berlin Rules: Perspectives on International Water Law." *Water Resources Development* 23 (2007): 625–40.

Salvemini, Gaetano. "Can Italy Live at Home?" *Foreign Affairs* 14 (1936): 243–58.

Salvemini, Gaetano. *Le origini del fascismo in Italia: Lezioni di Harvard.* Universale Economica Feltrinelli, 2015.

Salzman, James. *Drinking Water: A History.* New York: Harry N. Abrams, 2017.

Salzman, Michele R. "Apocalypse Then? Jerome and the Fall of Rome in 410." In *Maxima Debetur Magistro Reverentia. Essays on Rome and the Roman Tradition in Honor of Russell T. Scott.* Edited by P. B. Harvey, Jr., and C. Conybeare, 175–92. Biblioteca di Athenaeum, 54 (2009).

Samson, Fred B., F. L. Knopf, and W. R. Ostlie. "Great Plains Ecosystems: Past, Present, and Future." *Wildlife Society Bulletin* 32 (March 2004): 6–15.

Sapelli, Giulio. *Storia Economica Contemporanea.* Milan: Bruno Mondadori Editore, 2008.

Scarritt, James R., and S. M. Nkiwane. "Friends, Neighbors, and Former Enemies: The Evolution of Zambia-Zimbabwe Relations in a Changing Regional Context." *Africa Today* 43 (January–March 1996): 7–31.

Schiffrin, Harold Z. *Sun Yat-sen and the Origins of the Chinese Revolution.* Berkeley: University of California Press, 1968.

Schneider, Tapio, P. A. O'Gorman, and X. J. Levine. "Water Vapor and the Dynamics of Climate Changes." *Reviews of Geophysics* 48 (July 2010): RG3001.

Schramm, Gunter. "The Effects of Low-Cost Hydro Power on Industrial Location." *Canadian Journal of Economics* 2 (May 1969): 210–29.

Schwartz, Joel. "Illuminating Charles Darwin's Morality: Slavery, Humanity's Origin and Unity, and Darwin's Evolutionary Theory." *Evolution: Education and Outreach* 2 (April 2009): 334–37.

Scott, James C. *Seeing like a State: How Certain Schemes to Improve the Human Condition Have Failed.* New Haven, CT: Yale University Press, 1999.

Scott, Samuel P. *The Civil Law.* Cincinnati: Central Trust Co., 1932.

Scott, William H. "Demythologizing the Papal Bull 'Inter Caetera.'" *Philippine Studies* 35 (1987): 348–56.

Sekhar, C. S. C. "Surge in World Wheat Prices: Learning from the Past." *Economic and Political Weekly* 43 (May 2008): 12–14.

Severnini, Edson R. *The Power of Hydroelectric Dams: Agglomeration Spillovers.* Germany: IZA Discussion Paper No. 8082, 2014.

Shaw, Denis J. B. "Mastering Nature Through Science: Soviet Geographers and the Great Stalin Plan for the Transformation of Nature, 1948–53." *Slavonic and East European Review* 93 (January 2015): 120–46.

Shokr, Ahmad. "Hydropolitics, Economy, and the Aswan High Dam in Mid-Century Egypt." *Arab Studies Journal* 17 (Spring 2009): 9–31.

Showers, Kate B. "Beyond Mega on a Mega Continent: Grand Inga on Central Africa's Congo River." In *Engineering Earth,* edited by S. D. Brunn, 1651–79. Springer Science and Business Media, 2011.

Sial, Safdar, 2014. "The China-Pakistan Economic Corridor: An Assessment of Potential Threats and Constraints." *Conflict and Peace Studies, Pak Institute for Peace Studies* 6 (December 2014): 11–40.

Sima Qian. *The First Emperor.* Translated by R. Dawson. Oxford: Oxford World's Classics, 1994.

Singer, Itamar, 2000. "New Evidence on the End of the Hittite Empire." In *The Sea Peoples and Their Worlds: A Reassessment.* Edited by E. D. Oren, 21–33. Philadelphia: University of Pennsylvania Press, 2000.

Skinner, Quentin. *The Foundations of Modern Political Thought.* Cambridge: Cambridge University Press, 1978.

Skinner, Quentin. "Machiavelli's Discorsi and the Pre-humanist Origins of Republican Ideas." In *Machiavelli and Republicanism,* edited by G. Bock, Q. Skinner, and M. Viroli, 121–43. Cambridge: Cambridge University Press, 1990.

Smil, Vaclav. *Energy and Civilization: A History.* Cambridge, MA: MIT Press, 2017.

Smil, Vaclav. "Energy in the Twentieth Century: Resources, Conversions, Costs, Uses, and Consequences." *Annual Review of Energy and the Environment* 25 (November 2000): 21–51.

Smith, Adam. *The Wealth of Nations, Books I–III.* London: Penguin Classics, 1999.

Smith, Carl. *City Water, City Life: Water and the Infrastructure of Ideas in Urbanizing Philadelphia, Boston, and Chicago.* Chicago: Chicago University Press, 2013.

Smith, George O. *World Atlas of Commercial Geology. Part II: Water Power of the World.* Washington, DC: Department of the Interior, U.S. Geological Survey, 1921.

Smith, Vernon L. "The Primitive Hunter Culture, Pleistocene Extinction, and the Rise of Agriculture." *Journal of Political Economy* 83 (August 1975): 727–55.

Smolinski, Leon. "The Scale of Soviet Industrial Establishments." *American Economic Review* 52 (May 1962): 138–48.

Snarey, John. "The Natural Environment's Impact upon Religious Ethics: A Cross-Cultural Study." *Journal for the Scientific Study of Religion* 35 (June 1996): 85–96.

Sneddon, Christopher. *Concrete Revolution: Large Dams, Cold War Geopolitics, and the US Bureau of Reclamation.* Chicago: University of Chicago Press, 2015.

Snir, Ainit, D. Nadel, I. Groman-Yaroslavski, Y. Melamed, M. Sternberg, O. Bar-Yosef, and E. Weiss. "The Origin of Cultivation and Proto-Weeds, Long Before Neolithic Farming." *PLOS One* 10 (July 2015): e0131422.

Snow, John. *On the Mode of Communication of Cholera.* London: John Churchill, Princes Street, 1849.

Snowden, Frank M. *The Conquest of Malaria: Italy, 1900–1962.* New Haven, CT: Yale University Press, 2006.

Squatriti, Paolo. *Water and Society in Early Medieval Italy, AD 400–1000.* Cambridge: Cambridge University Press, 1998.

Stanković, Emilija. "Diocletian's Military Reforms." *Acta Univ. Sapientiae, Legal Studies* 1 (2012): 129–41.

Sternberg, Troy. "Chinese Drought, Wheat, and the Egyptian Uprising: How a Localized Hazard Became Globalized." In *The Arab Spring and Climate Change: A Climate and Security Correlations Series,* edited by C. E. Werrell and F. Femia, 7–14. Washington, DC: Stimson Center for American Progress, 2013.

Straub, Stéphane, C. Vellutini, and M. Warlters. *Infrastructure and Economic Growth in East Asia.* World Bank, Policy Research Paper No. 4589, 2008.

Subcommittee of the Committee on Appropriations. *Special Hearing, July 22, 2010—Washington, DC. Testimony Concerning Lessons from the 2010 Tennessee Flood.* Washington, DC: U.S. Senate, 111th Congress, 2nd Session, 2011.

Sun, Jimin. "Provenance of Loess Material and Formation of Loess Deposits on the Chinese Loess Plateau." *Earth and Planetary Science Letters* 203 (2002): 845–59.

Sun Yat-sen. *The International Development of China.* Shanghai: Commercial Press, 1920.

Taboulet, Georges. “Aux origines du canal de Suez. Le conflit entre F. de Lesseps et les Saint-Simoniens (1 re-partie).” *Revue Historique* 240 (1968): 89–114.

Taddese, Girma, K. Sonder, and D. Peden. “The Water of the Awash River Basin a Future Challenge to Ethiopia.” International Livestock Research Institute/International Water Management Institute Working Paper, 2004.

Tanzi, Vito, and L. Schuknecht. *Public Spending in the 20th Century: A Global Perspective.* Cambridge: Cambridge University Press, 2000.

Teichmann, Christian. “Canals, Cotton, and the Limits of De-colonization in Soviet Uzbekistan, 1924–1941.” *Central Asian Survey* 26 (May 2008): 499–519.

Temin, Peter. “The Economy of the Early Roman Empire.” *Journal of Economic Perspectives* 20 (Winter 2006): 133–51.

Temporelli, Giorgio, and N. Cassinelli. *Gli acquedotti genovesi.* Milan: Franco Angeli/Fondazione Famga, 2007.

Thomas, Harold A., Jr. “Roger Revelle: President-Elect, 1973.” *Science* 179 (February 1973): 818–20.

Thomas, Robert, and A. Wilson. “Water Supply for Roman Farms in Latium and South Etruria.” *Papers of the British School at Rome* 62 (November 1994): 139–96.

Thornton, Archibald P. “British Policy in Persia, 1858–1890. I.” *English Historical Review* 69 (October 1954): 554–79.

Thucydides. *The Peloponnesian Wars.* Translated by M. Hammonds. Oxford, UK: Oxford World’s Classics, 2009.

Tignor, Robert L. *W. Arthur Lewis and the Birth of Development Economics.* Princeton, NJ: Princeton University Press, 2006.

Tischler, Julia. *Light and Power for a Multiracial Nation: The Kariba Dam Scheme in the Central African Federation.* London: Palgrave Macmillan, 2013.

Tolmazin, David. “Recent Changes in Soviet Water Management: Turnabout of the ‘Project of the Century.’” *GeoJournal* 15 (October 1987): 243–58.

Torelli, Pietro. *Regesta Chartarum Italiae.* Vol. 1, *Regesto Mantovano.* Rome: Loescher & Co., 1914.

Totman, Conrad. *Early Modern Japan.* Berkeley: University of California Press, 1993.

Trenberth, Kevin E., and J. T. Fasullo. “Climate Extremes and Climate Change: The Russian Heat Wave and Other Climate Extremes of 2010.” *Journal of Geophysical Research* 117 (September 2012): D17103.

Trew, Alex. *Infrastructure Finance and Industrial Takeoff in England.* SIRE Discussion Papers 2010–25, Scottish Institute for Research in Economics (SIRE), 2010.

Tsomondo, Micah S. “From Pan-Africanism to Socialism: The Modernization of an African Liberation Ideology.” *Ufahamu: A Journal of African Studies* 6 (1975): 39–46.

Twain, Mark. *Life on the Mississippi.* New York: Harper and Brothers Publishers, 1917.

Twitchell, Karl S. *Saudi Arabia with an Account of the Development of Its Natural Resources.* Princeton, NJ: Princeton University Press, 1974.

Ulbrich, Uwe, P. Lionello, D. Belušić, J. Jacobeit, P. Knippertz, G. Kuglitsch, G. C. Leckebusch, J. Luterbacher, M. Maugeri, P. Maheras, K. M. Nissen, V. Pavan, J. G. Pinto, H. Saaroni, S. Seubert, A. Toreti, E. Xoplaki, and B. Ziv. “Climate of the Mediterranean: Synoptic Patterns, Temperature, Precipitation, Winds and Their Extremes.” In *The Climate of the Mediterranean Region: From the Past to the Future*, edited by P. Lionello, 301–46. London: Springer, 2012.

U.S. Army Corps of Engineers. *Cumberland and Duck River Basins: May 2010 Post Flood Technical Report.* Nashville, TN: USACE Nashville District, February 15, 2012.

Valla, François R., H. Khalaily, N. Samuelian, and F. Bocquentin. "What Happened in the Final Natufian?" *Journal of the Israel Prehistoric Society* 40 (2010): 131–48.

van Bath, B. H. Slicher. *Agrarian History of Western Europe, A.D. 500–1850.* London: Arnold, 1963.

Van Ingen, William B. "The Making of a Series of Murals at Panama." *Art World* 3 (October 1917): 17–19.

van Steenbergen, Frank, and W. Oliemans. "A Review of Policies in Groundwater Management in Pakistan, 1950–2000." *Water Policy* 4 (January 2002): 323–44.

Verhulst, Adriaan. *The Carolingian Economy.* Cambridge: Cambridge University Press, 2002.

Vials, Chris. "The Popular Front in the American Century: 'Life' Magazine, Margaret Bourke-White, and Consumer Realism, 1936–1941." *American Periodicals* 16 (March 2006): 74–102.

Vincent, Carol H., L. A. Hanson, and L. F. Bermejo. *Federal Land Ownership: Overview and Data.* Washington, DC: Congressional Research Service, 2020.

Vitalis, Robert. "The 'New Deal' in Egypt: The Rise of Anglo-American Commercial Competition in World War II and the Fall of Neocolonialism." *Diplomatic History* 20 (April 1996): 211–39.

Viviroli, Daniel, R. Weingartner, and B. Messerli. "Assessing the Hydrological Significance of the World's Mountains." *Mountain Research and Development* 23 (February 2003): 32–40.

Walder, Andrew G. Review of *The Science of Society: Toward an Understanding of the Life and Work of Karl August Wittfogel,* by G. L. Ulmen. *Journal of Asian Studies* 39 (May 1980): 535–38.

Wallimann, Isidor, H. Rosenbaum, N. Tatsis, and G. Zito. "Misreading Weber: The Concept of 'Macht.'" *Sociology* 14 (May 1980): 261–75.

Wang, Kaicun, and R. E. Dickinson. "A Review of Global Terrestrial Evapotranspiration: Observation, Modeling, Climatology, and Climatic Variability." *Reviews of Geophysics* 50 (May 2012): 1–54.

Wanner, Heinz, L. Mercolli, M. Grosjean, and S. P. Ritz. "Holocene Climate Variability and Change: A Data-Based Review." *Journal of the Geological Society* 172 (March 2015): 254–63.

Warner, John M., and J. T. Scott. "Sin City: Augustine and Machiavelli's Reordering of Rome." *Journal of Politics* 73 (July 2011): 857–71.

Waterbury, John. *The Egypt of Nasser and Sadat: The Political Economy of Two Regimes.* Princeton, NJ: Princeton University Press, 1983.

Webster, Peter J., V. E. Toma, and H.-M. Kim, 2011. "Were the 2010 Pakistan Floods Predictable?" *Geophysical Research Letters* 38 (February 2011): L04806.

Weiss, Barry. "The Decline of Late Bronze Age Civilization as a Possible Response to Climatic Change." *Climatic Change* 4 (June 1982): 173–98.

Weiss, Harvey. "4.2 ka BP Megadrought and the Akkadian Collapse." In *Megadrought and Collapse. From Early Agriculture to Angkor,* edited by H. Weiss, 93–159. Oxford: Oxford University Press, 2017.

Weiss, Harvey, M.-A. Courty, W. Wetterstrom, F. Guichard, L. Senior, R. Meadow, and A. Curnow. "The Genesis and Collapse of Third Millennium North Mesopotamian Civilization." *Science* 261 (August 1993): 995–1004.

Weiss, Harvey. "Megadrought, Collapse, and Causality." In *Megadroughts and Collapse: From Early Agriculture to Angkor,* edited by H. Weiss, 1–31. Oxford: Oxford University Press, 2017.

Weithman, Paul. "Augustine's Political Philosophy." In *The Cambridge Companion to Augus-*

tine, edited by E. Stump and N. Kretzmann, 234–52. Cambridge: Cambridge University Press, 2006.

Wells, Audrey. *The Political Thought of Sun Yat-sen: Development and Impact.* New York: Palgrave Macmillan, 2001.

Wells, Herbert G. *The War That Will End War*. New York: Duffield & Company, 1914.

Wells, Peter S. "Creating an Imperial Frontier: Archaeology of the Formation of Rome's Danube Borderland." *Journal of Archaeological Research* 13 (March 2005): 49–88.

Welton, George. *The Impact of Russia's 2010 Grain Export Ban.* Oxfam Research Report, 2011.

Westenholz, Aage. "The Old Akkadian Period: History and Culture." In *Mesopotamien: Akkade-Zeit und Ur III-Zeit*, edited by W. Sallaberger and A. Westenholz, 17–117. Göttingen: Vandenhoeck & Ruprecht, 1999.

Westenholz, Joan G. *Legends of the Kings of Akkade: The Texts.* Winona Lake, IN: Eisenbrauns, 1997.

Westermann, Andrea, 2015. "Geology and World Politics: Mineral Resource Appraisals as Tools of Geopolitical Calculation, 1919–1939." *Historical Social Research* 40 (2015): 151–73.

White, Gilbert F. "The Mekong River Plan." *Scientific American* 208 (April 1963): 49–59.

Wickham, Chris. *Framing the Early Middle Ages: Europe and the Mediterranean, 400–800.* Oxford: Oxford University Press, 2005.

Wickham, Chris. *The Inheritance of Rome. Illuminating the Dark Ages, 400–1000.* New York: Viking, 2009.

Wickham, Chris. *Sleepwalking into a New World.* Princeton, NJ: Princeton University Press, 2015.

Widell, Magnus, C. Hritz, J. A. Ur, and T. J. Wilkinson. "Land Use of the Model Communities." In *Models of Mesopotamian Landscapes*, edited by T. J. Wilkinson, M. Gibson, and M. Widell, 66–73. Oxford, UK: BAR International Series 2442, 2013.

Wijffels, Alain. "Flanders and the Scheldt Question: A Mirror of the Law of International Relations and Its Actors." *Sartoniana* 15 (2002): 213–80.

Wilkinson, T. J. "Hydraulic Landscapes and Irrigation Systems of Sumer." In *The Sumerian World*, edited by H. Crawford, 33–54. New York: Routledge, 2013.

Wilkinson, T. J., J. A. Ur, and C. Hritz. "Settlement Archaeology of Mesopotamia." In *Models of Mesopotamian Landscapes*, edited by T. J. Wilkinson, M. Gibson, and M. Widell, 34–55. Oxford, UK: BAR International Series 2442, 2013.

Williams, Geoffrey J. "The Changing Electrical Power Industry of the Middle Zambezi Valley." *Geography* 69 (June 1984): 257–61.

Williams, Stephen. *Diocletian and the Roman Recovery.* New York: Routledge, 1997.

Willis, Katherine J., L. Gillson, and T. M. Brncic. "How 'Virgin' is the Virgin Rainforest?" *Science* 304 (April 2004): 402–3.

Wilson, Andrew. "Machines, Power and the Ancient Economy." *Journal of Roman Studies* 92 (November 2002): 1–32.

Wimmer, Andreas, and B. Min. "From Empire to Nation-State: Explaining Wars in the Modern World, 1816–2001." *American Sociological Review* 71 (December 2006): 867–97.

Winterton, W. R. "The Soho Cholera Epidemic 1854." *History of Medicine* 8 (March/April 1980): 11–20.

Wirszubski, Chaim. *Libertas as a Political Idea at Rome During the Late Republic and Early Principate.* Cambridge: Cambridge University Press, 1968.

Wisser, Dominik, S. Frolking, S. Hagen, and M. F. P. Bierkens. "Beyond Peak Reservoir Storage? A Global Estimate of Declining Water Storage Capacity in Large Reservoirs." *Water Resources Research* 49 (August 2013): 5732–39.

Wittfogel, Karl A. *Oriental Despotism.* New York: Vintage, 1981.

Woertz, Eckart. *Oil for Food: The Global Food Crisis & the Middle East.* Oxford: Oxford University Press, 2013.

Wolf, Aaron, and J. Ross. "The Impact of Scarce Water Resources on the Arab-Israeli Conflict." *Natural Resources Journal* 32 (Winter 1992): 919–58.

Wolman, Abel, and W. H. Lyles. "John Lucian Savage, 1879–1967." *National Academy of Sciences Biographical Memoirs* 49 (1978): 225–38.

Wood, Adrian P. "Regional Development in Ethiopia." *East African Geographical Review* 15 (June 1977): 89–106.

Woodhouse, Connie A., S. T. Gray, and D. M. Meko. "Updated Streamflow Reconstructions for the Upper Colorado River Basin." *Water Resources Research* 42 (May 2006): W05415.

Woodhouse, Connie A., K. E. Kunkel, D. R. Easterling, and E. R. Cook. "The Twentieth-Century Pluvial in the Western United States." *Geophysical Research Letters* 32 (April 2005): L07701.

Woodhouse, Connie A., and J. T. Overpeck. "2000 Years of Drought Variability in the Central United States." *Bulletin of the American Meteorological Society* 79 (December 1998): 2693–2714.

Yergin, Daniel. *The Prize: The Epic Quest for Oil, Money & Power.* New York: Free Press, 2008.

Yin, Xungang, and S. E. Nicholson. "The Water Balance of Lake Victoria." *Hydrological Sciences Journal* 43 (March 1998): 789–811.

Yoffee, Norman. "Political Economy in Early Mesopotamian States." *Annual Review of Anthropology* 24 (1995): 281–311.

Zhang, Bin, and C. A. Carter. "Reforms, the Weather, and Productivity Growth in China's Grain Sector." *American Journal of Agricultural Economics* 79 (November 1997): 1266–77.

Zhang, David D., H. F. Lee, C. Wang, B. Li, Q. Pei, J. Zhang, and Y. An. "The Causality Analysis of Climate Change and Large-Scale Human Crisis." *Proceedings of the National Academy of Sciences* 108 (October 2011): 17296–301.

Zhang, David D., J. Zhang, H. F. Lee, and Y.-q. He. "Climate Change and War Frequency in Eastern China over the Last Millennium." *Human Ecology* 35 (April 2007): 403–14.

Zhang, Q., C.-Y Xu, T. Yang, and Z.-C. Hao. "The Historical Evolution and Anthropogenic Influences on the Yellow River from Ancient to Modern Times." In *Rivers and Society: From Early Civilizations to Modern Times,* edited by T. Tvedt and R. Coopey, 145–64. London: I. B. Tauris, 2010.

Zhuang, Yijie. "State and Irrigation: Archeological and Textual Evidence of Water Management in Late Bronze Age China." *Wiley Interdisciplinary Reviews: Water* 4 (April 2017): e1217.

Index